Studies in Systems, Decision and Control

Volume 240

Series Editor

Janusz Kacprzyk, Systems Research Institute, Polish Academy of Sciences, Warsaw, Poland

The series "Studies in Systems, Decision and Control" (SSDC) covers both new developments and advances, as well as the state of the art, in the various areas of broadly perceived systems, decision making and control–quickly, up to date and with a high quality. The intent is to cover the theory, applications, and perspectives on the state of the art and future developments relevant to systems, decision making, control, complex processes and related areas, as embedded in the fields of engineering, computer science, physics, economics, social and life sciences, as well as the paradigms and methodologies behind them. The series contains monographs, textbooks, lecture notes and edited volumes in systems, decision making and control spanning the areas of Cyber-Physical Systems, Autonomous Systems, Sensor Networks, Control Systems, Energy Systems, Automotive Systems, Biological Systems, Vehicular Networking and Connected Vehicles, Aerospace Systems, Automation, Manufacturing, Smart Grids, Nonlinear Systems, Power Systems, Robotics, Social Systems, Economic Systems and other. Of particular value to both the contributors and the readership are the short publication timeframe and the world-wide distribution and exposure which enable both a wide and rapid dissemination of research output.

** Indexing: The books of this series are submitted to ISI, SCOPUS, DBLP, Ulrichs, MathSciNet, Current Mathematical Publications, Mathematical Reviews, Zentralblatt Math: MetaPress and Springerlink.

More information about this series at http://www.springer.com/series/13304

Edmundo Capelas de Oliveira

Solved Exercises
in Fractional Calculus

Springer

Edmundo Capelas de Oliveira
IMECC
Universidade Estadual de Campinas
Campinas, São Paulo, Brazil

ISSN 2198-4182   ISSN 2198-4190   (electronic)
Studies in Systems, Decision and Control
ISBN 978-3-030-20526-3   ISBN 978-3-030-20524-9   (eBook)
https://doi.org/10.1007/978-3-030-20524-9

This Springer imprint is published by the registered company Springer Nature Switzerland AG
The registered company address is: Gewerbestrasse 11, 6330 Cham, Switzerland

Dedicated to Ivana,
my pretty woman.

Foreword I

The fractional calculus is an old topic born soon after the fundamental papers by Leibniz and Newton on integer-order calculus at the end of the seventeenth century. Indeed, we find contributions involving names of renowned researchers, among them, Euler, Fourier, Abel, Liouville, and Riemann. After the first international conference held in 1974, there was a significant increase in technical papers, and starting since the 1990s in textbooks. Nowadays, we can say that consolidated applications are found in diverse areas of science, but fractional calculus is still not considered a regular discipline in most universities. Faced with this scenario, the book of Prof. Capelas, written in a simple and objective way, aims to contribute to fill this gap. The book, after a brief historical introduction, exhibits an extensive list of solved and proposed exercises presented in a tutorial way in 4 chapters, along with a final chapter of applications. The reader can surely appreciate the way on which the solutions are presented: first the statements, then suggestions, and finally the detailed answers are found. May I suggest this book to both beginners and advanced scholars in fractional calculus.

Bologna, Italy   Francesco Mainardi
                                                                               University of Bologna

Foreword II

Is it possible to extend the order of the derivatives and integrals to any number irrational, fractional, or complex? Indeed, Gottfried Leibniz conceived that concept in 1695! This initial brilliant idea remained in the area of pure mathematics for centuries. However, recently fractional calculus (FC) attracted the attention of many mathematicians, physicists, and engineers for the developments of practical implementations. In fact, during the recent years FC emerged as an important scientific area, stimulated by the findings and applications, not only in physics and engineering, but also in areas such as finance, economy, or biology. This progress motivated the development of a large volume of studies and publishing of many books addressing distinct aspects of FC, going from pure mathematics up to research in specific topics of applied sciences. In spite of these efforts toward the dissemination of FC to a wider audience, we verify that there is still the need for educational works that guide readers without experience in the field. The new book entitled "Solved exercises in fractional calculus" is the first work purely dedicated to this exercise, that is, to introduce researchers, or even the general public, to the surprising tools and shining results produced under the umbrella of FC. This area, simultaneously old and new, with three hundred years, but rediscovered only recently, will become certainly more accessible when following this new educational work that establishes a compromise both between basic theory and advanced research, and mathematics and applications. This book will certainly enlighten and delight readers.

Porto, Portugal

José A. Tenreiro Machado
Institute of Engineering, Polytecnic of Porto

Foreword III

Fractional calculus has received considerable interest in recent years. It is natural therefore that students and researchers may be interested in the subject and wish to enter this research area. However, to advance in the study of something new in mathematics, nothing is more useful than solving problems. This book, by Prof. Edmundo Capelas de Oliveira, a leading specialist in fractional calculus, is undoubtedly the book that every beginner in an area would like to find. The book covers the main themes of fractional calculus with a smart strategy: after an introduction to a given topic, problems are proposed, followed by suggestions for solving problems, succeeded by a section with the solution of the problems, and ended with proposed problems. No doubt this is a very valuable book in the subject of fractional calculus.

Campinas, Brazil

Jayme Vaz Jr
University of Campinas

Preface

Arbitrary order calculus or fractional calculus is of considerable interest, as it opened new and fascinating horizons to science, freeing it from the dependence of the sensible world, limited to the integer order calculus. Although we cannot physically imagine the true appearance of fractional derivatives, their study deserves attention because it sharpens our intuition and imagination.[1]

The future of fractional calculus as a discipline is assured.[2]

It is common in almost all higher level courses in exact sciences and technology that the student, when facing for the first time the discipline calculus, will find it difficult to follow, as most students arrive at the university without the knowledge and maturity necessary to study it. For this reason, many authors write books with solved and proposed exercises by means of which they try to overcome such lack of basis by providing the basic knowledge necessary for the student to be able to learn the new subject. However, one must keep in mind that this is not enough. The student must do his part, as such books are usually just a kind of guide to be used in parallel with an adequate textbook on the discipline.

It is not different here. This is a book dedicated to solved and proposed exercises, but with the one peculiarity that discipline fractional calculus has yet to be a fully consolidated part of standard curricula. Fractional calculus is as old as the traditional, integer-order calculus proposed independently by Newton and Leibniz at the end of the seventeenth century. However, for one reason or another, fractional calculus had its development accelerated only from the second half of the twentieth century, after the first international conference on the subject. Nowadays, fractional calculus is completely consolidated and is a subject of research in several fields of knowledge, particularly in problems in which the memory effect is present, as in the cases of anomalous diffusion, polymer studies, and image processing, just to mention a few examples.

---

[1] Adapted from: G. Arcidiacono, *Spazio, Iperspazi, Frattali*, Di Renzo Editore, Roma, (1993).

[2] Miklós Mikolás, *On the recent trends in the development theory and applications of fractional calculus and its applications*, Lectures Notes in Mathématices, **457**, 357–375, Springer-Verlag, (1974).

With these facts in mind, we have elaborated this work. It does not intend to replace books dedicated to the theory of fractional calculus or its specificities, nor is it a compendium on the subject. It does have the intention of being an introductory text that we deem necessary for accompanying a possible course on fractional calculus. Besides, it can also be used to introduce students to this fascinating subject, aiming at new studies and research, as its wide range of applications provide young people a new and motivating field of work.

We believe that it is important for teachers to be able to answer the classic question posed by (mainly first year) students when one starts teaching a new subject: "What's the use?" or "Where will I need this?" Thus, we have tried to make it clear, for each of the topics addressed, where it can be used or applied—in the developments presented in later chapters, in one or more proposed problems or, in some cases, by making specific reference to works in which that topic played an auxiliary or more fundamental role.

After these few lines justifying the work, we can make a brief list of the topics addressed. In passing, we mention that our choices were made with a view to a future textbook for a discipline of fractional calculus to which this book might be a companion.

The historical introduction occupies the entire first chapter. In the second chapter, we present the gamma function as a generalization of the factorial function, together with a few other important functions: the incomplete gamma, beta, error, complementary error, hypergeometric and confluent hypergeometric functions. We also introduce Meijer's G-function, the generalized Wright function and Fox's H-function. The third chapter is dedicated to the Mittag-Leffler function, known as the queen of functions of fractional calculus, as well as to its variations, with emphasis on the Wright and Mainardi functions. The fourth chapter deals with integral transforms, with emphasis on Laplace, Fourier, and Mellin transforms. Then, in the fifth chapter, we introduce the concepts of fractional integral and fractional derivative, highlighting the Riemann–Liouville, Caputo and Hadamard versions and mentioning some possible generalizations. In the sixth chapter, by means of real problems and applications, we discuss some fractional differential, integral, and integro-differential equations and present some new results not included in previous chapters.

As our main objective is the solution of exercises, chapters two–five bring a list of proposed exercises. At first, we present only the statements, as we believe that students should try to discern the best way to address a given problem. After that, we present suggestions for solving each exercise and then their complete solutions. With this method, we try to make it clear that there are three distinct phases to pass through in order to obtain any particular solution. Moreover, at the end of chapters two–five, we present a further list of exercises which, we believe, can be solved by anyone who could understand the preceding developments. An appendix involving the essentials of Mellin–Barnes integral closes the book.

I would say that this is a small part of the studies in the past 10 years, but very important. A thank you to my former students, all of them teachers in various places in Brazil and abroad. Also, I would like to thank my colleagues

W. A. Rodrigues Jr. (in memoriam), J. Vaz Jr., M. J. Menon, B. Max Pimentel, J. M. Rosário, E. K. Lenzi, M. Lazo, F. Mainardi, J. A. Tenreiro Machado, M. D. Ortigueira, D. F. M. Torres, R. Almeida, J. Rezende, C. Braumann, Dr. José Emílio Maiorino, and Dr. Quintino Augusto Souza for several discussions and suggestions that improved the text as well as the figures.

Campinas, Brazil   Edmundo Capelas de Oliveira

Contents

Chapter 1
A Bit of History

Differently from ordinary, integer order calculus created by Newton and Leibniz, it can be said that fractional calculus had a start date. Several authors consider a letter sent from l'Hôpital to Leibniz on September 30, 1695, as the beginning of fractional calculus. In that letter, for the first time, l'Hôpital questioned Leibniz about the possibility and meaning of a derivative of order 1/2, that is, a derivative of fractionary order. The investigation that followed gave rise to the first results of what we presently call fractional calculus, also known by the names non-integer order calculus and arbitrary order calculus.

The purpose of this chapter is to sketch in chronological order the development of the fractional calculus, from its origins until recent times.

This chronology is based on several books and articles,[1] with a few additions which, as far as we know, did not yet appear in the literature on the history of fractional calculus.

For the sake of brevity, we do not present the complete references of the original works mentioned in this chronology, with the exception of those books and articles which do not appear in the references listed in footnote 1.

---

[1] K. B. Oldham and J. Spanier, *The Fractional Calculus* Academic Press, Inc., San Diego, (1974); K. S. Miller and B. Ross, *An Introduction to the Fractional Calculus and Fractional Differential Equations*, John Wiley and Sons, Inc., New York, (1993); J. A. Tenreiro Machado, V. Kiryakova and F. Mainardi, *A poster about the recent hystory of fractional calculus*, Fract. Calc. & Appl. Anal., **13**, 329–334, (2010); J. A. Tenreiro Machado, V. Kiryakova and F. Mainardi, *A poster about the old hystory of fractional calculus*, Fract. Calc. & Appl. Anal., **13**, 447–454, (2010); J. A. Tenreiro Machado, V. Kiryakova and F. Mainardi, *Recent history of fractional calculus*, Commun. Nonl. Sci. Num. Simul., **16**, 1140–1153, (2011); R. Figueiredo Camargo and E. Capelas de Oliveira, *Fractional Calculus*, Editora Livraria da Física, São Paulo, (2015); J. A. Tenreiro Machado and V. Kiryakova, *Historical Survey: The Chronicles of Fractional Calculus*, Fract. Calc. & Appl. Anal., **20**, 307–336, (2017); G. Sales Teodoro, D. S. Oliveira and E. Capelas de Oliveira, *On the fractional derivatives* (in Portuguese), Rev. Bras. Ens. Fis., **40**, (2), e2307 (2018).

© Springer Nature Switzerland AG 2019
E. Capelas de Oliveira, *Solved Exercises in Fractional Calculus*, Studies in Systems, Decision and Control 240,
https://doi.org/10.1007/978-3-030-20524-9_1

1.1 Historical Survey

Our chronology is divided in four periods: 1695–1975, 1976–1993, 1994–2015 and from 2015 on.

1.1.1 Period from 1695 to 1975

This subsection covers the period between the letter of September 30, 1695 and the date of the first scientific event, dedicated exclusively to the topic of fractional calculus, which occurred in June 1974, the annals of which were published in 1975, two hundred and eighty years after the missive already mentioned.

1695. In a letter of September 30, from G. A. l'Hôpital to G. W. Leibniz, l'Hôpital questioned Leibniz about the possibility and meaning of a derivative of order $1/2$. In his answer, Leibniz wrote, in a prophetic tone: "Thus, it follows that $d^{\frac{1}{2}}x$ will be equal to $x\sqrt[2]{dx : x}$, an apparent paradox, of which one day important consequences will be drawn." In another letter, addressed to J. Bernoulli, Leibniz mentioned the term derivative of the "general order."

1697. In a letter to J. Wallis, G. W. Leibniz discussed the infinite product of Wallis for π and used the notation $d^{\frac{1}{2}}y$ to denote the derivative of order ½.

1730. L. Euler wrote: "When n is a positive integer, the ratio of $d^n p$, p being a function of x, dx^n can always be expressed algebraically." He even suggested what might happen if instead of n we had a fraction and threw out the possibility of using interpolation of series.

1772. J. Lagrange developed the exponent law (indices) for differential operators of integer orders.

1812. Using integrals, P. S. Laplace wrote expressions for a derivative of non-integer order.

1819. In a treatise on differential and integral calculus, S. F. Lacroix obtained formally the derivative of order ½, arriving at the expression $(d^{\frac{1}{2}}/dx^{\frac{1}{2}})x = 2\sqrt{x}/\sqrt{\pi}$. Here it appeared the first mention to the name derivative of arbitrary order.

1822. J. B. J. Fourier published the classic "Analytical Theory of Heat" and obtained an integral formula for the derivative of order u (positive or negative). In 1835, Liouville came to suggest a more suitable way to obtain Fourier's result.

1823. H. N. Abel studied the problem of tautochrone using definite integrals. This result is considered to be the first application of fractional calculus.

1832. J. Liouville published the first major study on fractional calculus, starting the study on complementary solutions. In 1835 he suggested an expression for the fractional derivative by means of infinite series. In 1855 he proposed a series of definitions for the fractional derivative. In 1873 he discussed the integration of differential equations with derivatives of fractional order. It is worth mentioning that Liouville worked for more than forty uninterrupted years on the theme fractional calculus. Some researchers consider Liouville the father of the fractional calculus.

1833. G. Peacock published his "Principle of Permanence of Equivalent Forms", but he made a mistake in asserting its validity for all symbolic operations.

1841. D. F. Gregory solved the heat equation using symbolic operators.

1842. A. de Morgan initiated a polemic claiming that neither the formulation of Peacock nor that of Liouville for the derivative of non-integer order were correct.

1846. P. Kelland admitted the validity of the "Principle of Permanence of Equivalent Forms" for all symbolic operations, including the derivative.

1847. B. Riemann introduced an expression for the fractional integral which, with two changes, namely, eliminating the complementary function and taking the index less than zero, comes to be exactly the expression still used today. The publication of the important results of Riemann took place posthumously, in 1892.

1848. C. J. Hargreave discusses, for the first time, a generalization of the Leibniz rule.

1848. W. Center discussed the fractional derivative of a constant writing $1 = x^0$. In the three years that followed, he published several articles on fractional derivatives.

1867. A. K. Grünwald discussed the inversion of an integral equation.

1868. A. V. Letnikov proposed the sum of orders in the product of fractional derivatives and discussed the works of Liouville, Peacock and Kelland on fractional derivatives. In 1872 he discussed the generalization of Cauchy's integral formula and used fractional derivatives to address differential equations.

1868. H. Holmgren discussed the integration of an ordinary, linear and second-order differential equation with non-constant coefficients.

1869. N. Ya Sonin published an important work on arbitrary differentiation.

1880. A. Cayley discussed Riemann's work of 1847; in particular, the difficulty of interpreting the complementary function.

1884. H. Laurent generalized Leibniz's rule for the product of two functions, presenting the result in integral form.

1892. J. Hadamard studied functions given in terms of Taylor-type expansions.

1893. O. Heaviside discussed operator theory in Mathematical Physics.

1903. G. Mittag-Leffler introduced the function that bears his name and is of fundamental importance in fractional calculus. Mainardi coined it the queen of functions of fractional calculus.

1917. H. K. H. Weyl discussed non-integer order differentiation, proposing a different integral to what is now known as the Riemann-Liouville integral.

1919. E. Post discussed two problems that appeared in the American Mathematical Monthly using the Riemann definition of fractional derivative with the lower bound of the integral equal to zero.

1921. T. J. I'a Bromwich proved the effectiveness of the Heaviside operators.

1922. G. H. Hardy discussed the properties of integrals of fractional order. He continued this study in 1925 in collaboration with J. E. Littlewood.

1923. P. Levy considered a fractional derivative of a complex exponential.

1924. H. T. Davis discussed the application of fractional operators to a class of Volterra type integral equations. In 1927, he applied fractional operators to functional equations. In 1931 he discussed the properties of an operator involving the logarithm. In 1936 he published the book *The Theory of Linear Operators*.

1927. A. Marchaud proposes a derivative that currently bears its name.

1935. A. Zygmund discussed Weyl's fractional integration for the trigonometric systems. In 1945 he presented a theorem on fractional derivatives.

1938 A. Gemant discusses fractional differentials.

1939. A. Erdélyi used fractional integration in integrals with a hypergeometric function in the kernel.

1940. H. Kober extended the results of Hardy-Littlewood (1925); he also used the Mellin transform to solve integral equations.

1941. D. V. Widder correlated the Laplace transform with fractional integrals.

1948. A. N. Gerasimov generalizes laws of the linear deformation.

1949. M. Riesz discussed the Riemann-Liouville integral by means of the Cauchy integral.

1949. G. W. Scott Blair publishes the book *Survey of General and Applied Rheology*, where fractional modelling is presented.

1954. Publication of the classical *Bateman Manuscript Project*. The subject index of the series had twenty entries related to fractional integrals, without references. In 1960, Erdélyi, together with I. N. Sneddon, discussed dual integral equations. In 1962 he discussed the ½–order differential operator. In 1964 he discussed an integral equation containing Legendre functions.

1964. I. M. Gel'fand and G. E. Shilov presented functions that can be written as derivatives of arbitrary order of elementary functions.

1964. R. G. Buschman demonstrated several relationships involving integral operators by means of a Mellin-type convolution product.

1965. J. Cooke generalized the Erdélyi-Kober integral operators.

1965. V. A. Ditkin and A. P. Prudnikov published *Integral Transforms and Operational Calculus*, today a classic.

1966. M. M. Dzhrbashian published the book *Integral Transform and Representation of Functions in the Complex Domain* (in Russian).

1967. M. S. L. Higgins used fractional operators to discuss differential equations.

1967. M. Caputo discusses linear models of dissipation.

1968. M. M. Dzhrbashian and A. B. Nersesyan published an important paper on fractional differential equations. They proposed a particular differential operator similar to the Caputo operator.

1968. S. G. Samko discussed Abel's equation using fractional integral operators.

1969. M. Caputo published *Elasticità e Dissipazione*, in Italian. In this text, a new form is introduced for the fractional derivative.

1969. Yu N. Rabotnov published *Creep Problems in Structural Members*.

1970. K. B. Oldham and J. Spanier discussed Fick's law using the ½-order derivative, which they called semidifferentiation.

1971. T. J. Osler generalized Cauchy's integral formula in order to address the Leibniz rule associated with the product of two functions.

1971.   E. R. Love presented the fractional derivative of imaginary order.

1971.   M. Caputo and F. Mainardi present linear models of dissipation in anelastic solids and generalize the model of dissipation based on memory, introduced by Caputo.

1972.   T. R. Prabhakar discussed integral equations containing hypergeometric functions with two independent variables, through fractional integration. In this work the Mittag-Leffler function with three parameters was introduced.

1974.   G. W. Scott Blair published the book *An Introduction to Biorheology*.

1974.   Oldham and Spanier published the first book devoted exclusively to fractional calculus, *The Fractional Calculus*, now a classic. This was also the year of the first international conference dedicated exclusively to fractional calculus, in New Haven. In 2006 the book was reprinted by Dover.

1974.   B. Ross published a brief history exposing the fundamentals of fractional calculus.

1975.   B. Ross (ed.): *Fractional Calculus and its Applications*, Proceedings of the International Conference on Fractional Calculus and Applications, University of New Haven, West Haven, Conn., June 1974; Springer-Verlag, New York. Since it was the first conference, we mention here the name of the participants who contributed with works to the *Proceedings*: Bertram Ross, West Haven; Ian Naismith Sneddon, Glasgow; Kenneth S. Miller, New York; Mohammed Ali Al-Bassam, State of Kuwait; Richard Askey, Madison; Paul L. Butzer, Aachen; Ursula Westphal, Aachen; J. A. Donaldson, Whashington; Arthur Erdélyi, Edinburgh; Marvin C. Gaer, Delaware; Lee A. Rubel, Urbana; George Gasper, Evanstone; Theodore Parker Higgins, Seattle; Peter D. Johnson, Jr., Atlanta; Hikosaburo Komatsu, Tokyo; Juan Kucera, Pullman; Andre G. Laurent, Detroit; E. Russell Love, Parkville; Douglas E. Hatch, Schenectady; J. Richard Shanebrook, Schenectady; Raimond A. Struble, Raleigh; Willian L. Wainwright, Boulder; Stephen J. Wolfe, Newark; David H. Wood, Nato Saclant ASW Research Center; Jean L. Lavoie, Québec; R. Tremblay, Québec; Thomas Joseph Osler, Glassoro and Mikós Mikolás, Budapest.

1.1.2   Period from 1976 to 1993

This subsection refers to the period between the date of the first international conference on fractional calculus, in June 1974, and the publication in 1993 of Miller and Ross's book, approximately twenty years after that conference.

1978.   B. Ross and F. Northover used the complex-order derivative in fractional calculus.

1979.   B. Ross used fractional calculus methodology to obtain Legendre polynomials with arbitrary indexes.

1979.   A. C. McBride published a book on fractional calculus and integral transforms of generalized functions.

1979.   R. L. Bagley, directed by P. J. Torvik, presented what is usually considered the first doctoral thesis on fractional calculus, *Applications of Generalized Derivative to Viscoelasticity*, Air Force Institute of Technology. It should be noted, however, that in 1971 F. Mainardi, guided by M. Caputo, also defended a doctoral thesis on the same subject. Even more so, in 1970 Yu A. Rossikhin, in Russia, defended his doctoral thesis using tools of fractional calculus, according to Mainardi [1].

1979.   M. Stiassnie published an application of fractional calculus to the formulation of viscoelastic models.

1980.   M. Saigo discussed Hölder's continuity for integrals and fractional derivatives.

1981.   B. Ross discussed the Riemann-Liouville transformations.

1984.   K. Nishimoto published the first volume of his *Fractional Calculus*. The second volume appeared in 1987, the third in 1989 and the fourth in 1991. These last volume was devoted to fractional ordinary and partial differential equations.

1984.   L. M. B. C. Campos discussed the derivative of complex order applied to classical special functions.

1985.   A. C. McBride and G. F. Roach (eds.), *Fractional Calculus*, University of Stratchclyde, Glasgow, August 1984, Pitman Advanced Publishing Program. Collection of papers presented at the second Conference on Fractional Calculus and Applications. Contributed to this event, among others, P. Heywood, S. Kalla, W. Lamb, J. S. Lowndes (discussed fractional integration operators), A. C. McBride (used the Mellin transform in the fractional calculus), K. Nishimoto, P. G. Rooney and H. M. Srivastava.

1986.   R. Bagley and P. J. Torvik used fractional calculus to discuss problems in viscoelasticity.

1987.   S. Samko, O. Marichev and A. Kilbas published the classic book *Fractional Integrals and Derivatives and Some of Their Applications*, in Russian.

1990.   K. Nishimoto (ed.), *Fractional Calculus and its Applications*, Proceedings of the International Conference on Fractional Calculus and Applications, College of Engineering, Nihon University, may-june, Tokyo, 1989. Some of the participants were: M. Al-Bassam, R. Bagley, Y. A. Brichkov, L. M. B. C. Campos, R. Gorenflo, J. M. C. Joshi, S. Kalla, E. R. Love, M. Mikolás, K. Nishimoto, S. Owa, A. P. Prudnikov, B. Ross, S. Samko and H. M. Srivastava.

1990.   S. L. Kalla and V. S. Kiryakova published an article on Fox's *H* function and generalized fractional calculus.

1991.   R. Gorenflo and S. Vessella published the book *Abel Integral Equations*.

1991.   K. S. Miller and B. Ross discussed the fractional Green function.

1991.   S. Kamath discussed the problem of relativistic fractional tautochrone.

1991.   T. F. Nonnenmacher and W. G. Glöckle publish a fractional model for mechanical stress relaxation.

1992.   First specialized magazine: Journal of Fractional Calculus. Editor: K. Nishimoto.

1992.   N. Engheta discuss the role of non-integral (fractional) calculus in electrodynamics.

1993. First edition in English of the book by S. Samko, O. Marichev and A. Kilbas, *Fractional Integrals and Derivatives and Some of Their Applications*.

1993. K. Miller and B. Ross published the book *An Introduction to the Fractional Calculus and Fractional Differential Equations*.

1993. G. K. Watugala publishes Sumudu transform, an integral transform to solve differential equations.

1993. A. M. Mathai publishes the book *A Handbook of Generalized Special Functions for Statistical and Physical Sciences*.

It is clear that around the end of the 1970s, immediately after the 1974 conference, the number of researchers involved with fractional calculus experienced a growing trend so that any chronology will unfortunately become more excluding.

Note that we mention the names of some researchers who participated in the second (1984) and third (1989) congresses, exclusively because several continue to publish until the present day.

1.1.3 Period from 1994 to 2015

This subsection covers the period immediately following the date of publication of Miller and Ross's book in 1993 until the publication of Camargo and Capelas's book in 2015, a period of approximately two decades.

1994. S. Dugowson publishes his Ph.D. thesis: *Les différentielles métaphysiques*: *histoire et philosophie de la généralisation de l'ordre de dérivation*.

1994. V. S. Kiryakova published the book *Generalized Fractional Calculus and Applications*.

1994. G. Adomian published the book *Solving Frontier Problems of Physics*: *The Decomposition Method*.

1995. H. Schiessel, R. Metzler, A. Blumen and T. F. Nonnenmacher present generalized viscoelastic models with fractional equations.

1996. B. Rubin published the book *Fractional Integrals and Potentials*.

1996. M. Caputo publishes the Green function of the difusion of fluids in porous media with memory.

1996. F. Mainardi publishes fractional relaxation-oscillation and fractional differential wave phenomena.

1996. K. M. Kolwankar and A. D. Gangal introduced a new derivative which gave rise to local fractional calculus.

1997. A. Carpinteri and F. Mainardi edited the book *Fractal and Fractional Calculus in Continuous Mechanics*.

1998. Launched the specialized magazine *Fractional Calculus and Applied Analysis*. Editor: V. S. Kiryakova.

1998. S. Kempfle discussed viscous damped oscillations using fractional differential operators.

1998. C. F. Lorenzo and T. T. Hartley discussed the interesting problem of initialization, related to the lower limit of the fractional integral.

1999. I. Podlubny published the book *Fractional Differential Equations*, Mathematics in Science and Engineering.

2000. R. Hilfer edited the book *Applications of Fractional Calculus in Physics*. This was probably the first time someone cared to present a reason for the absence of M. Caputo's book (1969) in either of the two previous important historical surveys by Oldham and Spanier (1975) and Miller and Ross (1993): "Probably because it was written in Italian". Also worthy of note is the definition of the Hilfer fractional derivative, which contains, as particular cases, the fractional derivatives of Riemann-Liouville and Caputo.

2000. FRACALMO website was launched. The name is an acronym for **Fra**ctional **Cal**culus **Mo**delling.

2000. R. Metzler and J. Klafter present the random walk's guide to anomalous diffusion.

2000. R. Gorenflo, Y. Luchko and F. Mainardi presented solutions of the diffusion-wave equations in terms of the Wright function.

2000. N. Laskin began the study of fractional quantum mechanics and of Lévy's path integral.

2001. R. Gorenflo, F. Mainardi, E. Scalas and M. Raberto publish fractional calculus and continuous-time finance.

2002. I. Podlubny provided a possible interpretation, both geometrical and physical, of fractional integrals and derivatives.

2003. B. J. West, M. Bologna and P. Grigolini published the book *Physics of Fractal Operators*.

2003. E. K. Lenzi, R. S. Mendes, L. C. Malacarne and I. T. Pedron discussed anomalous diffusion using a fractional non-linear differential equation.

2004. K. Diethelm published the book *The Analysis of Fractional Differential Equations*.

2004. A. A. Kilbas and M. Saigo published the book *H-Transforms: Theory and Applications*.

2005. G. M. Zaslavsky published the book *Chaos and Fractional Dynamics*.

2005. M. N. Berberan-Santos publishes analytical inversion of the Laplace transform without contour integration.

2005. V. V. Novikov, K. W. Wojciechowski, O. A. Komkova and T. Thiel discussed anomalous relaxation in dietectrics with the help of equations with fractional derivatives.

2006. N. Heymans and I. Podlubny proposed an interpretation of the initial conditions for a problem with fractional derivative given in the Riemann-Liouville sense.

2006. A. A. Kilbas, H. M. Srivastava and J. J. Trujillo published the book *Theory and Applications of Fractional Differential Equations*.

2006. R. L. Magin published the book *Fractional Calculus in Bioengineering*.

2007. J. Sabatier, O. P. Agrawal and J. A. Tenreiro Machado edited the book *Advances in Fractional Calculus: Theoretical Developments and Applications in Physics and Engineering*.

2007. K. H. Lundberg, H. R. Miller and D. L. Trumper discussed what is known as *origin problem*, an issue involving initial conditions, generalized functions and the integral transform methodology.

2007. F. Mainardi and R. Gorenflo published an interesting survey involving time-fractional derivatives in relaxation processes.

2008. Shantanu Das published the book *Functional Fractional Calculus for System Identification and Controls*.

2008. V. S. Kiryakova published a text on the contribution of S. L. Kalla, to generalized fractional calculus. Kalla was the first to use a Fox H-function in the kernel of a fractional operator.

2008. S. Z. Rida, H. M. El-Sherbiny and A. A. M. Arafa discussed the fractional non-linear Schrödinger equation.

2009. J. A. Tenreiro Machado suggested an interpretation of the fractional derivative by means of the concept of probability.

2009. D. del-Castillo-Negrete discussed a superdifusive front propagation with Lévy flights.

2009. V. V. Uchaikin published the book *Methods of Fractional Derivatives*, in Russian. In 2013 the English version was published in two volumes: *Fractional Derivatives for Physicists and Engineers*, Vol. I (Background and Theory) and Vol. II (Applications).

2009. R. Figueiredo Camargo, with the guidance of E. Capelas de Oliveira, published his Ph.D. thesis that received an honorable mention from the Brazilian Society of Applied and Computational Mathematics.

2010. Launched a new specialized magazine: Fractional Dynamic Systems. Editors: J. Pečarić and Y. Zhou.

2010. F. Mainardi published the book *Fractional Calculus and Waves in Linear Viscoelasticity*.

2010. R. E. Gutiérrez, J. M. Rosário and J. A. Tenreiro Machado discussed basic concepts of fractional calculus with applications in engineering.

2010. R. Caponetto, G. Dongola, L. Fortuna and I. Petras published the book *Fractional Orders Sistems: Modelling and Control Applications*.

2010. A. M. Mathai, R. K. Saxena and H. J. Haulbod published the book *The H-Function: Theory and Applications, today a classic*.

2010 J. A. Tenreiro Machado, V. Kiryakova and F. Mainardi published two posters, available for download at http://www.diogenes.bg/fcaa/volume13, in which they list the most important highlights about old history and the recent history of fractional calculus.

2010. FDA'10—4th IFAC Workshop on Fractional Differentiation and Its Applications, Badajoz, Spain.

2011. V. E. Tarasov published the book *Fractional Dynamics: Applications of Fractional Calculus to Dynamics of Particles, Fields and Media*.

2011. U. N. Katugampola introduced a new fractional integral.

2011.   R. Herrmann published the book *Fractional Calculus: An Introduction for Physicists*.

2011.   X. J. Yang published the book *Local Fractional Functional Analysis and its Applications*.

2011.   M. D. Ortigueira published the book *Fractional Calculus for Scientists and Engineers*.

2011.   E. Capelas de Oliveira and J. Vaz Jr. published Tunneling in Fractional Quantum Mechanics.

2011.   M. M. Meerschaert and A. Sikorskii published the book *Stochastic Models for Fractional Calculus*.

2012.   X. J. Yang published the book *Advanced Local Fractional Calculus and its Applications*.

2012.   P. Perdikaris and G. E. Karniadakis proposed a fractional order viscoelasticity in one-dimensional models for blood flow.

2012.   A. B. Malinowska and D. F. M. Torres published the book *Introduction to the Fractional Calculus of Variations*.

2012.   D. Baleanu, K. Diethelm, E. Salas and J. J. Trujillo published the book *Fractional Calculus, Models and Numerical Methods*.

2012.   FDA'12—5th Symposium on Fractional Differentiation and Applications, Nanjing, China.

2013.   V. Daftardar-Gejji edited the book *Fractional Calculus Theory and Applications*.

2013.   Yu. Luchko discussed the fractional Schödinger equation for a particule moving in a potential well.

2014.   D. Valério, J. A. Tenreiro Machado, and V. Kiryakova, published the paper *Some Pioneers of the Applications of Fractional Calculus*.

2014.   R. Gorenflo, A. A. Kilbas, F. Mainardi and S. V. Rogosin published the book *Mittag-Leffler Functions, Related Topics and Applications*.

2014.   E. Capelas de Oliveira and J. A. Tenreiro Machado published an article collecting the various definitions of fractional derivative.

2014.   A. Oustaloup published the book *Diversity and Non-Integer Differentiation for Systems Dynamics*.

2014.   J. Creson edited the book *Fractional Calculus in Analysis, Dynamics & Optimal Control*.

2014.   R. Khalil, M. al Horani, A. Yousef and M. Sababheh introduced the conformal derivative.

2014.   T. Abdeljawad proposed the conformal fractional calculus and discussed the fractional versions of the chain rule and integration by parts, among other subjects.

2014.   Jumarie published the book *Fractional Differential Calculus for Non-differentiable Functions: Mechanics, Geometry, Stochastics, Information Theory*.

2014.   S. Abbas published the book *Topics in Fractional Differential Equations*.

2014.   H. M. Srivastava, R. K. Raina, Xiao-Jun Yang, published the book: *Special Functions in Fractional Calculus and Related Fractional Differintegral Equations*.

2014.   FDA'14—6th International Conference on Fractional Differentiation and its Applications, Catania, Italy.

2014.   D. Valério, José Tenreiro Machado, and Virginia Kiryakova, published a paper presenting some pioneers of the applications of fractional calculus.

2014.   S. Rogosin and F. Mainardi published a paper where they discuss Scott Blair's pioneerism in the use of fractional calculus in rheology.

2014.   Y. Zhou published the book *Basic Theory of Fractional Differential Equations*.

2015.   M. Caputo and M. Fabrizio published an article in which they introduced a new fractional derivative, with non-singular kernel.

2015.   M. Ortigueira and J. A. Tenreiro Machado proposed a criterion, more restrictive than the criterion proposed by Ross (1974), to determine if a given derivative can be considered a fractional derivative.

2015.   X.-J. Yang, D. Baleanu and H. M. Srivastava published the book *Local Fractional Integral Transforms and their Applications*.

2015.   R. Figueiredo Camargo and E. Capelas de Oliveira published the book *Fractional Calculus* (in Portuguese), the first book on this subject written in Portuguese.

It is noticeable that, with the possibility of publishing in specialized magazines and the facility introduced by the internet, the number of researchers involved with fractional theme continues to grow.

It is worth highlighting, in this period, the proposals of two new formulations of fractional derivatives, the local derivatives and those with a nonsingular function in the kernel. Two decades after [2] and after the publication of [3], there is still an issue concerning the use of the term fractional for these two formulations, since they do not satisfy the criterion proposed by [4, 5]. In addition to these two formulations there are still others that do not satisfy any of the two criteria mentioned above. In spite of this, two new fronts of research were initiated.

1.1.4 Period After 2015

For this period, we included the complete reference for each work mentionied; this seems pertinent as we believe that they were not mentioned in previous chronologies.

2016.   V. E. Tarasov published an article in which he justifies the statement "For a derivative to be considered fractional, Leibniz's rule must not apply" [6].

2016.   O. S. Iyiola and E. R. Nwaeze discussed some results of conformal fractional calculus [7].

2016.   A. Atangana and D. Baleanu introduced a fractional derivative with a non-local and non-singular kernel [8].

2016.   FDA'16—7th International Conference on Fractional Differentiation and its Applications, Novi Sad, Serbia.

2017.   R. Almeida proposed a fractional derivative with respect to a function [9].

2017.   A. M. Mathai and J. H. Haubold published the book *Matrix Methods and Fractional Calculus* [10].

2017.   J. A. Tenreiro Machado and V. Kiryakova published an interesting historical survey of fractional calculus [11].

2017.   D. Zhao and M. Luo introduced a general conformable fractional derivative and its physical interpretation [12].

2017.   D. S. Oliveira and E. Capelas de Oliveira proposed a generalization of the Hadamard derivative via Katugampola integral and, in 2018, proposed the (k, ρ)-fractional derivative, depending on two parameters [13, 14].

2017.   N. A. Sheikh, F. Ali, M. Saqib, I. Khan, S. A. A. Jan, A. S. Alshomrani and S. S. Alghamdi discussed the pros and cons of fractional versions as introduced by Atangana-Baleanu and Caputo-Fabrizio [15].

2017.   Y. Yan, Z.-Zhong Sun and J. Zhang proposed a rapid numerical method to discuss second order differential equations with derivative of the Caputo type [16].

2017.   M. Kaplan and A. Bekir constructed exact solutions for fractional differential equations in space and time [17].

2017   X.-J. Yang and J. A. Tenreiro Machado discussed anomalous diffusion using a new fractional operator of variable order [18].

2018.   J. Vanterler da C. Sousa and E. Capelas de Oliveira proposed the ψ-Hilfer derivative which contains, as particular cases, more than twenty distinct formulations of fractional derivative [19].

2018.   M. Kaplan and A. Akbulut discussed the wave equation by means of the conformal fractional derivative [20].

2018.   S. Rogosin and M. Dubatovskaya published a survey on Letnikov vs. Marchaud [21].

2018.   X. Liang, F. Gao, C-B. Zhou, Z. Wang and X.-J. Yang proposed a model for anomalous diffusion based on Mittag-Leffler and Wiman functions [22].

2018.   F. Ferrari discusses the Marchaud and Weyl derivatives [23].

2018.   L. R. Evangelista and E. K. Lenzi published the book Fractional Diffusion Equations and Anomalous Diffusion [24].

2018.   6th Workshop on Fractional Calculus, Probability and Non-Local Operators: Applications and Recent Developments, Bilbao, Spain.

2018.   M. D. Ortigueira and J. A. Tenreiro Machado published, as invited paper, "On fractional vectorial calculus" [25].

2018.   X.-J. Yang, F. Gao and H. M. Srivastava introduced a computational methodology to solve fractional nonlinear differential equations expressed by a local derivative [26].

2018.   P. Agarwal and A. A. El-Sayed introduced an unconventional finite difference methodology to solve fractional diffusion equations [27].

2018.   FDA'18—8th The International Conference on Fractional Differentiation and its Applications, Amman, The Hashemite Kingdom of Jordan.

2018. B. Cuahutenango-Barro, M. A. Taneco-Hernández and J. F. Gómez-Aguilar discussed the fractional-time wave equation via Atangana-Baleanu fractional derivative [28].

2018. A. M. F. Andrade, E. G. Lima and C. A. Dartora published an introduction to fractional calculus with applications in electric circuits [29].

2018. H. G. Sun, Y. Zhang, D. Baleanu, W. Chen and Y. Q. Chen presented a collection of real world applications of fractional calculus [30].

2018. T. Akman, B. Yildiz and D. Baleanu proposed a discretization of Caputo-Fabrizio derivative [31].

2019. M. D. Ortigueira, D. Valério, and J. A. T. Machado presented a discussion of variable order fractional systems [32].

2019. D. Zhao and M. Luo introduced the general fractional derivatives with memory effects [33].

2019. E. Capelas de Oliveira, S. Jarosz, and J. Vaz Jr. published Fractional calculus via Laplace transform and its application in relaxation processes [34].

2019. Workshop on Nonlinear Fractional Operator, Sapienza University of Rome, Rome.

2019. G. Sales Teodoro, J. A. Tenreiro Machado, and E. Capelas de Oliveira, published A review of definitions of fractional derivatives and other operators [35].

2019. A. R. Gómez Plata and E. Capelas de Oliveira published the book *Introducción al cálculo fraccionario*, Editorial Neogranadina, Bogotá, 2019.

Here, too, found in more recent references and because we understood that it would be worth doing the proper registration.

It is clear that as in the end of last century, the number of researchers involved with the fractional theme is still growing. It is also noteworthy the growth in methodologies involving numerical and computational procedures, aiming at the solution of fractional (ordinary and partial) differential equations.

We believe that this growth is due to several facts, among which we mention: congresses and symposia have occurred more frequently and regularly throughout the world; there is presently a huge number of textbooks on fractional calculus; the increase in the number of journals that publish specific articles involving fractional calculus and its correlated areas in all branches of science, whether it is classified as pure or applied or technological and, undoubtedly, the facility provided by the internet with its sites and blogs.

We conclude this brief introduction by mentioning three important papers resulting from the last three conferences on fractional differentiation and applications, all containing discussions and opinions of several researchers in the area: J. Machado, F. Mainardi, V. Kiryakova, *Fractional Calculus: Quo Vadimus? (Where Are We Going?)* Contributions to Round Table Discussion held at International Conference "Fractional Differentiation and Applications (ICFDA 2014), Fract. Cal. Appl. Anal., **18**, 495–526, (2015); José António Tenreiro Machado, Francesco Mainardi, Virginia Kiryakova, Teodor Atanackovic, *Fractional Calculus: D'où Venons-Nous? Que Sommes-Nous? Où Allons-Nous?* (Contributions to Round Table Discussion held at ICFDA 2016), Fract. Cal. Appl. Anal., **19**, 1074–1104, (2016); J. A. Tenreiro Machado, Virginia Kiryakova, Francesco Mainardi, Shaher Momani, *Fractional Calculus's Adventures in Wonderland* (Round Table Held at ICFDA 2018), Fract. Cal. Appl. Anal., **21**, 1151–1155, (2018).

References

1. Mainardi, F.: Fractional Calculus and Waves in Linear Viscoelasticity. An Introduction to Mathematical Models. Imperial College, London (2010)
2. Kolwankar, K.M., Gangal, A.D.: Fractional differentiability of nowhere differentiable functions and dimensions. Chaos. **6**, 505–513 (1996)
3. Caputo, M., Fabrizio, M.: A new definition of fractional derivative without singular kernel. Prog. Frac. Differ. Appl. **2**, 73–85 (2015)
4. Katugampola, U.N.: Correction to "what is a fractional derivative?" by Ortigueira and Machado [J. Comp. Phys. **293**, 4–13 (2015), Special issue on Fractional PDEs]. J. Comp. Phys. **321**, 1255–1257 (2016)
5. Ortigueira, M.D., Tenreiro Machado, J.A.: What is a fractional derivative? J. Comp. Phys. **293**, 4–13 (2015)
6. Tarasov, V.E.: On chain rule for fractional derivatives. Commun. Nonlinear Sci. Numer. Simul. **30**, 1–4 (2016)
7. Iyiola, O.S., Nwaeze, E.R.: Some new results on the new conformable fractional calculus with application using D'Álembert approach. Progr. Frac. Diff. Appl. **2**, 115–122 (2016)
8. Atangana, A., Baleanu, D.: New fractional derivatives with nonlocal and non-singular kernel: theory and application to heat transfer model (2016). arXiv:1602.03408v1 [math.GM]
9. Almeida, R.: A Caputo fractional derivative of a function with respect to another function. Commun. Nonlinear Sci. Numer. Simulat. **44**, 460–481 (2017)
10. Mathai, A.M., Haubold, H.J.: Matrix Methods and Fractional Calculus. World Scientific, New Jersey (2017)
11. Tenreiro Machado, J.A., Kiryakova, V.: Historical survey: the chronicles of fractional calculus **20**, 307–336 (2017)
12. Zhao, D., Luo, M.: General conformable fractional derivative and its physical interpretation. Calcolo **54**, 903–917 (2017)
13. Oliveira, D.S., Capelas de Oliveira, E.: Hilfer-Katugampola fractional derivative (online). Comp. Appl. Math. 1–19 (2017)
14. Oliveira, D.S., Capelas de Oliveira, E.: On the generalized (k, ρ)-fractional derivative. Progr. Fract. Differ. Appl. **2**, 133–145 (2018)
15. Sheikh, N.A., Ali, F., Saqib, M., Khan, I., Jan, S.A.A., Alshomrani, A.S., Alghamdi, S.S.: Comparison and analysis of the Atangana-Baleanu and Caputo-Fabrizio fractional derivatives for generalized Casson fluid model with heat generation and chemical reaction. Results Phys. **7**, 789–800 (2017)
16. Yan, Y., Sun, Z.-Z., Zhang, J.: Fast evaluation of the Caputo fractional derivative and its applications to fractional diffusion equations: a second-order scheme. Commun. Comput. Phys. **22**, 1028–1048 (2017)
17. Kaplan, M., Bekir, A.: Construction of exact solutions to the space-time fractional differential equations via new approach. Optik **132**, 1–8 (2017)
18. Yang, X.-J., Tenreiro Machado, J.A.: A new fractional operator of variable order: Application in the description of anomalous diffusion. Phys. A Stat. Mech. Appl. **481**, 276–283 (2017)
19. Sousa, J.V.C., Capelas de Oliveira, E.: On the ψ-Hilfer fractional derivative. Commun. Nonlinear Sci. Numer. Simulat. **60**, 72–91 (2018)

20. Kaplan, M., Akbulut, A.: Application of two different algorithms to the approximate long water wave equation with conformable fractional derivative. Arab. J. Basic Appl. Sci. **25**, 77–84 (2018)
21. Rogosin, S., Dubatovskaya, M.: Letnikov vs. Marchaud: A survey on two prominent constructions of fractional derivatives. Mathematics **6**, 3 (2018). https://doi.org/10.3390/math6010003
22. Liang, X., Gao, F., Zhou, C-B., Wang, Z., Yang, X-J.: An anomalous diffusion model based on a new general fractional operator with the Mittag-Leffler function of Wiman type. Adv. Diff. Equat. Open Acess (2018)
23. Ferrari, F.: Weyl and Marchaud derivatives: a forgotten history. Mathematics **6**, 6 (2018). https://doi.org/10.3390/math6010006
24. Evangelista, L.R., Lenzi, E.K.: Fractional diffusion equations and anomalous diffusion. Cambridge University Press, Cambridge (2018)
25. Ortigueira, M.D., Tenreiro Machado, J.A.: On fractional vectorial calculus. Bull. Pol. Ac. Tec. **66**, 399–402 (2018)
26. Yang, X.-J., Gao, F., Srivastava, H.M.: A new computational approach for solving nonlinear local fractional PDEs. J. Comput. Appl. Math. **339**, 285–296 (2018)
27. Agarwal, P., El-Sayed, A.A.: Non-standard finite difference and Chebyshev collocation methods for solving fractional diffusion equation. Phys. A Stat. Mech. Appl. **500**, 40–49 (2018)
28. Cuahutenango-Barro, B., Taneco-Hernández, M.A., Gómez-Aguilar, J.F.: The fractional-time wave equation via Atangana-Baleanu fractional derivative. Chaos, Solitons and Fractals **115**, 283–299 (2018)
29. Andrade, A.M.F., Lima,E.G., Dartora, C.A.: An Introduction to fractional calculus and its applications in electric circuit. Rev. Bras. Ens. Fis. **40**, e3314 (2018)
30. Sun, H.G., Zhang, Y., Baleanu, D., Chen, W., Chen, Y.Q.: A new collection of real world applications of fractional calculus in science and engineering. Commun. Nonlinear Sci. Numer. Simulat. **64**, 213–231 (2018)
31. Akman, T., Yildiz, B., Baleanu, D.: New discretization of Caputo-Fabrizio derivative. Comput. Appl. Math. **37**, 3307–3333 (2018)
32. Ortigueira, M.D., Valério, D., Tenreiro Machado, J.A.: Variable order fractional systems. Commun. Nonlinear Sci. Num. Simulat. **71**, 231–243 (2019)
33. Zhao, D., Luo, M.: Representations of acting processes and memory effects; general fractional derivative and its applications to theory of heat condition with finite wave speeds. Appl. Math. Comp. **346**, 531–544 (2019)
34. Capelas de Oliveira, E., Vaz Jr, J., Jarosz, S.: Fractional calculus via Laplace transform and its application in relaxation processes. Commun. Nonlinear Sci. Numer. Simulat. **69**, 58–72 (2019)
35. Sales Teodoro, G., Tenreiro Machado, J.A., Capelas de Oliveira, E.: A review of definitions of fractional derivatives and other operators. J. Comput. Phys. **388**, 195–208 (2019)

Chapter 2
Special Functions

In analogy to integer order calculus, fractional calculus also presents classes of functions that can be called special functions. In addition to the Mittag-Leffler function and its particular cases, which will be presented in the third chapter, other functions emerge naturally, among them the incomplete gamma function and the error function, all of which are considered to be particular cases of Meijer's G-function, and their general case, Fox's H-function, which are related with the Mellin-Barnes integral, as we will discuss in the Appendix.

The study of the classical special functions is part of the mathematical analysis and it is all directed to the hypergeometric function and the confluent hypergeometric function, whose differential equations present, respectively, three and two singular points, as well as particular cases, among others we mention, the Jacobi polynomials, depending on three parameters; the Laguerre polynomials and the Whittaker functions, depending on two parameters and the Bessel functions and the Legendre polynomials, depending on a parameter.

Moreover, since these special functions are solutions of differential equations, we consider the classical gamma functions, a generalization of the concept of factorial, and beta, depending on two parameters, as being part of this class of functions, although they are not solution of a differential equation.

We start by presenting the *function*, as introduced by Gel'fand-Shilov, due to its importance in the definition associated with the product of Laplace convolution, as well as in the fractional integral, to later introduce the concepts of gamma function, Pochhammer symbol, due to the importance in the simplification of the notation, incomplete gamma function, error function, complementary error function and beta function. Furthermore, we present the hypergeometric function, the confluent hypergeometric function, and we only mention, as adequate parameter choices, as special cases, other special functions.

Finally, we justify this chapter taking into account that the confluent hypergeometric function is the function that is related to a particular Mittag-Leffler function, the

© Springer Nature Switzerland AG 2019
E. Capelas de Oliveira, *Solved Exercises in Fractional Calculus*, Studies in Systems, Decision and Control 240,
https://doi.org/10.1007/978-3-030-20524-9_2

most important special function of the fractional calculus, which will be discussed in the next chapter. We must mention that all these functions are particular cases of Meijer's G-function as, in general, of Fox's H-functions, both presented at the end of section four.

In summary, the hypergeometric function is for the integer order calculus, just as Fox's H-function is for the fractional calculus. Once again, we would like to emphasize that this chapter serves to introduce the necessary tools for the good progress of the other chapters, as well as providing a brief review of the classic special functions.

2.1   Functions and Pochhammer Symbol

In this section, we will present the Gel'fand-Shilov *function*, the gamma function, the Pochhammer symbol, the incomplete gamma function, the error function, the complementary error function, and the beta function.

Definition 2.1.1 (*The Gel'fand-Shilov function*) Let $n \in \mathbb{N}$. We define the Gel'fand-Shilov *function*, denoted by $\phi_n(t)$, as

$$
\phi_n(t) := \begin{cases} \dfrac{t^{n-1}}{(n-1)!} & \text{if } t \geq 0, \\[2mm] 0 & \text{if } t < 0. \end{cases}
$$

Note that we emphasize, with another typeface, the word *function*, because, in fact, it is not a function in the classic sense of the definition. From this observation, always bearing in mind that it is not a function in the classical sense, we no longer use different style of letter.

Before introducing the concept of gamma function, understood as a generalization of the concept of factorial, it is adequate to mention that there are different ways of introducing this concept. For example, Weierstrass studied it using an infinite product, while Gauss studied it through an adequate limiting process. Here, we chose to introduce the gamma function in the manner proposed by Euler, through an improper integral associated with the factorial. This integral is known as the Euler integral of the second kind. The improper integral, defining the concept of factorial, is given by

$$
n! := \int_0^\infty e^{-\xi} \xi^n \, d\xi \, ,
$$

being $n = 0, 1, 2, \ldots$ From this integral, we replace the integer n with a number, in principle complex, $z = x + iy$, with $x, y \in \mathbb{R}$ so, we get

$$
\int_0^\infty e^{-\xi} \xi^z \, d\xi \, .
$$

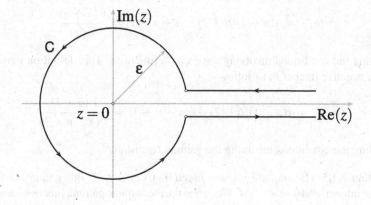

Fig. 2.1 Hankel contour, C

Definition 2.1.2 (*The gamma function*) We define the gamma function, denoted by $\Gamma(z)$, through the improper integral

$$\Gamma(z) := \int_0^\infty e^{-\xi} \xi^{z-1} \, d\xi \tag{2.1}$$

with the condition $\mathrm{Re}(z) > 0$. For the purpose of this book, we consider $x \in \mathbb{R}$, other than zero and a negative integer, where the poles of the gamma function are located. Further, integrating by parts Eq. (2.1), we can show the following functional relationship

$$\Gamma(x + 1) = x \, \Gamma(x)$$

interpreted as a generalization of the identity $n! = n(n - 1)!$, as already mentioned.

Due to the importance, we present an integral representation for the inverse of the gamma function, which, like the Mittag-Leffler expansion, makes explicit the poles of the gamma function, as well as plays an important role in obtaining several integral representations, specifically, for the Mittag-Leffler function.

Definition 2.1.3 (*Integral representation for the inverse gamma function*) Let μ be nonzero and a negative integer. The following relationship is valid

$$\frac{1}{\Gamma(\mu)} = \frac{1}{2\pi i} \int_{\mathrm{Ha}} e^z z^{-\mu} \, dz$$

where Ha is the called Hankel contour, as we will see in Fig. 2.1. Note that this representation, with the constraint in the μ parameter, shows the singularities of the gamma function, that is, they are at points $z = 0$ and $z = -n$ with $n \in \mathbb{N}$.

Definition 2.1.4 (*The Pochhammer symbol*) Let n be a non-negative integer and $\alpha \in \mathbb{R}$. We define the Pochhammer symbol, denoted by $(\alpha)_n$, from the following quotient involving gamma functions

$$(\alpha)_n := \alpha(\alpha+1)(\alpha+2)\cdot(\alpha+n-1) = \frac{\Gamma(\alpha+n)}{\Gamma(\alpha)}$$

respecting the conditions involving the gamma functions. This definition may also contain negative integer n, as follows

$$(\alpha)_{-n} := \alpha(\alpha-1)(\alpha-2)\cdots(\alpha-n+1) = \frac{\Gamma(\alpha+1)}{\Gamma(\alpha-n+1)}$$

respecting the conditions involving the gamma functions.

Definition 2.1.5 (*Incomplete gamma functions*) Let $\mu \in \mathbb{R}$, other than zero or a negative integer, and $0 < x < \infty$. We define the incomplete gamma function, denoted by $\gamma(\mu, x)$, and the complementary incomplete gamma function, denoted by $\Gamma(\mu, x)$, from the integrals

$$\gamma(\mu, x) = \int_0^x e^{-\xi}\xi^{\mu-1}\,d\xi \qquad \text{and} \qquad \Gamma(\mu, x) = \int_x^{\infty} e^{-\xi}\xi^{\mu-1}\,d\xi$$

respectively. From the definition of the incomplete gamma functions, it is immediate to verify that the sum of the two results is the gamma function, that is,

$$\gamma(\mu, x) + \Gamma(\mu, x) = \Gamma(\mu).$$

Definition 2.1.6 (*Error functions*) Let $x \in \mathbb{R}$ be non-negative. We define the error function, denoted by $\text{erf}(x)$, and the complementary error function, denoted by $\text{erfc}(x)$, from the integrals

$$\text{erf}(x) = \frac{2}{\sqrt{\pi}}\int_0^x e^{-\xi^2}\,d\xi \qquad \text{and} \qquad \text{erfc}(x) = \frac{2}{\sqrt{\pi}}\int_x^{\infty} e^{-\xi^2}\,d\xi$$

respectively. From the definition of the error functions, it is immediate to verify that the sum of the two results is unity, that is,

$$\text{erf}(x) + \text{erfc}(x) = 1,$$

which justifies the name of one being complementary to the other.

In analogy to the gamma function, we introduce the beta function, involving two parameters, also known as the Euler integral of first kind, as well as presenting a symmetry relation, obtained through a simple change of variable. It is not excessive to remember that, although the definition applies to complexes $p, q \in \mathbb{C}$, with the restrictions $\text{Re}(p) > 0$ and $\text{Re}(q) > 0$. In this text, we consider p and q real numbers.

Definition 2.1.7 (*Beta function*) Let $p, q \in \mathbb{R}$ be such that $p > 0$ and $q > 0$, which guarantee the existence of the integral. We define the beta function, denoted by $B(p, q)$, from the following integral

$$B(p, q) := \int_0^1 \xi^{p-1}(1 - \xi)^{q-1}\, d\xi\,. \tag{2.2}$$

From the change of variable $\xi = 1 - \eta$, it is immediate to show that

$$B(p, q) = B(q, p)$$

a symmetry relation involving the parameters. Finally, an important relation involving the beta and gamma functions is given by the following quotient

$$B(x, y) = \frac{\Gamma(x)\Gamma(y)}{\Gamma(x + y)}$$

respecting the conditions of existence of the gamma functions.

Just to mention, in analogy to the gamma functions, we can introduce the called incomplete beta functions and their complementary function simply by separating the integration interval, that is, from zero to μ and from μ to one, with $0 < \mu < 1$.

2.2 Hypergeometric Function and Particular Cases

As already mentioned, confluent hypergeometric function, a function involving two parameters, is related to a particular Mittag-Leffler function. Since confluent hypergeometric function is obtained, with an adequate limiting process, from the hypergeometric function, a function involving three parameters, we will present it and mention the most relevant particular cases. Moreover, since the hypergeometric function is a solution of an ordinary differential equation containing three singular points, we will consider only the expansion in terms of a single point, that is, we introduce the hypergeometric function through a series expansion in the neighborhood of the point $z = 0$.

Definition 2.2.1 (*Hypergeometric function*) Let $x \in \mathbb{R}$ and the parameters $a, b, c \in \mathbb{R}$. We define the hypergeometric function, denoted by $_2F_1(a, b; c; x)$, from the following expansion

$$_2F_1(a, b; c; x) = \sum_{k=0}^{\infty} \frac{(a)_k (b)_k}{(c)_k} \frac{x^k}{k!} \tag{2.3}$$

which is a convergent series imposing the condition $|x| < 1$, provided that the conditions imposed on the gamma functions, derived from the Pochhammer symbols, are satisfied. Similar expressions can be obtained for expansions around the other two singular points [1].

Considering particular cases of the parameters a, b and c, we will list the most relevant ones. Let the parameters be in the form $a = -n$, with $n = 0, 1, 2, \ldots b = \alpha + \beta + n + 1$ and $c = \alpha + 1$ such that $\alpha > -1$ and $\beta > -1$. In this case, we have the called Jacobi polynomials, denoted by $P_n^{(\alpha,\beta)}(x)$ and given by the following expression

$$P_n^{(\alpha,\beta)}(x) = \frac{\Gamma(\alpha+n+1)}{\Gamma(\alpha+1)\,n!} {}_2F_1\left(-n, \alpha+\beta+n+1; \alpha+1; \frac{1-x}{2}\right).$$

Let us now consider the parameters $\alpha = \beta = \lambda - 1/2$ with $\lambda > -1/2$. With this we obtain the Gegenbauer polynomials, denoted by $C_n^\lambda(x)$ and $\lambda \neq 0$, given by the expression

$$C_n^\lambda(x) = \frac{\Gamma(2\lambda+n)}{\Gamma(2\lambda)\,n!} {}_2F_1\left(-n, n+2\lambda; \lambda+1/2; \frac{1-x}{2}\right).$$

From the Gegenbauer polynomials, with $\lambda = 1$, we have the second kind Chebyshev polynomials, denoted by $U_n(x)$ and given, in terms of the hypergeometric function, by the expression

$$U_n(x) = (n+1)\,{}_2F_1\left(-n, n+2; 3/2; \frac{1-x}{2}\right).$$

In the case where $\lambda = 0$, we obtain the first kind Chebyshev polynomials, denoted by $T_n(x)$ and given, in terms of the hypergeometric function, by the expression

$$T_n(x) = {}_2F_1\left(-n, n+1; 1; \frac{1-x}{2}\right).$$

It should be noted that, in this case, we take into account the following limit $C_n^0(x) = \lim_{\lambda \to 0} \frac{1}{\lambda} C_n^\lambda(x)$.

In the case where $\lambda = 1/2$, we obtain the classic Legendre polynomials, denoted by $P_n(x)$ and given, in terms of the hypergeometric function, by

$$P_n(x) = {}_2F_1\left(-n, n; 1/2; \frac{1-x}{2}\right).$$

It is important to note that all these polynomials are solutions of an linear ordinary second-order differential equation and therefore admit another linearly independent solution, but this solution will not be a polynomial, but rather a function.

2.3 Confluent Hypergeometric Function and Particular Cases

In order to define the confluent hypergeometric function, we introduce the change of variable $x \to x/b$ in Eq. (2.3), so that you can write

$$_2F_1(a, b; c; x/b) = \frac{\Gamma(c)}{\Gamma(a)} \sum_{k=0}^{\infty} \frac{\Gamma(a+k)}{\Gamma(c+k)} \Lambda(b, k) \frac{x^k}{k!}$$

where we use the definition of the Pochhammer symbol and denote

$$\Lambda(b, k) = \frac{\Gamma(b+k)}{\Gamma(k) b^k},$$

of which, taking the $b \to \infty$ limit, provides

$$\lim_{b \to \infty} \Lambda(b, k) = \lim_{b \to \infty} \frac{b(b+1) \cdots (b-k-1)}{b^k}$$

$$= \left(1 + \frac{1}{b}\right)\left(1 + \frac{2}{b}\right) \cdots \left(1 + \frac{n-1}{b}\right) = 1.$$

Definition 2.3.1 (*Confluent hypergeometric function*) Let $x \in \mathbb{R}$ and the parameters $a, c \in \mathbb{R}$. We define the confluent hypergeometric function, denoted by $_1F_1(a; c; x)$, from the following expansion

$$\lim_{b \to \infty} {}_2F_1(a, b; c; x) \equiv {}_1F_1(a; c; x) = \sum_{k=0}^{\infty} \frac{(a)_k}{(c)_k} \frac{x^k}{k!} \qquad (2.4)$$

which is a convergent series imposing the condition $|x| < 1$, provided that the conditions imposed on the gamma functions arising from Pochhammer symbols, be satisfied.

Note that this function, dependent on two parameters, is also the solution of a linear, ordinary, second-order differential equation with two singular points. In analogy, to the hypergeometric function, we will consider only two particular cases of the parameters, mentioning only the relation with the confluent hypergeometric function, namely: Whittaker functions, denoted by $M_{v,\mu}(x)$ and Laguerre polynomials, denoted by $L_n(x)$.

Considering the parameters $a = \mu - v + 1/2$ and $c = 1 + 2\mu$ in confluent hypergeometric function, we obtain

$$M_{v,\mu}(x) = e^{-x/2} x^{\mu+1/2} {}_1F_1(\mu - v + 1/2; 1 + 2\mu; x)$$

with the restriction $\mu \neq -1/2, -3/2, \ldots$ This function, also with two parameters, is only a distinct way of presenting one of the solutions of the respective ordinary differential equation obtained from an adequate dependent variable change and the parameters as defined above.

On the other hand, in order to mention a particular case involving a parameter, the called classical Laguerre polynomials, we use the following limit

$$\lim_{\beta \to \infty} P_n^{(0,\beta)} \left(1 - \frac{2}{\beta}x\right) \equiv L_n(x)$$

where $P_n^{(0,\beta)} \left(1 - \frac{2}{\beta}x\right)$ are particular Jacobi polynomials. In short, we can write the Laguerre polynomial in terms of a confluent hypergeometric function,

$$L_n(x) = \frac{1}{n!} {}_1F_1(-n; 1; x)$$

with $n = 0, 1, 2, \ldots$ Finally, from the confluent hypergeometric function we can obtain, as particular cases of the parameters, other functions and polynomials, among which we mention: the Hermite polynomials, the Weber functions and the Bessel functions.

We conclude the section by pointing out that these special functions have, in addition to a purely mathematical study, in particular studying orthogonality and general functions, a vast class of applications in problems arising from mathematical physics, among which we mention only the simple pendulum, harmonic oscillator, and the hydrogen atom. Moreover, these functions can be generalized, in a natural way, by admitting p terms in the numerator and q terms in the denominator, with notation

$$_pF_q(a_1, \ldots, a_p; b_1, \ldots, b_q; x) = {}_pF_q((a_p); (b_q); x) = \sum_{k=0}^{\infty} \frac{(a_1)_k \cdots (a_p)_k}{(b_1)_k \cdots (b_q)_k} \frac{x^k}{k!}$$

where $(a_j)_k$ and $(b_j)_k$ are the Pochhammer symbols, as we will see next.

2.4 Generalized Hypergeometric Functions

As already mentioned, the hypergeometric function, depending on three parameters, containing several functions as particular cases, is the solution of an ordinary differential equation of the second order. There are three ways of thinking about a possible generalization of hypergeometric function, namely: inserting a larger number of parameters, increasing the number of independent variables or both. In this book, we will only be concerned with the insertion of the number of parameters in order to introduce the generalized hypergeometric function with several parameters that, as a particular case, recover the classical hypergeometric function.

There are three classes of generalized hypergeometric functions which, due to their particular use, deserve to be highlighted, in particular, by the (Bessel–Wright, Fox–Wright, among others); Meijer's function, also known as Meijer's G-function and the Fox function, also known as Fox's H-function.

It is important to note that this work is not intended to compute all the generalized hypergeometric functions, but to mention those that are closer to the fractional calculus, as will be seen in Chap. 3 in particular, dedicated exclusively to the Mittag-Leffler functions and some applications, in Chap. 6.

Definition 2.4.1 (*Generalized hypergeometric function*) Let p and q be non-negative integers; a_i with $i = 1, 2, \ldots, p$; b_j and $j = 1, 2, \ldots, q$ parameters with none of b_j a non-positive integer. The generalized hypergeometric function, denoted by one of the following forms

$$ {}_pF_q, \quad {}_pF_q\left(\begin{matrix} \mathbf{a} \\ \mathbf{b} \end{matrix}; z\right), \quad {}_pF_q(\mathbf{a}, \mathbf{b}; z), \quad {}_pF_q(a_1, \ldots, a_p; b_1, \ldots, b_q; z) $$

is defined by the series

$$ {}_pF_q\left(\begin{matrix} a_1, a_2, \ldots, a_p \\ b_1, b_2, \ldots, b_q \end{matrix}; z\right) = \sum_{k=0}^{\infty} \frac{(a_1)_k (a_2)_k \cdots (a_p)_k}{(b_1)_k (b_2)_k \cdots (b_q)_k} \frac{z^k}{k!} \tag{2.5} $$

with $(\cdot)_k$ denoting the Pochhammer symbols.

Note that in the case where $p = 2$ and $q = 1$ we recover the classical hypergeometric function. For a detailed study, in particular, involving the convergence of the series and properties, we suggest [2] where a vast list of references can be found.

2.4.1 Wright Functions

In this section devoted to Wright's functions, we will only present what is meant by Wright's generalized function [3], also known as Fox–Wright function, as well as its relation to the classical hypergeometric function for, then present the classic Wright function.

Definition 2.4.2 (*Generalized Wright function*) We define the generalized Wright function, denoted by ${}_p\Psi_q\left[z \middle| \begin{matrix} (a_i, \alpha_i)_{1,p} \\ (b_j, \beta_j)_{1,q} \end{matrix}\right]$, through the following sum,

$$ {}_p\Psi_q\left[z \middle| \begin{matrix} (a_i, \alpha_i)_{1,p} \\ (b_j, \beta_j)_{1,q} \end{matrix}\right] = \sum_{k=0}^{\infty} \frac{\displaystyle\prod_{i=1}^{p} \Gamma(a_i + \alpha_i k)}{\displaystyle\prod_{j=1}^{q} \Gamma(b_i + \beta_j k)} \frac{z^k}{k!} $$

for $z \in \mathbb{C}$, with $p, q \in \mathbb{N}$, $a_i, b_j \in \mathbb{C}$ and $\alpha_i, \beta_i \in \mathbb{R}^*$ such that $i = 1, 2, \ldots, p$ and $j = 1, 2, \ldots, q$.

As we have already mentioned, the classical hypergeometric function, also known as the Gauss function or Gauss hypergeometric function, denoted by $_2F_1(a, b; c; z)$ depends on three parameters, is a particular case. Considering $p = 2$, $q = 1$ and $\alpha_1 = \alpha_2 = \beta_1 = 1$ in the generalized Wright function, we can write the relation between Wright's function and the hypergeometric function, maintaining the notation used for the generalized Wright function,

$$_2F_1(a_1, a_2; b_1; z) = \frac{\Gamma(b_1)}{\Gamma(a_1)\Gamma(a_2)} {}_2\Psi_1 \left[z \left| \begin{array}{l} (a_1, 1), (a_2, 1) \\ (b_1, 1) \end{array} \right. \right]$$

for $|z| < 1$.

The classic Wright function is also called the Bessel–Wright function, since the most important particular case is a Bessel function and, by all accounts, it was a way of generalizing the Bessel function. In this book we mention only the simplest Wright function, denoted by $\phi(\alpha, \beta; z)$ and defined by the power series

$$\phi(\alpha, \beta; z) = \sum_{k=0}^{\infty} \frac{1}{\Gamma(\alpha k + \beta)} \frac{z^k}{k!} \tag{2.6}$$

with the parameters $\alpha, \beta \in \mathbb{C}$ and $z \in \mathbb{C}$. We mention that for $\alpha = 1$, $\beta = \mu + 1$ and $z = \pm x^2/4$ we obtain Bessel and modified Bessel functions, both of first kind. Further, identifying with Eq. (2.5), we find that we do not have parameters in the numerator.

Finally, another way of presenting Wright's function is by means of an integral in the complex plane, as a Mellin-Barnes type (Appendix A), that is,

$$\phi(\alpha, \beta; z) = \frac{1}{2\pi i} \int_\gamma \frac{\Gamma(s)}{\Gamma(\beta - \alpha s)} (-z)^{-s} \, ds \tag{2.7}$$

where the integration path, γ, separates all poles $s = -k$ with $k \in \mathbb{N}^*$ leaving them to the left of the straight line $\mathrm{Re}(s) = \gamma$.

Just to mention, after the concept of integral transforms, in particular the Mellin and Laplace transforms, we will recover the expression

$$\mathcal{M}[\phi(\alpha, \beta; t)](s) = \frac{\Gamma(s)}{\Gamma(\beta - \alpha s)}$$

with $\mathrm{Re}(s) > 0$ which shows the Eq. (2.7) is the respective inverse Mellin transform, that is, recovers the Wright function, as well as, after introducing the Mittag-Leffler function in Chap. 3, the following relation

$$\mathcal{L}[\phi(\alpha, \beta; t)](s) = \frac{1}{s} E_{\alpha, \beta}(1/s) \tag{2.8}$$

for $\alpha > -1$, $\beta \in \mathbb{C}$ and $\text{Re}(s) > 0$ being $E_{\alpha,\beta}(\cdot)$ a Mittag-Leffler function with two parameters. Moreover, s is the respective parameter associated with the integral transform, not necessarily the same, despite the letter, for Mellin and Laplace transforms.

2.4.2 Meijer's G-Function

Meijer's G-function is also a particular case of generalized hypergeometric function and was introduced as a possible generalization of the classical hypergeometric function, in particular, an adequate constant, involving gamma functions, multiplying the generalized hypergeometric function.

Definition 2.4.3 (*Meijer G-function*) We define Meijer's G-Meijer, denoted by $G_{p,q}^{m,n}\left[z \left| \begin{array}{c} a_1, \ldots, a_p \\ b_1, \ldots, b_q \end{array} \right. \right]$, with alternative notations

$$G_{p,q}^{m,n}\left[z \left| \begin{array}{c} a_1, \ldots, a_p \\ b_1, \ldots, b_q \end{array} \right. \right] \equiv G_{p,q}^{m,n}\left[z \left| \begin{array}{c} (a_p) \\ (b_q) \end{array} \right. \right] = G_{p,q}^{m,n}(z)$$

with $m, n, p, q \in \mathbb{N}$, such that

$$G_{p,q}^{m,n}\left[-z \left| \begin{array}{c} 1 - a_1, \ldots, 1 - a_p \\ 0, 1 - b_1, \ldots, 1 - b_q) \end{array} \right. \right] = \frac{\Gamma(a_1) \cdots \Gamma(a_p)}{\Gamma(b_1) \cdots \Gamma(b_q)} \, _pF_q(\mathbf{a}, \mathbf{b}; z) \quad (2.9)$$

or, in analogy to Wright's function, in terms of a Mellin-Barnes integral

$$G_{p,q}^{m,n}\left[z \left| \begin{array}{c} a_1, \ldots, a_p \\ b_1, \ldots, b_q \end{array} \right. \right] = \frac{1}{2\pi i} \int_\gamma \mathscr{G}_{p,q}^{m,n}(s) \, z^s \, \mathrm{d}s$$

with $z \neq 0$ and $\mathscr{G}_{p,q}^{m,n}(s)$ given by (here, also, through an integral in the complex plane, as a quotient of gamma functions)

$$\mathscr{G}_{p,q}^{m,n}(s) = \frac{\displaystyle\prod_{i=1}^{m} \Gamma(b_i - s) \cdot \prod_{j=1}^{n} \Gamma(1 - a_j + s)}{\displaystyle\prod_{i=m+1}^{q} \Gamma(1 - b_i + s) \cdot \prod_{j=n+1}^{p} \Gamma(a_j - s)}.$$

In this expression we admit an empty product equal to the unit and the parameters satisfying the inequalities $1 \leq m \leq q$ and $0 \leq n \leq p$. The complex parameters a_j and b_i are such that none of the poles of $\Gamma(b_i - s)$ with $i = 1, \ldots, m$ coincide with the poles of $\Gamma(1 - a_j + s)$ with $j = 1, \ldots, n$.

It is important to mention that, according to the particularity of the parameters, we can obtain several relations between the Meijer's function and the generalized hypergeometric function. Here we only mention the relation between a particular Meijer's function [4]

$$\frac{\displaystyle\prod_{j=1}^{r}\Gamma(a_j)}{\displaystyle\prod_{k=1}^{t}\Gamma(b_k)}\, _rF_t(a_1,\ldots,a_r;b_1,\ldots,b_t;z) = G^{1,r}_{r,t+1}\left(-z\left|\begin{array}{l}1-a_1,\ldots,1-a_r \\ 0,1-b_1,\ldots,1-b_t\end{array}\right.\right).$$

We conclude this subsection by mentioning that Meijer's function was exhaustively discussed in a series of works by Meijer himself[1] as well as, most recently, by Mathai [5], where one can find a discussion involving the conditions to be imposed on the parameters. In analogy to Wright's function, there are expressions involving the Mellin and Laplace transforms, due to the similarity between these (inverses) and Mellin-Barnes integrals [6–11].

2.4.3   Fox's H-Function

This section intends to introduce Fox's H-function [12] from the definition and retrieve the Meijer's G-function as a particular case. Moreover, this function contains, as particular cases, also the hypergeometric function, as well as the Mittag-Leffler functions that will be presented in Chap. 3, since they play a fundamental role in the fractional calculus, among others.

Definition 2.4.4 (*Fox's H-function*) The Fox function or Fox's H-function, denoted by $H^{m,n}_{p,q}(z)$ or, alternatively, by

$$H^{m,n}_{p,q}\left[z\left|\begin{array}{l}(a_1,\alpha_1),\ldots,(a_p,\alpha_p) \\ (b_1,\beta_1),\ldots,(b_q,\beta_q)\end{array}\right.\right] \equiv H^{m,n}_{p,q}\left[z\left|\begin{array}{l}\{(a_p,\alpha_p)\} \\ \{(b_q,\beta_q)\}\end{array}\right.\right] = H^{m,n}_{p,q}(z)$$

is defined, in analogy to the Meijer's G-function, through an integral in the complex plane, the called Mellin-Barnes integrals,

$$H^{m,n}_{p,q}\left[z\left|\begin{array}{l}(a_1,\alpha_1),\ldots,(a_p,\alpha_p) \\ (b_1,\beta_1),\ldots,(b_q,\beta_q)\end{array}\right.\right] = \frac{1}{2\pi i}\int_{\gamma}\mathscr{H}^{m,n}_{p,q}(s)\,z^s\,\mathrm{d}s$$

with $0 \le n \le p$, $1 \le m \le q$, $\alpha_1,\alpha_2,\ldots,\alpha_p \in \mathbb{R}_+$ and $\beta_1,\beta_2,\ldots,\beta_q \in \mathbb{R}_+$ and a_j and b_j, with $j=1,2,\ldots$, which can be complex numbers. The contour in the complex plane, denoted by γ, separates the poles from $\Gamma(1-a_j+\alpha_j s)$ with $j=1,2,\ldots,n$ and the poles from $\Gamma(b_j-\beta_j s)$ with $j=1,2,\ldots,m$, with $\mathscr{H}^{m,n}_{p,q}(s)$

---

[1]C. S. Meijer, *On the G function*, Proc. Nederl. Akad. Wetensch., **49**, 227–236, 344–356, 457, 469, 632–641, 765–772, 936–943, 1062–1072, 1165–1175 (1946).

given by

$$\mathcal{H}_{p,q}^{m,n}(s) = \frac{A(s)B(s)}{C(s)D(s)}$$

with

$$A(s) = \prod_{j=1}^{m} \Gamma(b_j - \beta_j s) \quad \text{and} \quad B(s) = \prod_{j=1}^{n} \Gamma(1 - a_j + \alpha_j s),$$

$$C(s) = \prod_{j=m+1}^{q} \Gamma(1 - b_j + \beta_j s) \quad \text{and} \quad D(s) = \prod_{j=n+1}^{p} \Gamma(a_j - \alpha_j s).$$

Note that $\alpha_j = 1 = \beta_j$ for all $j = 1, 2, \ldots$ the Fox's H-function is reduced to a Meijer's G-function, with notation

$$H_{p,q}^{m,n}\left[z \left| \begin{matrix} (a_1, 1), \ldots, (a_p, 1) \\ (b_1, 1), \ldots, (b_q, 1) \end{matrix} \right.\right] = G_{p,q}^{m,n}\left[z \left| \begin{matrix} a_1, \ldots, a_p \\ b_1, \ldots, b_q \end{matrix} \right.\right].$$

In analogy to the Meijer's G-function, an empty product is equal to unity, as exemplified, in the following cases

$$n = 0 \Leftrightarrow B(s) = 1, \quad m = q \Leftrightarrow C(s) = 1, \quad n = p \Leftrightarrow D(s) = 1.$$

It is important to note that, due to its generality, the study of convergence and properties, was and is the reason for several approaches, but is beyond the scope of this work. We mention, in addition to the work by Fox [12], the book [13] where a discussion associated with the conditions of existence of the integral can be found, among others.

We conclude this subsection, in analogy to the Meijer functions, mentioning that there are expressions involving the Mellin and Laplace transforms of the Fox H-function, due to the similarity between these integral transforms and the Mellin-Barnes integrals [6–10], as well as playing fundamental role, in particular in the resolution of differential equations, which is also beyond the scope of this book, intended as an introduction.

2.5 Exercises

We separate this section into four subsections. The first one contains only the statements of the exercises (Exercise list), the second account with a suggestion (Suggestions) for the respective solution, while the third contains the resolutions (Solutions) themselves. The fourth subsection presents exercises (Proposed exercises) to be solved by the students, most of them similar to those discussed in the text.

2.5.1 Exercise list

1. Show that $\Gamma(x + 1) = x\Gamma(x)$ for $x \in \mathbb{R}$ different from zero and a negative integer. Discuss the case $x = n$ with $n = 1, 2, 3, \ldots$
2. Show that

$$\Gamma(x) = \sum_{k=0}^{\infty} \frac{(-1)^k}{k!(x+k)} + \int_1^{\infty} e^{-\xi}\xi^{x-1}\, d\xi$$

   called the Mittag-Leffler expansion that evidences the gamma function poles, located at $x = -k$ with $k = 0, 1, 2, \ldots$, that is, they are in zero and in the negative integers.
3. Show that the residues of the gamma function at the poles $x = -n$ with $n = 0, 1, 2, \ldots$ are given by $(-1)^n/n!$.
4. Show the Legendre duplication formula

$$\Gamma(x)\Gamma(x + 1/2) = \sqrt{\pi}\, 2^{1-2x}\Gamma(2x)$$

   provided that the gamma functions are defined.
5. Let $m, n \in \mathbb{N}$. Show that $(n + m)! = n!(n + 1)_m$.
6. Let $a \neq 1, 2, 3, \ldots, k$ and $k = 1, 2, 3, \ldots$ Show that

$$(a)_{-k} = \frac{(-1)^k}{(1-a)_{-k}}.$$

7. Let $a \neq 0, 1, 2, \ldots$ Show that $\gamma(a, x) = e^{-x}\sum_{k=0}^{\infty} \frac{x^{a+k}}{(a)_{k+1}}.$
8. Show that $\gamma(1, x) = 1 - e^{-x}$.
9. Let $a > 0$ and $a + b > 0$. Show that

$$\int_0^{\infty} \xi^{a-1}\Gamma(b, \xi)\, d\xi = \frac{\Gamma(a+b)}{a}.$$

10. Express the error function in terms of an incomplete gamma function, that is,

$$\mathrm{erf}\,(x) = \frac{1}{\sqrt{\pi}}\gamma\left(\frac{1}{2}, x^2\right).$$

11. Show that $\mathrm{erf}\,(x) = \dfrac{2}{\sqrt{\pi}}\displaystyle\sum_{k=0}^{\infty} \dfrac{(-1)^k}{k!}\dfrac{x^{2k+1}}{2k+1}.$
12. Obeying the restrictions of the gamma functions, show that the relation between them and the beta function is valid

$$B(p, q) = \frac{\Gamma(p)\Gamma(q)}{\Gamma(p+q)}.$$

13. Show that the relation is valid

$$B(p, q) = 2 \int_0^{\pi/2} \cos^{2p-1} \theta \, \sin^{2q-1} \theta \, d\theta.$$

14. Let $x > \eta$. Evaluate the integral

$$\Omega(\eta, x) = \int_\eta^x (x - \xi)^{\alpha-1} (\xi - \eta)^{\beta-1} \, d\xi$$

expressing it in terms of a beta function.
15. Get an integral representation for the hypergeometric function

$$_2F_1(a, b; c; x) = \frac{\Gamma(c)}{\Gamma(c - b)\Gamma(b)} \int_0^1 t^{b-1} (1 - t)^{c-b-1} (1 - xt)^{-a} \, dt \quad (2.10)$$

admitting $c > b > 0$.
16. Evaluate the integral

$$\int_0^\infty {}_2F_1(a, b; c; -t) \, t^{-x-1} \, dt$$

for $c \neq 0, -1, -2, \ldots, x < 0, a + x > 0$ and $b + x > 0$.
17. Discuss the particular case of the preceding one, considering the $a = b = c = 1$.
18. Let $c - a - b > 0$. Show that $_2F_1(a, b; c; 1) = \dfrac{\Gamma(c)\Gamma(c - a - b)}{\Gamma(c - b)\Gamma(c - a)}$.
19. Let $\{x \in \mathbb{R} : -1 \leq x \leq 1\}$. The Jacobi polynomials are expressed in terms of the hypergeometric functions from the expression

$$P_n^{(\alpha,\beta)}(x) = \frac{\Gamma(n + \alpha + 1)}{n!\Gamma(\alpha + 1)} \, {}_2F_1\left(-n, \alpha + \beta + n + 1; \alpha + 1; \frac{1 - x}{2}\right).$$

Evaluate $P_n^{(\alpha,\beta)}(1)$ and $P_n^{(\alpha,\beta)}(-1)$.
20. Use the generating function, denoted by $G(x, t)$, for the Legendre polynomials, $P_k(x)$,

$$G(x, t) := (1 - 2xt + t^2)^{-1/2} = \sum_{k=0}^\infty P_k(x) t^k$$

in order to show the recurrence relation

$$P_k(x) = P'_{k+1}(x) - 2x P'_k(x) + P'_{k-1}(x)$$

valid for $k \geq 1$ with the line denoting derivative in relation to x.
21. Use the Rodrigues formula, given by the following expression,

$$P_k(x) = \frac{1}{2^k k!} \frac{d^k}{dx^k}(x^2 - 1)^k$$

with $k = 0, 1, 2, \ldots$ to write the first three Legendre polynomials.

22. Express the incomplete gamma function, $\gamma(\mu, x)$, in terms of the confluent hypergeometric function, $_1F_1(a; c; x)$.

23. Show the relationship between confluent hypergeometric functions

$$_1F_1(a; c; x) = e^x \, _1F_1(c - a; c; -x)$$

respecting the conditions of existence.

24. Let $x > 0$, $x > k$ and $\mu > 0$. Show that

$$\int_0^\infty t^{\mu-1} e^{-xt} \, _1F_1(a; c; kt)\, dt = \Gamma(\mu)x^{-\mu} \, _2F_1(a, \mu; c; k/x).$$

This expression can be interpreted as being the Laplace transform of the function $t^{\mu-1} \, _1F_1(a; c; kt)$, as we will see in Chap. 4.

25. Let $\mu > -1/2$. Show that

$$\Gamma\left(\mu + \frac{1}{2}\right) J_\mu(x) = \frac{2}{\sqrt{\pi}} \left(\frac{x}{2}\right)^\mu \int_0^{\pi/2} \cos(x \sin\theta)(\cos\theta)^{2\mu}\, d\theta,$$

an integral representation for the Bessel function, the called Poisson integral.

26. A way of defining the Laguerre polynomials, denoted by $L_k(x)$ with $k = 0, 1, 2, \ldots$, is through the Rodrigues formula, given by

$$L_k(x) = \frac{e^x}{k!} \frac{d^k}{dx^k}\left(e^{-x} x^k\right).$$

Obtain the explicit form of the first three Laguerre polynomials.

27. Another way to present the Laguerre polynomials is by means of the series

$$L_n(x) = \sum_{k=0}^\infty (-1)^k \frac{n!}{(n-k)! k!} \frac{x^k}{k!}.$$

Note that for a polynomial of degree n the sum goes to n. Express $L_n(x)$ in terms of confluent hypergeometric functions, that is, obtain the expression

$$L_n(x) = \, _1F_1(-n; 1; x).$$

28. We define the Hermite polynomials, denoted by $H_n(x)$ with $n = 0, 1, 2, \ldots$ from an expansion in power series

$$H_n(x) = n! \sum_{k=0}^{[n/2]} \frac{(-1)^k}{k!} \frac{(2x)^{n-2k}}{(n-2k)!}$$

with $[\xi]$ denotes the largest integer in ξ. Show that the following relation is valid

$$\frac{d}{dx} H_n(x) = 2n H_{n-1}(x)$$

for $n = 0, 1, 2, \ldots$

29. Let $k \in \mathbb{N}$ and $\phi(\alpha, \beta; x)$ be the Wright function. Show that

$$\frac{d^k}{dx^k} \phi(\alpha, \beta; x) = \phi(\alpha, k\alpha + \beta; x).$$

30. (Symmetry) For $a_1 = b_q$ show that

$$G_{p,q}^{m,n} \left[z \left| \begin{array}{c} a_1, \ldots, a_p \\ b_1, \ldots, b_q \end{array} \right. \right] = G_{p-1,q-1}^{m,n-1} \left[z \left| \begin{array}{c} a_2, \ldots, a_p \\ b_1, \ldots, b_{q-1} \end{array} \right. \right].$$

31. Show that

$$G_{2,2}^{1,2} \left[z \left| \begin{array}{c} 1, 1 \\ 1, 0 \end{array} \right. \right] = \ln(1 + z)$$

for $|z| < 1$.

32. Using the above, show that $\ln(1 + z) = z \, {}_2F_1(1, 1; 2; -z)$, with ${}_2F_1(a, b; c; x)$ the classic hypergeometric function.

33. Show that

$$H_{1,1}^{1,0} \left[z \left| \begin{array}{c} (\alpha + \beta + 1, 1) \\ (\alpha, 1) \end{array} \right. \right] = \frac{z^\alpha (1 - z)^\beta}{\Gamma(\beta + 1)}.$$

34. Consider $|x| < 1$. Show that

$$G_{2,2}^{1,1} \left[x \left| \begin{array}{c} 0, 1/2 \\ 0, 1/2 \end{array} \right. \right]$$

is a multiple of the sum of the terms of a geometric (infinite) progression of first term equal to unity and ratio equal to x.

35. Let $\gamma(\mu, x)$ be the incomplete gamma function. Show that

$$\gamma(\mu, x) = G_{1,2}^{1,1} \left[x \left| \begin{array}{c} 1 \\ \mu, 0 \end{array} \right. \right].$$

36. Show that

$$\frac{d}{dx} H_{1,2}^{1,1} \left[x \left| \begin{array}{c} (a, \alpha) \\ (a, \alpha), (0, 1) \end{array} \right. \right] = H_{1,2}^{1,1} \left[x \left| \begin{array}{c} (a - \alpha, \alpha) \\ (a - \alpha, \alpha), (0, 1) \end{array} \right. \right].$$

37. Explain Hankel's contour to show that it is worth the integral representation for the inverse gamma function, according to Definition 2.1.3.

38. Let $z \in \mathbb{C}$ and $\mu \in \mathbb{R}$. Denoting by $I_\mu(z)$ the modified Bessel function of first kind, as in Eq. (6.27), obtain the following integral representation

$$I_\mu(z) = \frac{(z/2)^\mu}{\Gamma(\mu + 1/2)\sqrt{\pi}} \int_{-1}^{1} e^{-z\xi}(1 - \xi^2)^{\mu-1/2}\,d\xi \cdot$$

39. Let $n \in \mathbb{N}$ and $\mu \in \mathbb{C}$. Show that

$$\frac{(-\mu)_n}{n!} = (-1)^n \binom{\mu}{n}$$

with $(\mu)_n$ being the Pochhammer symbol,

$$(\mu)_n = \mu(\mu + 1)\cdots(\mu + n - 1) = \frac{\Gamma(\mu + n)}{\Gamma(\mu)}$$

and $\binom{\mu}{n}$ the binomial coefficient.

40. Let $n \in \mathbb{N}$ and $\alpha, \beta \in \mathbb{R}$. Show that the sum involving the product of two binomial coefficients is given by

$$\sum_{i=0}^{n} \binom{\alpha}{i}\binom{\beta}{n-i} = \binom{\alpha+\beta}{n} \cdot$$

2.5.2 Suggestions

1. Use the definition of gamma function and integration by parts.
2. Use the definition of gamma function and separate it into two integrals, from zero to one and from one to infinity. In the first integral use a Maclaurin expansion for the exponential, since the second integral is analytic throughout the complex plane, that is, an integer function.
3. Multiply the Mittag-Leffler expansion by $x + k$ and evaluate the limit $x \to -k$.
4. Using the definition of $\Gamma(x + y)$ and enter the variable change $\xi = s(1 + p)$ with $p > -1$. Multiply the result by $p^{x-1}(1 + p)^{y-1}$ and integrate in the variable p to obtain the expression

$$\Gamma(x + y) \int_{0}^{\infty} \frac{p^{x-1}}{(1 + p)^{x+y}}\,dp = \Gamma(x)\Gamma(y) \cdot$$

Finally, enter another variable change $p = t/(1-t)$, consider $y = x$ and evaluate the remaining integral.

5. Direct from the definition.

6. Use the definition of the Pochhammer symbol, the reflection formula involving the functions and expression that provides the sine of difference.

7. From the definition of incomplete gamma function, multiply by e^x and expand the exponential in a Maclaurin series. In the resulting integral take an adequate variable changes, use the result given by Eq. (2.12) and the Pochhammer symbol.

8. Direct from the definition.

9. Replace the definition of complementary function and change the integration order. Calculate the integrals, one in terms of the gamma function and the other is immediate.

10. Take the variable change $\xi^2 = u$ and use the definition of incomplete gamma function.

11. Expand the exponential function and integrate term by term.

12. Direct from Eq. (2.12).

13. Use the change of variable $\xi = \cos^2 \theta$.

14. Use the change of variable $\xi = \eta + (x - \eta)t$.

15. Consider the integral

$$J = \int_0^1 t^{b-1}(1-t)^{c-b-1}(1-xt)^{-a}\, dt$$

for $c > b > 0$. Admit $|x| < 1$, expand $(1-xt)^{-a}$ into a uniformly convergent series and use the definition of beta function.

16. Use the integral representation for the hypergeometric function, change the integration order, and use the definition of beta function.

17. Simply replace $a = b = c = 1$ and simplify.

18. Use the integral representation for the hypergeometric function and the definition of beta function.

19. For $x = 1$ show that $_2F_1(a, b; c; 0) = 1$ and for $x = -1$ use the result of Exercise (18) and the result of Proposed exercise (3).

20. Derive in relation to x. Obtain the following first order partial differential equation

$$\frac{\partial}{\partial x} G(x, t) = \frac{t}{1 - 2xt + t^2} G(x, t).$$

Replace $G(x, t)$ with the power series and equal coefficients of the same power.

21. Just substitute $k = 0, 1, 2$ and evaluate the derivatives.

22. Use the series expansion for the gamma function by writing the coefficients in terms of gamma functions and compare with the confluent hypergeometric series.

23. Enter the variable change $t = 1 - \xi$ in the integral representation.

24. Use the series representation for the hypergeometric function and confluent hypergeometric function.

25. Use the expansion for the Bessel function and the Legendre duplication formula. From the integral defining the beta function, written in terms of trigonometric functions, and the Maclaurin series of the cosine function, conclude the result.
26. Just substitute $k = 0, 1, 2$ and evaluate the derivatives.
27. Write the factorial in terms of the Pochhammer symbol, use the reflection formula, and compare with the series expansion of the confluent hypergeometric function.
28. Simply derive in relation to x and compare with the series expansion.
29. Use Eq. (2.6) by deriving it k times. Derive the first time, reset the index, then derive again and reset again to show the recurrence.
30. Separate from the products the term a_1, simplify and rearrange.
31. Identify m, n, p and q. Substitute in the expression for $\mathscr{G}_{p,q}^{m,n}(s)$ and determine the coefficients a_k, b_k. Use the expansion for the logarithm, valid for $-1 < \ln(1 + x) \leq 1$,

$$\ln(1 + x) = \sum_{k=0}^{\infty}(-1)^k \frac{x^{k+1}}{k+1}.$$

32. Use the series representation for the hypergeometric function and compare with the result obtained in the previous one.
33. In this case, like all unitary α_i and β_i, this is also a Meijer function. Identify to get m, n, p and q. Replace in the expression for $\mathscr{G}_{p,q}^{m,n}(s)$ and determine the coefficients a_k and b_k. Use the expression

$$\sum_{j=0}^{\infty}\binom{q}{j}x^j \equiv \sum_{j=0}^{\infty}\frac{\Gamma(q+1)}{j!\Gamma(q+1-j)}x^j = (1+x)^q.$$

34. Identify to get m, n, p and q. Replace in the expression for $\mathscr{G}_{p,q}^{m,n}(s)$ and determine the coefficients a_k and b_k. Use the result

$$\Gamma\left(\frac{1}{2}+z\right)\Gamma\left(\frac{1}{2}-z\right) = \frac{\pi}{\cos(\pi z)}.$$

35. Identify to get m, n, p and q. Replace in expression for $\mathscr{G}_{p,q}^{m,n}(s)$ and determine the coefficients a_k and b_k. Use the result of Exercise (2).
36. After identifying the indices and parameters, write the representation of Mellin-Barnes type, use the functional relation to the gamma function, take a variable change and re-identify with the representation of the Fox's H-function.
37. Evaluate the integral $\int_C e^{-z}z^\mu \, dz$ with $\mathrm{Re}(\mu) > -1$ and C an adequate contour in the complex plane, and use the definition of gamma function.
38. Express the exponential on the second member in a power series, take a variable change, and use the definition of beta function. Justify an adequate change of index in order to get the result.

39. Direct from the definitions of the Pochhammer symbol and the binomial coefficient.
40. Express the binomial coefficients in terms of an adequate gamma function and use the relation between gamma and beta functions. Use the integral representation of the beta function and the sum of the terms of a geometric progression.

2.5.3 Solutions

1. Consider the definition of the gamma function for $x \to x + 1$. Then, we can write

$$\Gamma(x + 1) = \int_0^\infty e^{-\xi} \xi^x \, d\xi.$$

Integrating by parts: $e^{-\xi} d\xi = dv \Rightarrow v = -e^{-\xi}$ and $\xi^x = u \Rightarrow du = x\xi^{x-1} d\xi$, we get

$$\Gamma(x + 1) = \xi^x \cdot (-e^{-\xi})\big|_0^\infty - \int_0^\infty (-e^{-\xi}) \cdot x \cdot \xi^{x-1} \, d\xi.$$

For $x > -1$ and nonzero, we obtain

$$\Gamma(x + 1) = x \int_0^\infty e^{-\xi} \cdot \xi^{x-1} \, d\xi$$

which, from the definition of the gamma function, allows write

$$\Gamma(x + 1) = x\Gamma(x)$$

which is the desired result. In the case where $x = n$ we have $\Gamma(n + 1) = n\Gamma(n) = n!$ with $\Gamma(1) = 0! = 1$. We see that the gamma function generalizes the concept of factorial   ◇

2. Using the definition of gamma function, we get

$$\Gamma(x) = \int_0^1 e^{-\xi} \xi^{x-1} \, d\xi + \int_1^\infty e^{-\xi} \xi^{x-1} \, d\xi$$

for x nonzero and non-negative integer. In the first parcel, we introduced the Maclaurin expansion for the exponential then, treating it separately, we have

$$\Lambda \equiv \int_0^1 e^{-\xi} \xi^{x-1} \, d\xi = \int_0^1 \sum_{k=0}^\infty \frac{(-\xi)^k}{k!} \xi^{x-1} \, d\xi$$

or by exchanging the sum signal with the integral signal (the exponential is uniformly convergent at any compact interval), in the form

$$\Lambda = \sum_{k=0}^{\infty} \frac{(-1)^k}{k!} \int_0^1 \xi^{k+x-1} \, d\xi$$

whose integration furnishes

$$\Lambda = \sum_{k=0}^{\infty} \frac{(-1)^k}{k!(x+k)}$$

from which the desired result follows, that is, the Mittag-Leffler expansion which evidences the poles of the gamma function   ◇

3. Multiplying the Mittag-Leffler expansion by $x + k$ and taking the limit we have

$$\operatorname*{Res}_{x=-k} \Gamma(x) = \lim_{x \to -k} (x+k)\Gamma(x)$$

$$= \lim_{x \to -k} \left\{ (x+k) \sum_{k=0}^{\infty} \frac{(-1)^k}{k!(x+k)} + (x+k) \int_1^{\infty} e^{-\xi} \xi^{x-1} \, d\xi \right\}.$$

The limit of the second parcel is zero and, using the result

$$\operatorname*{Res}_{x=-k} \Gamma(x) = \sum_{k=0}^{\infty} \frac{(-1)^k}{k!} \tag{2.11}$$

that, for a fixed $k = n$, the residue is given by

$$\operatorname*{Res}_{x=-k} \Gamma(x) = \frac{(-1)^n}{n!}$$

which is the desired result   ◇

4. Let x and y be positive real numbers. Using the definition of gamma function, we have

$$\Gamma(x+y) = \int_0^{\infty} e^{-\xi} \xi^{x+y-1} \, d\xi.$$

Consider the change of variable $\xi = s(1+p)$ with $p > -1$ which, replaced in the previous one, provides

$$\Gamma(x+y) = (1+p)^{x+y} \int_0^{\infty} e^{-s(1+p)} s^{x+y-1} \, ds.$$

Multiplying this expression by $p^{x-1}(1+p)^{-x-y}$ and integrating in the variable p we can write

$$\Gamma(x+y) \int_0^\infty p^{x-1}(1+p)^{-x-y}\,\mathrm{d}p = \int_0^\infty e^{-s}s^{x+y-1}\left[\int_0^\infty e^{-ps}p^{x-1}\,\mathrm{d}p\right]\mathrm{d}s.$$

Introducing the change $ps = u$ in the integral between brackets, rearranging and using the definition of gamma function, we obtain

$$\Gamma(x+y) \int_0^\infty \frac{p^{x-1}}{(1+p)^{x+y}}\,\mathrm{d}p = \Gamma(x)\Gamma(y).$$

We will manipulate the remaining integral. For this, we introduce the change of variable $p = t/(1-t)$ from where it follows

$$\Gamma(x)\Gamma(y) = \Gamma(x+y) \int_0^1 t^{x-1}(1-t)^{y-1}\,\mathrm{d}t. \tag{2.12}$$

Considering the particular case $y = x$, we have

$$\Gamma(x)\Gamma(x) = \Gamma(2x) \int_0^1 t^{x-1}(1-t)^{x-1}\,\mathrm{d}t$$

which, with the change of variable $t = (1 + xi)/2$, provides

$$\Gamma(x)\Gamma(x) = 2^{1-2x}\Gamma(2x) \int_{-1}^1 (1-\xi^2)^{x-1}\,\mathrm{d}\xi.$$

Finally, since the resulting integral has an even function in the integraand and the integration interval is symmetric, we introduce the change of variable $\xi^2 = u$, then

$$\Gamma(x)\Gamma(x) = 2^{1-2x}\Gamma(2x) \int_0^1 u^{-1/2}(1-u)^{x-1}\,\mathrm{d}u$$

which, compared to Eq. (2.12), allows to write

$$\Gamma(x)\Gamma(x+1/2) = \sqrt{\pi}2^{1-2x}\Gamma(2x)$$

which is the desired result   ◇

5. Using the definition of Pochhammer symbol, we get

$$(n+1)_m = \frac{\Gamma(n+1+m)}{\Gamma(n+1)} = \frac{(n+m)!}{n!}$$

where the latter equality is due to the fact that $m, n \in \mathbb{N}$.
Rearranging, we can write

$$(n + m)! = n!(n + 1)_m$$

which is the desired result   ◇

6. Using the definition of Pochhammer symbol, we get

$$(a)_{-k} = \frac{\Gamma(a - k)}{\Gamma(a)}.$$

From the reflection formula involving the gamma functions, we can write

$$\Gamma(a - k)\Gamma(1 - a + k) = \frac{\pi}{\sin \pi(a - k)} \quad \text{and} \quad \Gamma(a)\Gamma(1 - a) = \frac{\pi}{\sin \pi(a)}$$

whose quotient allows us to write

$$\frac{\Gamma(a - k)\Gamma(1 - a + k)}{\Gamma(a)\Gamma(1 - a)} = \frac{\sin \pi a}{\sin \pi(a - k)}$$

or using the definition of the Pochhammer symbol as follows

$$(a)_{-k}(1 - a)_k = \frac{\sin \pi a}{\sin \pi(a - k)}. \tag{2.13}$$

Consider the expression for the sine of difference

$$\sin \pi(a - k) = \sin \pi a \cos \pi k - \sin \pi k \cos \pi a = \sin \pi a \, (-1)^k$$

where the second equality is justified, since $k \in \mathbb{N}$. Substituting this result into Eq. (2.13) and rearranging we obtain

$$(a)_{-k} = \frac{(-1)^k}{(1 - a)_k}$$

which is the desired result   ◇

7. Multiplying the two members of the definition of incomplete gamma function by e^x and rearranging, we can write

$$\gamma(a, x) = e^{-x} \int_0^x e^{-(\xi - x)} \xi^{a-1} \, d\xi.$$

Expanding the exponential in a Maclaurin series and simplifying, we have, already inverting the order of the integral with the summation

$$\gamma(a, x) = e^{-x} \sum_{k=0}^{\infty} \frac{(-1)^k}{k!} \int_0^x (\xi - x)^k \xi^{a-1} \, d\xi.$$

Introducing the change of variable $\xi = xt$ in the preceding and rearranging, we have

$$\gamma(a, x) = e^{-x} \sum_{k=0}^{\infty} \frac{x^{k+a}}{k!} \int_0^1 (1 - t)^k t^{a-1} \, dt.$$

Using the result given by Eq. (2.12) we can write

$$\gamma(a, x) = e^{-x} \sum_{k=0}^{\infty} x^{k+a} \frac{\Gamma(a)}{\Gamma(k + a + 1)}.$$

Finally, from the definition of the Pochhammer symbol, we obtain

$$\gamma(a, x) = e^{-x} \sum_{k=0}^{\infty} \frac{x^{a+k}}{(a)_{k+1}}$$

which is the desired result ◇

8. This is a particular case of the previous exercise. Consider $a = 1$ in the definition of the incomplete gamma function

$$\gamma(1, x) = \int_0^x e^{-\xi} \, d\xi$$

whose integration provides directly

$$\gamma(1, x) = 1 - e^{-x}$$

which is the desired result ◇

9. From the definition of the complementary gamma function we obtain

$$\int_0^{\infty} \xi^{a-1} \Gamma(b, \xi) \, d\xi = \int_0^{\infty} \xi^{a-1} \left[\int_{\xi}^{\infty} e^{-t} t^{b-1} \, dt \right] d\xi.$$

Introducing the change of variable $t = \xi u$ in the integral between brackets and rearranging, already changing the order of the integrations, we can write

$$\int_0^{\infty} \xi^{a-1} \Gamma(b, \xi) \, d\xi = \int_1^{\infty} u^{b-1} \, du \int_0^{\infty} e^{-\xi u} \xi^{a+b-1} \, d\xi.$$

From another change of variable $\xi u = t$ and the definition of gamma function, we get

$$\int_0^{\infty} \xi^{a-1} \Gamma(b, \xi) \, d\xi = \Gamma(a + b) \int_1^{\infty} u^{-a-1} \, du.$$

whose integration allows us to write

$$\int_0^\infty \xi^{a-1} \Gamma(b, \xi)\, d\xi = \frac{\Gamma(a+b)}{a}$$

which is the desired result   ◇

10. Introducing the change of variable $\xi^2 = u \Rightarrow d\xi = \frac{u^{-1/2}}{2}\, du$ in the integral of the definition of error function and simplifying, we obtain

$$\mathrm{erf}\,(x) = \frac{1}{\sqrt{\pi}} \int_0^{x^2} e^{-u} u^{-1/2}\, du.$$

Identifying with the parameter of the incomplete gamma function, we conclude that

$$\mathrm{erf}\,(x) = \frac{1}{\sqrt{\pi}} \gamma\left(\frac{1}{2}, x^2\right)$$

which is the desired result   ◇

11. Expanding the exponential, in the definition of the error function, in a Maclaurin series, permuting the order of integration with the summation and simplifying, we have

$$\mathrm{erf}\,(x) = \frac{2}{\sqrt{\pi}} \sum_{k=0}^\infty \frac{(-1)^k}{k!} \int_0^x \xi^{2k}\, d\xi.$$

Integrating and rearranging, we obtain

$$\mathrm{erf}\,(x) = \frac{2}{\sqrt{\pi}} \sum_{k=0}^\infty \frac{(-1)^k}{k!} \frac{x^{2k+1}}{2k+1}$$

which is the desired result   ◇

12. It is a direct consequence of Eq. (2.12). It is noteworthy that another way of showing this result is through the product of two gamma functions, however, not as done in the text, using polar coordinates in the plane   ◇

13. Introducing the change of variable $\xi = \cos^2 \theta \Rightarrow d\xi = -2\cos\theta \sin\theta\, d\theta$ in the definition of beta function and rearranging, we obtain

$$B(p, q) = 2 \int_0^{\pi/2} \cos^{2p-1}\theta \, \sin^{2q-1}\theta \, d\theta$$

which is the desired result   ◇

14. Noting that the extremes are exactly the constants that are found, sometimes summing up by subtracting the integration variable, we introduce the following change of variable $\xi = \eta + (x - \eta)t \Rightarrow d\xi = -\eta\, dt$ in the integral, whence it follows, after rearranging

$$\Omega(\eta, x) = (x - \eta)^{\alpha+\beta-1} \int_0^1 (1 - t)^{\alpha-1} t^{\beta-1} \, dt.$$

The remaining integral in the previous expression is nothing more than the definition of the beta function, according to Eq. (2.12), so we get

$$\Omega(\eta, x) = (x - \eta)^{\alpha+\beta-1} B(\alpha, \beta)$$

which is the desired result   ◇

15. Consider the integral

$$J = \int_0^1 t^{b-1}(1 - t)^{c-b-1}(1 - xt)^{-a} \, dt$$

with $c > b > 0$. With the condition $|x| < 1$ we can expand $(1 - xt)^{-a}$ in a uniformly convergent series, then

$$J = \int_0^1 t^{b-1}(1 - t)^{c-b-1} \left[\sum_{k=0}^{\infty} \frac{\Gamma(a + k)}{\Gamma(a)} \frac{(xt)^k}{k!} \right] dt$$

or, by changing integration with the summation, in the following way

$$J = \frac{1}{\Gamma(a)} \sum_{k=0}^{\infty} \frac{\Gamma(a + k)}{k!} x^k \int_0^1 t^{b-1}(1 - t)^{c-b-1} \, dt.$$

The remaining integral is given in terms of the beta function, that is, by identifying this integral with Eq. (2.12), we can write

$$J = \frac{1}{\Gamma(a)} \sum_{k=0}^{\infty} \frac{\Gamma(a + k)}{k!} B(b + k, c - b) x^k$$

or using the relation between the beta and gamma functions, in the form

$$J = \frac{\Gamma(c - b)\Gamma(b)}{\Gamma(c)} \sum_{k=0}^{\infty} \frac{(a)_k (b)_k}{(c)_k} \frac{x^k}{k!}$$

where $(\cdot)_k$ are the Pochhammer symbols. Using the integral representation for the hypergeometric function, we have

$$_2F_1(a, b; c; x) = \frac{\Gamma(c)}{\Gamma(c - b)\Gamma(b)} \int_0^1 t^{b-1}(1 - t)^{c-b-1}(1 - xt)^{-a} \, dt$$

which is the desired result   ◇

16. First, we introduce the notation

$$\Lambda(a, b, c, x) \equiv \int_0^\infty {}_2F_1(a, b; c; -t)\, t^{-x-1}\, dt.$$

From the integral representation for the hypergeometric function we obtain

$$\Lambda(a, b, c, x) = \frac{\Gamma(c)}{\Gamma(c-b)\Gamma(b)} \int_0^\infty \left[\int_0^1 \xi^{b-1}(1-\xi)^{c-b-1}(1+t\xi)^{-a}\, d\xi \right] t^{-x-1}\, dt.$$

By changing the order of integration and rearranging, we can write

$$\Lambda(a, b, c, x) = \frac{\Gamma(c)}{\Gamma(b-c)\Gamma(b)} \int_0^1 \xi^{b-1}(1-\xi)^{c-b-1}\, d\xi \int_0^\infty \frac{t^{-x-1}}{(1+t\xi)^a}\, dt.$$

Entering the change of variable $t\xi = u$ we have

$$\Lambda(a, b, c, x) = \frac{\Gamma(c)}{\Gamma(b-c)\Gamma(b)} \int_0^1 \xi^{x+b-1}(1-\xi)^{c-b-1}\, d\xi \int_0^\infty \frac{u^{-x-1}}{(1+u)^a}\, du.$$

We evaluate the integral in the variable u. Considering the following change of variable $u = \xi(1+u)$ and the definition of beta function, we obtain

$$\Lambda(a, b, c, x) = \frac{\Gamma(c)\Gamma(-x)\Gamma(a+x)}{\Gamma(b-c)\Gamma(b)\Gamma(a)} \int_0^1 \xi^{x+b-1}(1-\xi)^{c-b-1}\, d\xi.$$

The remaining integral is also calculated from the definition of beta function, then

$$\Lambda(a, b, c, x) = \frac{\Gamma(c)\Gamma(-x)\Gamma(a+x)}{\Gamma(b-c)\Gamma(b)\Gamma(a)} \frac{\Gamma(c-b)\Gamma(b+x)}{\Gamma(c+x)}.$$

Using the definition of the Pochhammer symbol and simplifying it, we have

$$\int_0^\infty {}_2F_1(a, b; c; -t)\, t^{-x-1}\, dt = \frac{(a)_x (b)_x}{(c)_x}\Gamma(-x)$$

which is the desired result   ◇

17. Substituting $a = b = c = 1$ into the previous expression, we obtain

$$\int_0^\infty {}_2F_1(1, 1; 1; -t)\, t^{-x-1}\, dt = \frac{(1)_x (1)_x}{(1)_x}\Gamma(-x)$$

which, by simplifying, provides

$$\int_0^\infty {}_2F_1(1, 1; 1; -t)\, t^{-x-1}\, dt = \Gamma(1+x)\Gamma(-x)$$

which is the desired result   ◇

18. Substituting $x = 1$ into the expression that gives the integral representation we have

$$\,_2F_1(a, b; c; 1) = \frac{\Gamma(c)}{\Gamma(c - b)\Gamma(b)} \int_0^1 t^{b-1}(1 - t)^{c-b-1}(1 - t)^{-a}\, dt,$$

or, in the following way

$$\,_2F_1(a, b, ; c; 1) = \frac{\Gamma(c)}{\Gamma(c - b)\Gamma(b)} \int_0^1 t^{b-1}(1 - t)^{c-b-a-1}\, dt.$$

Identifying the remaining integral with the definition of beta function, we get

$$\,_2F_1(a, b; c; 1) = \frac{\Gamma(c)}{\Gamma(c - b)\Gamma(b)} B(b, c - a - b)$$

which, expressed in terms of gamma functions, allows to write, already simplifying

$$\,_2F_1(a, b; c; 1) = \frac{\Gamma(c)\Gamma(c - a - b)}{\Gamma(c - b)\Gamma(c - a)}$$

which is the desired result   ◇

19. For $x = 1$ we have, directly replacing the relation involving the Jacobi polynomials and the hypergeometric function

$$P_n^{(\alpha,\beta)}(1) = \frac{\Gamma(n + \alpha + 1)}{n!\Gamma(\alpha + 1)}\, {}_2F_1\left(-n, \alpha + \beta + n + 1; \alpha + 1; 0\right).$$

Since ${}_2F_1(a, b; c; 0) = 1$ it follows

$$P_n^{(\alpha,\beta)}(1) = \frac{\Gamma(n + \alpha + 1)}{n!\Gamma(\alpha + 1)} = \frac{(\alpha + 1)_n}{n!}$$

which is the desired result. On the other hand, replacing $x = -1$ we obtain

$$P_n^{(\alpha,\beta)}(-1) = \frac{\Gamma(n + \alpha + 1)}{n!\Gamma(\alpha + 1)}\, {}_2F_1\left(-n, \alpha + \beta + n + 1; \alpha + 1; 1\right).$$

From the result of Exercise (18) and simplifying, we can write

$$P_n^{(\alpha,\beta)}(-1) = \frac{\Gamma(-\beta)}{n!\,\Gamma(-\beta - n)}.$$

Using the formula of reflection for the gamma function we have, already simplifying,

$$P_n^{(\alpha,\beta)}(-1) = \frac{(-1)^n}{n!} \frac{\Gamma(n+\beta+1)}{\Gamma(\beta+1)} = \frac{(-1)^n}{n!}(\beta+1)_n$$

which is the desired result   ◇

20. Deriving from the variable x the expression for $G(x,t)$ we obtain

$$\frac{\partial}{\partial x}G(x,t) = t(1-2xt+t^2)^{-3/2} = \sum_{k=0}^{\infty} P_k'(x)t^k$$

or, in the following way

$$\frac{\partial}{\partial x}G(x,t) = \frac{t}{1-2xt+t^2}G(x,t)$$

which is a first order partial differential equation in the variable x. Using the power series we obtain the identity

$$t\sum_{k=0}^{\infty} P_k(x)t^k = (1-2xt+t^2)\sum_{k=0}^{\infty} P_k'(x)t^k$$

which, using the distributive property, provides

$$\sum_{k=0}^{\infty} P_k(x)t^{k+1} = \sum_{k=0}^{\infty} P_k'(x)t^k - 2x\sum_{k=0}^{\infty} P_k'(x)t^{k+1} + \sum_{k=0}^{\infty} P_k'(x)t^{k+2}.$$

Considering in the first and third sums the change of index $k \to k-1$ and in the fourth sum $k \to k-2$, we can write

$$\sum_{k=1}^{\infty} P_{k-1}(x)t^k = \sum_{k=0}^{\infty} P_k'(x)t^k - 2x\sum_{k=1}^{\infty} P_{k-1}'(x)t^k + \sum_{k=2}^{\infty} P_{k-2}'(x)t^k.$$

Note that all parcels are with the same exponent, t^k, but with different index beginnings. We separate the terms $k=1$ in the first and third sums and $k=0$ and $k=1$ on the second sum so that all sums start at the $k=2$ index. So we have

$$P_0(x)t + \sum_{k=2}^{\infty} P_{k-1}(x)t^k =$$

$$P_0'(x) + P_1'(x)t + \sum_{k=2}^{\infty} P_k'(x)t^k - 2xP_0'(x)t - 2x\sum_{k=2}^{\infty} P_{k-1}'(x)t^k + \sum_{k=2}^{\infty} P_{k-2}'(x)t^k.$$

Then, by equating terms with the same exponent, the identities follow

$$
\begin{aligned}
&\text{(i) Independent } && P_0'(x) = 0 \\
&\text{(ii) Power of } t && P_0(x) = P_1'(x) - 2x P_0'(x) \\
&\text{(iii) } k \geq 2 && P_{k-1}(x) = P_k'(x) - 2x P_{k-1}'(x) + P_{k-2}'(x)
\end{aligned}
$$

Before proceeding, we can conclude that: (i) it follows that $P_0(x)$ must be a constant and (ii) that $P_1(x)$ is a polynomial of degree one. Introducing the change of index $k \to k + 1$ in (iii) we obtain

$$
P_k(x) = P_{k+1}'(x) - 2x P_k'(x) + P_{k-1}'(x)
$$

which is justified for $k \geq 1$ since for $k = 0$ in the previous one it provides exactly (ii), which is the desired result   ◇

21. Substituting $k = 0, 1, 2$ and deriving, according to the following table, we have

$$
\begin{aligned}
k = 0 \quad & P_0(x) = 1 \\
k = 1 \quad & P_1(x) = \frac{1}{2} \cdot \frac{d}{dx}(x^2 - 1) = x \\
k = 2 \quad & P_2(x) = \frac{1}{2^2 \cdot 2} \cdot \frac{d^2}{dx^2}(x^2 - 1)^2 = \frac{1}{2}(3x^2 - 1)
\end{aligned}
$$

Therefore, the first three Legendre polynomials are given by $P_0(x) = 1$, $P_1(x) = x$ and $P_2(x) = (3x^2 - 1)/2$ which is the desired result   ◇

22. Using the integral representation for the gamma function, expanding the exponential as a Maclaurin series and integrating, we obtain

$$
\gamma(\mu, x) = x^\mu \sum_{k=0}^{\infty} \frac{(-1)^k}{k + \mu} \frac{x^k}{k!},
$$

which can be written as follows

$$
\gamma(\mu, x) = x^\mu \sum_{k=0}^{\infty} \frac{\Gamma(k + \mu)}{\Gamma(k + \mu + 1)k!} (-x)^k.
$$

On the other hand, the confluent hypergeometric function is given by

$$
{}_1F_1(a; c; x) = \frac{\Gamma(c)}{\Gamma(a)} \sum_{k=0}^{\infty} \frac{\Gamma(a + k)}{\Gamma(c + k)k!} x^k
$$

which identifies with the expression for the incomplete gamma function, allows to write $a = \mu$ and $c = \mu + 1$, from where follows

$$\gamma(\mu, x) = \frac{\Gamma(\mu)}{\Gamma(\mu + 1)} x^{\mu} {}_1 F_1(\mu; \mu + 1; -x)$$

or after simplification of the gamma functions, in the form

$$\gamma(\mu, x) = \frac{x^{\mu}}{\mu} {}_1 F_1(\mu; \mu + 1; -x),$$

which is the desired result   ◇

23. Consider the integral representation for the confluent hypergeometric function

$$_1 F_1(a; c; x) = \frac{\Gamma(c)}{\Gamma(a)\Gamma(c - a)} \int_0^1 e^{xt} t^{a-1} (1 - t)^{c-a-1} dt \qquad (2.14)$$

satisfied the conditions of existence. By introducing the change of variable $t = 1 - \xi$ in the previous expression, we can write, by rearranging, the expression

$$\Gamma(c - a)\Gamma(a) {}_1 F_1(a; c; x) = e^x \Gamma(c) \int_0^1 e^{-x\xi} \xi^{c-a-1} (1 - \xi)^{a-1} d\xi.$$

Identifying with the integral representation, we have

$$\Gamma(c - a)\Gamma(a) {}_1 F_1(a; c; x) = e^x \Gamma(c) \frac{\Gamma(c - a)\Gamma(a)}{\Gamma(c)} {}_1 F_1(c - a; c; -x)$$

which, by simplifying, provides

$$_1 F_1(a; c; x) = e^x {}_1 F_1(c - a; c; -x),$$

which is the desired result   ◇

24. Introducing the series representation for the confluent hypergeometric function and changing the order of integration with the summation, we obtain

$$\int_0^{\infty} t^{\mu-1} e^{-xt} \sum_{n=0}^{\infty} \frac{(a)_n}{(c)_n} \frac{(kt)^n}{n!} dt = \sum_{n=0}^{\infty} \frac{(a)_n}{(c)_n} \frac{(k)^n}{n!} \int_0^{\infty} t^{\mu+n-1} e^{-xt} dt.$$

By changing the variable $xt = \xi$ into the integral of the second member and rearranging, we have

$$\int_0^{\infty} t^{\mu-1} e^{-xt} \sum_{n=0}^{\infty} \frac{(a)_n}{(c)_n} \frac{(kt)^n}{n!} dt = \sum_{n=0}^{\infty} \frac{(a)_n}{(c)_n} \frac{(k)^n}{n!} \frac{1}{x^{\mu+n}} \int_0^{\infty} \xi^{\mu+n-1} e^{-\xi} d\xi.$$

The resulting integral is nothing more than the gamma function $\Gamma(n + n)$ which, written in terms of the Pochhammer symbol, allows writing, already rearranging

$$\int_0^\infty t^{\mu-1} e^{-xt} \sum_{n=0}^\infty \frac{(a)_n}{(c)_n} \frac{(kt)^n}{n!}\, dt = \Gamma(\mu) x^{-\mu} \sum_{n=0}^\infty \frac{(a)_n}{(c)_n} \frac{(\mu)_n}{n!} \left(\frac{k}{x}\right)^n.$$

Since the summation in the second member is a hypergeometric function, we can write

$$\int_0^\infty t^{\mu-1} e^{-xt} \, {}_1F_1(a; c; kt)\, dt = \Gamma(\mu) x^{-\mu} \, {}_2F_1(a, \mu; c; k/x)$$

which is the desired result   ◇

25. From the series expansion of the Bessel function

$$J_\mu(x) = \sum_{k=0}^\infty \frac{(-1)^k}{\Gamma(\mu+k+1)k!} \left(\frac{x}{2}\right)^{\mu+2k}$$

and the Legendre duplication formula, written adequately in the form

$$\frac{1}{k!} = \frac{1}{\sqrt{\pi}} \frac{2^{2k}}{(2k)!} \Gamma(k+1/2)$$

we can write

$$J_\mu(x) = \frac{1}{\sqrt{\pi}} \left(\frac{x}{2}\right)^\mu \sum_{k=0}^\infty \frac{(-1)^k}{(2k)!} \frac{\Gamma(k+1/2)}{\Gamma(\mu+k+1)} x^{2k}.$$

Multiplying two members by $\Gamma(mu + 1/2)$ and rearranging, we have

$$\Gamma(\mu+1/2) J_\mu(x) = \frac{1}{\sqrt{\pi}} \left(\frac{x}{2}\right)^\mu \sum_{k=0}^\infty \frac{(-1)^k}{(2k)!} \underbrace{\frac{\Gamma(k+1/2)\Gamma(\mu+1/2)}{\Gamma(\mu+k+1)}} x^{2k}.$$

Note that the highlight in the previous equation is the integral that defines the beta function, written in terms of trigonometric functions

$$\Gamma(\mu+1/2)\Gamma(k+1/2) = 2\Gamma(\mu+k+1) \int_0^{\pi/2} (\cos\theta)^{2\mu}(\sin\theta)^{2k}\, d\theta$$

from where it follows, already changing the order of integral with summation and rearranging

$$\Gamma(\mu+1/2) J_\mu(x) = \frac{2}{\sqrt{\pi}} \left(\frac{x}{2}\right)^\mu \int_0^{\pi/2} (\cos\theta)^{2\mu}\, d\theta \sum_{k=0}^\infty \frac{(-1)^k}{(2k)!} (x\sin\theta)^{2k}.$$

The remaining sum is the series for $\cos(x \sin\theta)$ then

$$\Gamma\left(\mu+\frac{1}{2}\right)J_\mu(x)=\frac{2}{\sqrt{\pi}}\left(\frac{x}{2}\right)^\mu\int_0^{\pi/2}\cos(x\,\sin\theta)(\cos\theta)^{2\mu}\,d\theta,$$

which is the desired result   ◇

26. Substituting $k=0,1,2$ and deriving, according to the following table, we have

$$k=0\ \ L_0(x)=1$$
$$k=1\ \ L_1(x)=\frac{e^x}{1!}\cdot\frac{d}{dx}(e^{-x}x)=1-x$$
$$k=2\ \ L_2(x)=\frac{e^x}{2!}\cdot\frac{d^2}{dx^2}(e^{-x}\,x^2)=\frac{1}{2}(x^2-4x+2)$$

Thus, the first three Laguerre polynomials are given by $L_0(x)=1$, $L_1(x)=1-x$ and $L_2(x)=(x^2-4x+2)/2$ which is the desired result   ◇

27. We write the quotient, involving two factorials, $n!/(n-k)!$, which appears in the expansion, as follows

$$L_n(x)=\sum_{k=0}^\infty(-1)^k\frac{n!}{(n-k)!k!}\frac{x^k}{k!}$$

in terms of gamma functions, using the reflection formula, that is, we have

$$n!=-\frac{\pi}{\Gamma(-n)\sin\pi n}\quad\text{and}\quad(n-k)!=-\frac{\pi(-1)^k}{\Gamma(k-n)\sin\pi n}$$

from where we can write

$$L_n(x)=\sum_{k=0}^\infty\frac{(-n)_k}{k!k!}x^k.$$

From the particular Pochhammer symbol $(1)_k$ we have $(1)_k=\dfrac{\Gamma(k+1)}{\Gamma(1)}$, which can be written as, $k!=(1)_k$. Replacing in the series expansion we have

$$L_n(x)=\sum_{k=0}^\infty\frac{(-n)_k}{(1)_k}\frac{x^k}{k!}$$

which, compared with the series expansion for confluent hypergeometric function, leads us to

$$L_n(x)={}_1F_1(-n;1;x)$$

which is the desired result   ◇

28. Deriving the series expansion in relation to x we have

$$\frac{d}{dx}H_n(x) = 2n! \sum_{k=0}^{[n/2]} \frac{(-1)^k}{k!}(n-2k)\frac{(2x)^{n-2k-1}}{(n-2k)!}$$

for $n = 0, 1, 2, \ldots$ Using the relation $n! = n(n-1)!$ we can write

$$\frac{d}{dx}H_n(x) = 2n\,(n-1)! \underbrace{\sum_{k=0}^{[(n-1)/2]} \frac{(-1)^k}{k!}\frac{(2x)^{n-1-2k}}{(n-1-2k)!}}.$$

The expression that is pointed out is nothing more than the power series expansion for the Hermite polynomials of degree $n-1$,

$$\frac{d}{dx}H_n(x) = 2n\,H_{n-1}(x)$$

which is the desired result   ◇

29. Consider Eq. (2.6)

$$\phi(\alpha, \beta; z) = \sum_{k=0}^{\infty} \frac{1}{\Gamma(\alpha k + \beta)}\frac{z^k}{k!}.$$

Deriving both sides in relation to x we have

$$\frac{d}{dx}\phi(\alpha, \beta; x) = \sum_{k=1}^{\infty} \frac{k\,x^{k-1}}{k!\Gamma(\alpha k + \beta)} = \sum_{k=0}^{\infty} \frac{x^k}{k!\Gamma(\alpha k + \alpha + \beta)}$$

where, in the last passage we make the change of index $k \to k+1$.
Deriving once again we can write

$$\frac{d^2}{dx^2}\phi(\alpha, \beta; x) = \sum_{k=1}^{\infty} \frac{k\,x^{k-1}}{k!\Gamma(\alpha k + \alpha + \beta)} = \sum_{k=0}^{\infty} \frac{x^k}{k!\Gamma(\alpha k + 2\alpha + \beta)}$$

from which we conclude that the n-order derivative is given by

$$\frac{d^n}{dx^n}\phi(\alpha, \beta; x) = \sum_{k=0}^{\infty} \frac{x^k}{k!\Gamma(\alpha k + n\alpha + \beta)} = \phi(\alpha, n\alpha + \beta; x)$$

which is the desired result   ◇

30. We begin with the definition, writing only the quotient of the products involving the gamma functions, that is,

$$G_{p,q}^{m,n}(s) = \frac{\prod\limits_{i=1}^{m} \Gamma(b_i - s) \cdot \prod\limits_{j=1}^{n} \Gamma(1 - a_j + s)}{\prod\limits_{i=m+1}^{q} \Gamma(1 - b_j + s) \cdot \prod\limits_{j=n+1}^{p} \Gamma(a_j - s)}.$$

Separating the term a_1 we can write

$$G_{p,q}^{m,n}(s) = \Gamma(1 - a_1 + s) \frac{\prod\limits_{i=1}^{m} \Gamma(b_i - s) \cdot \prod\limits_{j=2}^{n} \Gamma(1 - a_j + s)}{\prod\limits_{i=m+1}^{q} \Gamma(1 - b_j + s) \cdot \prod\limits_{j=n+1}^{p} \Gamma(a_j - s)}.$$

Since $a_1 = b_q$, replacing in the previous one and separating the term b_q in the denominator and rearranging the indexes, we obtain

$$\frac{\Gamma(1 - b_q + s)}{\Gamma(1 - b_q + s)} \frac{\prod\limits_{i=1}^{m} \Gamma(b_i - s) \cdot \prod\limits_{j=1}^{n-1} \Gamma(1 - a_j + s)}{\prod\limits_{i=m+1}^{q-1} \Gamma(1 - b_j + s) \cdot \prod\limits_{j=n}^{p-1} \Gamma(a_j - s)}$$

which, after simplification and returning with the notation, allows to write

$$G_{p,q}^{m,n}\left[z \left| \begin{matrix} a_1, \ldots, a_p \\ b_1, \ldots, b_q \end{matrix} \right.\right] = G_{p-1,q-1}^{m,n-1}\left[z \left| \begin{matrix} a_2, \ldots, a_p \\ b_1, \ldots, b_{q-1} \end{matrix} \right.\right],$$

which is the desired result   ◇

31. From the definition we have $m = 1$ and $n = 2 = p = q$. Substituting such values into the expression for $\mathscr{G}_{p,q}^{m,n}(s)$ and remembering that an 'empty' product is unitary, we get

$$\mathscr{G}_{2,2}^{1,2}(s) = \frac{\Gamma(b_1 - s)\Gamma(1 - a_1 + s)\Gamma(1 - a_2 + s)}{\Gamma(1 - b_2 + s)}.$$

Identifying with the data we have $a_1 = a_2 = 1$, $b_1 = 1$ and $b_2 = 0$, from where it follows

$$\mathscr{G}_{2,2}^{1,2}(s) = \frac{\Gamma(1 - s)\Gamma(s)\Gamma(s)}{\Gamma(1 + s)} = \frac{\Gamma(1 - s)\Gamma(s)}{s}$$

where in the last passage we use the relation $\Gamma(z + 1) = z\Gamma(z)$ and simplify. Note that we can simplify, since $s \neq 0$. Returning to the expression providing $G_{2,2}^{2,1}(s)$ in terms of a Mellin-Barnes integral, we obtain

$$G_{2,2}^{1,2}(s) = \frac{1}{2\pi i} \int_\gamma \frac{\Gamma(1-s)\Gamma(s)}{s} (z)^s \, ds,$$

where the contour γ separates the poles of $\Gamma(1-s)$, leaving them to the left of the vertical axis, of those of $\Gamma(s)$, leaving them to the right of the vertical axis. Using the residue theorem we can write

$$\sum_{k=0}^{\infty} \lim_{s \to k+1} (k-s+1) \left[\frac{\Gamma(1-s)\Gamma(s)}{s} (z)^s \right].$$

Since the limit of the product is the product of the limit and using the result

$$\lim_{s \to k+1} (k-s+1)\Gamma(1-s) = \lim_{s \to k+1} \frac{(k-s+1)(k-s)\cdots(1-s)\Gamma(1-s)}{(k-s)\cdots(1-s)}$$

$$= \lim_{s \to k+1} \frac{\Gamma(k-s+2)}{(k-s)\cdots(1-s)} = \frac{(-1)^k}{k!}.$$

we can write

$$\sum_{k=0}^{\infty} \lim_{s \to k+1} (k-s+1) \left[\frac{\Gamma(1-s)\Gamma(s)}{s} (z)^s \right] = \sum_{k=0}^{\infty} \frac{(-1)^k}{k!} \frac{\Gamma(k+1)}{k+1} (z)^{k+1}.$$

Simplifying and identifying with the series expansion of the logarithm, provides

$$G_{2,2}^{1,2}\left[z \left| \begin{matrix} 1,1 \\ 1,0 \end{matrix} \right. \right] = \ln(1+z),$$

which is the desired result   ◇

32. From the result obtained in the previous exercise

$$\sum_{k=0}^{\infty} \frac{(-1)^k}{k!} \frac{\Gamma(k+1)}{k+1} (z)^{k+1}$$

we write, using the gamma function, $\Gamma(z+1) = z\Gamma(z)$ with $z = k+1$ from where it follows

$$\sum_{k=0}^{\infty} \frac{(-1)^k}{k!} \frac{\Gamma(k+1)}{k+1} (z)^{k+1} = z \sum_{k=0}^{\infty} \frac{\Gamma(k+1)\Gamma(k+1)}{\Gamma(k+2)} \frac{(-z)^k}{k!}.$$

We now use the definition of the Pochhammer symbol $\Gamma(a+k)/\Gamma(a) = (a)_k$, since defined, to write the quotient of gamma functions, then

$$\sum_{k=0}^{\infty} \frac{(-1)^k}{k!} \frac{\Gamma(k+1)}{k+1} (z)^{k+1} = z \sum_{k=0}^{\infty} \frac{(1)_k (1)_k}{(2)_k} \frac{(-z)^k}{k!}.$$

Finally, identifying the previous expression with the series representation of the classical hypergeometric function, we obtain

$$\ln(1+z) \equiv \sum_{k=0}^{\infty} \frac{(-1)^k}{k!} \frac{\Gamma(k+1)}{k+1} (z)^{k+1} = z \, _2F_1(1, 1; 2; -z)$$

which is the desired result   ◇

33. As already mentioned, this Fox's H-function is nothing more than a Meijer's G-function, since $\alpha_i = 1 = \beta_i$, for all $i \in \mathbb{N}$. Then, identifying with the Meijer's G-function, in order to obtain the parameters, we have $m = 1 = p = q$ and $n = 0$, from where follows

$$\mathscr{G}_{1,1}^{1,0}(s) = \frac{\prod\limits_{i=1}^{1} \Gamma(b_i - s) \cdot \prod\limits_{j=1}^{0} \Gamma(1 - a_j + s)}{\prod\limits_{i=2}^{1} \Gamma(1 - j + s) \cdot \prod\limits_{j=1}^{1} \Gamma(a_j - s)} = \frac{\prod\limits_{i=1}^{1} \Gamma(b_i - s)}{\prod\limits_{j=1}^{1} \Gamma(a_j - s)}$$

the second equality being valid, since the other two products are equal to unity. From the previous expression, we conclude that $a_1 = \alpha + \beta + 1$ and $b_1 = \alpha$, which allows to write the following Mellin-Barnes integral

$$H_{1,1}^{1,0}\left[z \left|\begin{array}{c} (\alpha+\beta+1, 1) \\ (\alpha, 1) \end{array}\right.\right] = \frac{1}{2\pi i} \int_\gamma \frac{\Gamma(\alpha - s)}{\Gamma(\alpha+\beta+1 - s)} z^s \, ds$$

whose poles of the integrand are at $\alpha - s = -k$ with $k = 0, 1, 2, \ldots$
In analogy to the previous ones, we will evaluate the sum of the residues at the poles, that is,

$$\sum_{k=0}^{\infty} \operatorname*{Res}_{s=k+\alpha} \left\{ \frac{\Gamma(\alpha - s)}{\Gamma(\alpha+\beta+1 - s)} z^s \right\} = \sum_{k=0}^{\infty} \lim_{s \to k+\alpha} \left\{ (-s + k + \alpha) \frac{\Gamma(\alpha - s)}{\Gamma(\alpha+\beta+1 - s)} z^s \right\}$$

from where it follows, already returning in the expression for the Fox's H-function and rearranging,

$$H_{1,1}^{1,0}\left[z \left|\begin{array}{c} (\alpha+\beta+1, 1) \\ (\alpha, 1) \end{array}\right.\right] = z^\alpha \sum_{k=0}^{\infty} \frac{(-z)^k}{k! \, \Gamma(\beta+1 - k)}.$$

Using the result

$$\sum_{j=0}^{\infty} \frac{\Gamma(q+1)}{j!\Gamma(q+1-k)} x^j = (1+x)^q,$$

we finally get,

$$H_{1,1}^{1,0}\left[z \left| \begin{array}{c} (\alpha+\beta+1, 1) \\ (\alpha, 1) \end{array} \right.\right] = \frac{z^{\alpha}(1-z)^{\beta}}{\Gamma(\beta+1)}$$

which is the desired result   ◇

34. We have, identifying, the parameters: $m = 1 = n$ and $p = 2 = q$, from where follows

$$\mathscr{G}_{2,2}^{1,1}(s) = \frac{\displaystyle\prod_{i=1}^{1}\Gamma(b_i - s) \cdot \prod_{j=1}^{1}\Gamma(1 - a_j + s)}{\displaystyle\prod_{i=2}^{2}\Gamma(1 - j + s) \cdot \prod_{j=2}^{2}\Gamma(a_j - s)} = \frac{\Gamma(b_1 - s)\Gamma(1 - a_1 + s)}{\Gamma(1 - b_2 + s)\Gamma(a_2 - s)}.$$

From this expression we conclude that $a_1 = 0 = b_1$ and $a_2 = 1/2 = b_2$, from where we can write to the integral

$$G_{2,2}^{1,1}\left[x \left| \begin{array}{c} 0, 1/2 \\ 0, 1/2 \end{array} \right.\right] = \frac{1}{2\pi i} \int_{\gamma} \frac{\Gamma(-s)\Gamma(1+s)}{\Gamma(s+1/2)\Gamma(1/2-s)} z^s \, ds,$$

where the contour γ separates the poles of $\Gamma(-s)$ from the poles of the function $\Gamma(1+s)$. Using the residue theorem we can write

$$\sum_{k=0}^{\infty} \operatorname*{Res}_{s=k}\left\{ \frac{\Gamma(-s)\Gamma(1+s)}{\Gamma(s+1/2)\Gamma(1/2-s)} z^s \right\} = \sum_{k=0}^{\infty} \lim_{s\to k}\left\{ (-s+k) \frac{\Gamma(-s)\Gamma(1+s)}{\Gamma(s+1/2)\Gamma(1/2-s)} z^s \right\}$$

which, by evaluating the limit, provides

$$G_{2,2}^{1,1}\left[x \left| \begin{array}{c} 0, 1/2 \\ 0, 1/2 \end{array} \right.\right] = \sum_{k=0}^{\infty} \frac{(-x)^k}{\Gamma(z+1/2)\Gamma(1/2-k)}.$$

In order to calculate this sum, we use the result involving the gamma functions

$$\Gamma\left(\frac{1}{2}+z\right)\Gamma\left(\frac{1}{2}-z\right) = \frac{\pi}{\cos(\pi z)}.$$

So, we can finally write

$$G_{2,2}^{1,1}\left[x \left| \begin{array}{c} 0, 1/2 \\ 0, 1/2 \end{array} \right.\right] = \frac{1}{\pi} \sum_{k=0}^{\infty} x^k = \frac{1/\pi}{1-x},$$

which is the desired result   ◇

35. The parameters are such that $m = n = p = 1$ and $q = 2$, from where follows

$$\mathscr{G}_{1,2}^{1,1}(s) = \frac{\displaystyle\prod_{i=1}^{1}\Gamma(b_i - s) \cdot \prod_{j=1}^{1}\Gamma(1 - a_j + s)}{\displaystyle\prod_{i=2}^{2}\Gamma(1 - j + s) \cdot \prod_{j=2}^{1}\Gamma(a_j - s)} = \frac{\Gamma(b_1 - s)\Gamma(1 - a_1 + s)}{\Gamma(1 - b_2 + s)}$$

where we use the fact that we have an 'empty' product. Comparing with what we are calculating, we conclude that: $a_1 = 1$, $b_1 = \mu$ and $b_2 = 0$, simplifying, we can write for the Mellin-Barnes type integral,

$$G_{1,2}^{1,1}\left[x \left|\begin{matrix}1\\\mu, 0\end{matrix}\right.\right] = \frac{1}{2\pi i}\int_{\mathcal{L}}\frac{\Gamma(\mu - s)}{s}z^s\,ds\,,$$

where the poles are at $s = k + \mu$ with $k = 0, 1, 2, \ldots$ Calculating the sum of the residues, we can write

$$\sum_{k=0}^{\infty}\operatorname*{Res}_{s=\mu+k}\left\{\frac{\Gamma(\mu - s)}{s}z^s\right\} = \sum_{k=0}^{\infty}\lim_{s\to\mu+k}\left\{(-s + \mu + k)\frac{\Gamma(\mu - s)}{s}z^s\right\}$$

or, in the following form

$$G_{1,2}^{1,1}\left[x \left|\begin{matrix}1\\\mu, 0\end{matrix}\right.\right] = x^\mu\sum_{k=0}^{\infty}\frac{(-x)^k}{k!(\mu + k)}$$

which, using the result of Exercise (2), allows us to write, finally

$$\gamma(\mu, x) = G_{1,2}^{1,1}\left[x \left|\begin{matrix}1\\\mu, 0\end{matrix}\right.\right],$$

which is the desired result   ◇

36. Identifying the indices $m = n = p = 1$ and $q = 2$, while the pair $a_1 = a = b_1$, $\alpha_1 = \alpha = \beta_1$, $b_2 = 0$ and $\beta_2 = 1$. We can then write the expression on the first member in terms of an integral of the Mellin-Barnes type

$$\frac{d}{dx}\left\{\frac{1}{2\pi i}\int_{\mathcal{L}}\frac{\Gamma(a + \alpha s)\Gamma(1 - a - \alpha s)}{\Gamma(1 - s)}x^{-s}\,ds\right\},$$

where the integration contour $\mathcal{L}$ separates the poles of $\Gamma(a + \alpha s)$ leaving them on the left, and $\Gamma(1 - a - \alpha s)$, leaving them to the right, these not contributing to the calculation of integral. By calculating the derivative, we obtain

$$\frac{1}{2\pi i} \int_{\mathcal{L}} \frac{\Gamma(a+\alpha s)\Gamma(1-a-\alpha s)}{\Gamma(1-s)}(-s)\, x^{-s-1}\, ds.$$

Using the functional relation $\Gamma(s+1) = s\Gamma(s)$ and effecting the change of variable $s \rightarrow s-1$, we can write

$$\frac{1}{2\pi i} \int_{\mathcal{L}} \frac{\Gamma(a-\alpha+\alpha s)\Gamma(1-a+\alpha-\alpha s)}{\Gamma(1-s)}(-s)\, x^{-s}\, ds.$$

From this expression, we identify the parameters $a_1 = a - \alpha = b_1$, $\alpha_1 = \alpha = \beta_1$, $b_2 = 0$ and $\beta_2 = 1$ which, returning in the representation for the Fox's H-function, provides

$$\frac{d}{dx}\, H_{1,2}^{1,1}\left[x \left|\begin{array}{l}(a,\alpha) \\ (a,\alpha),(0,1)\end{array}\right.\right] = H_{1,2}^{1,1}\left[x \left|\begin{array}{l}(a-\alpha,\alpha) \\ (a-\alpha,\alpha),(0,1)\end{array}\right.\right]$$

which is the desired result   ◇

37. Consider in the complex plane a contour C, oriented in the positive sense, consisting of a circle, centered at the origin and radius ε and two straight segments $\varepsilon < x < \infty$ (first quadrant) and $\infty > x > \varepsilon$ (fourth quadrant), that is, the point $z = 0$ (singularity of the integrand) is surrounded, that is, it is not inside the contour, as in Fig. 2.1. This is the called Hankel contour. We will evaluate the integral

$$\int_C e^{-z} z^\mu\, dz$$

with the condition $\mathrm{Re}\,(\mu) > -1$. This condition can be extended to all μ, except for zero and negative integers.
The integral of ∞ to ε when $\varepsilon \rightarrow 0$, is equal to $-\Gamma(\mu+1)$, since the phase of $z\, e^{2i\pi 0}$ is zero. The integral of ε through ∞ (fourth quadrant) is equal to $e^{2i\pi\mu}\Gamma(\mu+1)$ since, now, the phase is 2π. Now, taking the limit $\varepsilon \rightarrow 0$ we have that the integral in the circumference goes to zero, simply by parameterizing the circumference $z = \varepsilon\, e^{i\theta}$ with $0 < \theta < 2\pi$. Therefore, we can write, already rearranging,

$$\int_C e^{-z} z^\mu\, dz = \int_\infty^0 e^{-x} x^\mu\, dx + \int_0^\infty e^{-x}(e^{2i\pi}x)^\mu\, dx$$

$$= \left(e^{2i\pi\mu} - 1\right) \int_0^\infty e^{-x} x^\mu\, dx.$$

From the previous expression, for $\mu \neq 0$ and $\mu \neq -n$ with $n \in \mathbb{N}$, and using the definition of the gamma function follows an integral representation for the gamma function

$$\Gamma(\mu+1) = \frac{1}{e^{2i\pi\mu} - 1} \int_C e^{-z} z^\mu\, dz.$$

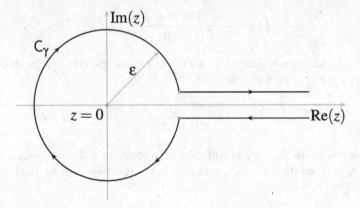

Fig. 2.2 C_γ is the symmetric, with respect to the origin, of the contour C

By introducing the changes $z \to -z$ and $\mu \to -\mu$ in the previous expression and rearranging, we can write

$$\Gamma(1-\mu) = \frac{1}{1-e^{-2i\pi\mu}} \int_{C_\gamma} e^z z^{-\mu} \, dz.$$

The contour C_γ is the symmetric, with respect to the origin, of the contour C, as in Fig. 2.2.

Using the reflection formula (complementarity) and simplifying, we obtain

$$\frac{1}{\Gamma(\mu)} = \frac{1}{2\pi i} \int_{C_\gamma} e^z z^{-\mu} \, dz. \tag{2.15}$$

This integral representation shows that the singularities of the gamma function lie at points such that $\mu = -n$ with $n = 0, 1, 2, \ldots$ which is the desired result $\diamond$

38. We start by denoting the second member for II_M. Substituting the exponential in the integrand for an expansion in power series and changing the order of integration with the summation, we obtain

$$II_M = \frac{(z/2)^\mu}{\Gamma(\mu+1/2)\sqrt{\pi}} \sum_{k=0}^{\infty} \frac{(-z)^k}{k!} \int_{-1}^{1} \xi^k (1-\xi^2)^{\mu-1/2} \, d\xi.$$

In order to calculate the resulting integral, we separate into two others, the first with the extremes of -1 to 0 and the second of 0 to $+1$. In the first one we change the variable $\xi \to -\xi$ from where it follows, already rearranging,

$$II_M = \frac{(z/2)^\mu}{\Gamma(\mu+1/2)\sqrt{\pi}} \sum_{k=0}^{\infty} \frac{1}{2}[1+(-1)^k]\frac{(-z)^k}{k!} \int_{0}^{1} \xi^k (1-\xi^2)^{\mu-1/2} \, d\xi.$$

The term in brackets ensures that only even values for the sum index contribute, hence considering $k \to 2k$ and simplifying, we have

$$II_M = \frac{(z/2)^\mu}{\Gamma(\mu + 1/2)\sqrt{\pi}} \sum_{k=0}^{\infty} \frac{z^{2k}}{(2k)!} \int_0^1 \xi^{2k}(1 - \xi^2)^{\mu - 1/2} \, d\xi.$$

By changing the variable $\xi^2 = t$ into the resulting integral, using the definition of the beta function and the relation between the beta and gamma functions, we obtain

$$II_M = \frac{(z/2)^\mu}{\sqrt{\pi}} \sum_{k=0}^{\infty} \frac{z^{2k}}{(2k)!} \frac{\Gamma(\frac{2k+1}{2})}{\Gamma(k + \mu + 1)}.$$

From the duplication formula for the gamma function

$$\sqrt{\pi}\Gamma(2k + 1) = 2^{2k}\Gamma(k + 1/2)\Gamma(k + 1)$$

and rearranging, we obtain

$$II_M = \sum_{k=0}^{\infty} \frac{(z/2)^{\mu + 2k}}{\Gamma(\mu + k + 1)\, k!},$$

the series expansion of powers of the μ-order first kind modified Bessel function, according to Eq. (6.27), from where follows

$$I_\mu(z) = \frac{(z/2)^\mu}{\Gamma(\mu + 1/2)\sqrt{\pi}} \int_{-1}^1 e^{-z\xi}(1 - \xi^2)^{\mu - 1/2} \, d\xi$$

which is the desired result   ◇

39. Using the definition of the Pochhammer symbol and starting from the first member, we can write

$$\frac{(-\mu)_n}{n!} = \frac{(-\mu)(-\mu + 1) \cdots (-\mu + n - 1)}{n!}.$$

Factoring (-1) in the numerator and writing in terms of the gamma function we have

$$\frac{(-\mu)_n}{n!} = \frac{(-1)^n \, \mu(\mu - 1) \cdots (\mu - n + 1)}{n!} = \frac{(-1)^n \Gamma(\mu + 1)}{n!\Gamma(\mu - n + 1)}$$

which, using the definition of the binomial coefficient, allows to write

$$\frac{(-\mu)_n}{n!} = (-1)^n \binom{\mu}{n},$$

which is the desired result (can be obtained from the reflection formula)   ◇
40. We start by writing the two binomial coefficients as follows:

$$\binom{\alpha}{i} = \frac{(-1)^i \Gamma(i - \alpha)}{i! \Gamma(-\alpha)} \quad \text{and} \quad \binom{\beta}{n-i} = \frac{(-1)^{n-i} \Gamma(n - i - \beta)}{(m - i)! \Gamma(-\beta)}$$

where, in both we made use, besides the definition, of the reflection formula.
Introducing these two results in the expression for the sum and rearranging, we
have

$$\sum_{i=0}^{n} \binom{\alpha}{i} \binom{\beta}{n-i} = \frac{(-1)^n}{\Gamma(-\alpha)\Gamma(-\beta)} \sum_{i=0}^{n} \frac{\Gamma(i - \alpha)\Gamma(n - i - \beta)}{i!(n - i)!}.$$

Using the relation between beta and gamma functions, the product of the two
gamma functions found in the numerator can be expressed in terms of a beta
function, so

$$\sum_{i=0}^{n} \binom{\alpha}{i} \binom{\beta}{n-i} = \frac{(-1)^n \Gamma(n - \alpha - \beta)}{\Gamma(-\alpha)\Gamma(\beta)} \sum_{i=0}^{n} \frac{B(i - \alpha, n - i - \beta)}{i!(n - i)!}$$

which, from the integral representation of the beta function, can be written in
the form

$$\sum_{i=0}^{n} \binom{\alpha}{i} \binom{\beta}{n-i} = \frac{(-1)^n \Gamma(n - \alpha - \beta)}{\Gamma(-\alpha)\Gamma(\beta)} \sum_{i=0}^{n} \frac{1}{i!(n - i)!} \int_0^1 \xi^{i-\alpha-1}(1 - \xi)^{n-i-\beta-1} \, d\xi.$$

Exchanging the order of integration with the sum and rearranging, we obtain

$$\sum_{i=0}^{n} \binom{\alpha}{i} \binom{\beta}{n-i} = \frac{(-1)^n \Gamma(n - \alpha - \beta)}{n! \Gamma(-\alpha)\Gamma(\beta)} \int_0^1 \xi^{-\alpha-1}(1 - \xi)^{n-\beta-1} \Omega(n, \xi) \, d\xi$$

where we introduce the notation

$$\Omega(n, \xi) = \sum_{i=0}^{n} \binom{n}{i} \left(\frac{\xi}{1 - \xi}\right)^i.$$

In order to sum, we use the result involving the sum of the terms of a geometric
progression of first unit term and ratio q, that is, the following expression

$$\sum_{i=0}^{n} \binom{n}{i} q^i = (1 + q)^n.$$

Considering $q = \xi/(1 - \xi)$ in the previous expression, we obtain $\Omega(n, \xi) = (1 - \xi)^{-n}$ where it goes, simplifying

$$\sum_{i=0}^{n} \binom{\alpha}{i}\binom{\beta}{n-i} = \frac{(-1)^n \Gamma(n - \alpha - \beta)}{n! \Gamma(-\alpha)\Gamma(\beta)} \int_0^1 \xi^{-\alpha-1}(1 - \xi)^{-\beta-1}\, d\xi.$$

The remaining integral is exactly the integral representation for the beta function, using the relation between the beta and gamma functions and simplifying, we get

$$\sum_{i=0}^{n} \binom{\alpha}{i}\binom{\beta}{n-i} = \frac{(-1)^n \Gamma(n - \alpha - \beta)}{n! \Gamma(-\alpha - \beta)}.$$

Finally, using the reflection formula, we obtain

$$\sum_{i=0}^{n} \binom{\alpha}{i}\binom{\beta}{n-i} = \binom{\alpha + \beta}{n}$$

which is the desired result   ◇

2.5.4 Proposed Exercises

1. Show that $\Gamma(1/2) = \sqrt{\pi}$.
2. Let $0 < x < 1$. Show that

$$\Gamma(x)\Gamma(1 - x) = \int_0^{\infty} \frac{\xi^{x-1}}{\xi + 1}\, d\xi.$$

3. Use the residue theorem to show that the integral of the Proposed exercise (2) is equal to $\pi/\sin(\pi x)$ in order to obtain the called reflection formula

$$\Gamma(x)\Gamma(1 - x) = \frac{\pi}{\sin \pi x}.$$

4. Let $m, n \in \mathbb{N}$ and $a \neq 0, 1, 2, \ldots$ Show that $(a)_{n+m} = (a)_n (a + n)_m$.
5. Let $n \in \mathbb{N}$. Show that
$$\frac{(2n - 1)!!}{(2n)!!} = \frac{(1/2)_n}{n!}$$

where $(\cdot)!!$ denotes the double factorial.
6. Set $a \neq 0, -1, -2, \ldots$ Show that

$$\Gamma(a, x) = \Gamma(a) - \sum_{k=0}^{\infty} \frac{(-1)^k}{k!} \frac{x^{a+k}}{a+k}.$$

7. Show that $\Gamma(1, x) = e^{-x}$.
8. Show the recurrence relation involving incomplete gamma function

$$\gamma(a + 1, x) = a\,\gamma(a, x) - x^a e^{-x}.$$

9. Show that $\text{erfc}\,(x) = \frac{1}{\sqrt{\pi}} \Gamma\left(\frac{1}{2}, x^2\right).$
10. Using the relation between the complementary error and error function with the respective incomplete gamma and complementary gamma functions, show that

$$\text{erf}\,(x) + \text{erfc}\,(x) = 1.$$

11. Show the symmetry relation $B(p, q) = B(q, p)$.
12. For values of p and q within the domain, show that

$$\frac{B(p + 1, q)}{B(p, q + 1)} = \frac{p}{q}.$$

13. Let $n \in \mathbb{N}$ and $\mu > -1$. Show that the integral

$$\int_{-1}^{1} \xi^{2n+1} (1 - \xi^2)^{\mu}\, d\xi$$

is zero, if n is an odd number and $B(n + 1, \mu + 1)$, if n is an even number.
14. Show that $_2F_1(a, b, c, 0) = 1$.
15. Use the integral representation for the hypergeometric function to show that

$$_2F_1(b; c; x) = (1 - x)^{-a}\, _2F_1\left(a, c - b; c; \frac{x}{x - 1}\right).$$

16. Use the series expansion of the logarithm $\ln(1 + x) = \sum_{k=1}^{\infty} \frac{(-1)^{k-1}}{k} x^k$ valid for
    $|x| < 1$ to show that

$$\ln(1 - x) = -x\, _2F_1(1, 1; 2; x).$$

17. Let $|x| < 1$. Use properties of the Pochhammer symbol to show that

$$_2F_1(a, b + 1; c; x) - _2F_1(a, b; c; x) = \frac{ax}{c}\, _2F_1(a + 1, b + 1; c + 1; x)$$

with the conditions of existence valid.

18. Let $P_n^{(\alpha,\beta)}(x)$ be the Jacobi polynomials. Show that

$$P_n^{(\alpha,\beta)}(-x) = (-1)^n P_n^{(\beta,\alpha)}(x).$$

19. Let $P_k(x)$ be the Legendre polynomials. Show that

$$k P_k(x) = x P_k'(x) - P_{k-1}'(x)$$

for $k \geq 1$.
20. We define the Kronecker delta symbol, denoted by δ_{nk}, through

$$\delta_{nk} = \begin{cases} 1 \text{ if } n = k, \\ 0 \text{ if } n \neq k. \end{cases}$$

Show the orthogonality relation of the Legendre polynomials

$$\int_{-1}^{1} P_n(x) P_k(x) \, dx = \frac{2}{2k+1} \delta_{nk}.$$

21. Show that the error function denoted by $\text{erf}(x)$, is given in terms of the confluent hypergeometric function, by

$$\text{erf}(x) = \frac{2x}{\sqrt{\pi}} \, {}_1F_1\left(\frac{1}{2}; \frac{3}{2}; -x^2\right).$$

22. Show that the relationship between confluent hypergeometric functions

$$\frac{d}{dx} \, {}_1F_1(a; c; x) = \frac{a}{c} \, {}_1F_1(a+1; c+1; x).$$

23. Let $c - a > 0$. Use the symmetry relation ${}_2F_1(a, b; c; x) = {}_2F_1(b, a; c; x)$ from the integral representation of the hypergeometric function, to obtain

$$ {}_1F_1(a; c; x) = \frac{\Gamma(c)}{\Gamma(a)\Gamma(c-a)} \int_0^1 e^{xt} t^{a-1}(1-t)^{c-a-1} dt$$

that is, an integral representation for confluent hypergeometric function.
24. Let $x - k > 0$ and $\mu > 0$. Show that

$$\int_0^\infty t^{\mu-1} e^{-xt} \, {}_1F_1(a; \mu; kt) \, dt = \Gamma(\mu) x^{-\mu} \left(1 - \frac{k}{x}\right)^{-a}.$$

25. Evaluate $_1F_1(a; c; 0)$ and $\lim_{b \to \infty} {}_2F_1\left(a, b; a; \dfrac{xt}{b}\right)$ being $_1F_1(a; c; x)$ and $_2F_1(a, b; c; x)$ the confluent hypergeometric function and hypergeometric function, respectively.

26. The Bessel function of order μ, denoted by $J_\mu(\cdot)$, is given by the expansion

$$J_\mu(x) = \sum_{k=0}^{\infty} \frac{(-1)^k}{\Gamma(\mu + k + 1)k!} \left(\frac{x}{2}\right)^{\mu + 2k}.$$

Use this expansion to show the following recurrence relation

$$J_{k-1}(x) = \frac{k}{x} J_k(x) + J_k'(x).$$

27. Get the integral representation for the zero order Bessel function

$$J_0(x) = \frac{2}{\pi} \int_0^{\pi/2} \cos(x \sin \theta) \, d\theta.$$

28. Show the orthogonality relation of the Laguerre polynomials

$$\int_0^{\infty} e^{-x} L_m(x) L_n(x) \, dx = \delta_{mn}$$

where δ_{mn} is the Kronecker symbol.

29. Show that it is worth the recurrence relation for the Laguerre polynomials

$$L_n(x) = \frac{d}{dx} \left[L_n(x) - L_{n+1}(x) \right]$$

with $n = 0, 1, 2, \ldots$

30. Let $H_n(x)$ be the Hermite polynomials of degree n. Show that

$$H_n(x) = \frac{n!(-1)^{-n/2}}{\left(\frac{n}{2}\right)!} \, {}_1F_1\left(-\frac{n}{2}; \frac{1}{2}; x^2\right) \qquad n = 0, 2, 4, \ldots$$

and

$$H_n(x) = 2x \frac{n!(-1)^{(1-n)/2}}{\left(\frac{n-1}{2}\right)!} \, {}_1F_1\left(-\frac{n-1}{2}; \frac{3}{2}; x^2\right) \qquad n = 1, 3, 5, \ldots$$

where $_1F_1(a; c; x)$ are the confluent hypergeometric functions. Verify that both satisfy the called Hermite equation

$$\frac{d^2}{dx^2} H_n(x) - 2x \frac{d}{dx} H_n(x) + 2n H_n(x) = 0$$

with $n = 0, 1, 2, \ldots$

31. Another way to introduce the Hermite polynomials is by means of the Rodrigues formula

$$H_n(x) = (-1)^n e^{x^2} \frac{d^n}{dx^n} \left(e^{-x^2} \right)$$

with $n = 0, 1, 2, \ldots$ Get the first three Hermite polynomials.

32. Let $|z| < 1$ and $\mathrm{Re}(\mu) > 0$. Show that

$$G_{1,1}^{1,1} \left[-z \left| \begin{matrix} 1 - \mu \\ 0 \end{matrix} \right. \right] = \Gamma(\mu) (1 + z)^{-\mu}.$$

33. Since the parameters are as defined in the text and $\mu \in \mathbb{R}$, show that

$$z^\mu H_{p,q}^{m,n} \left[z \left| \begin{matrix} (a_1, \alpha_1), \ldots, (a_p, \alpha_p) \\ (b_1, \beta_1), \ldots, (b_q, \beta_q) \end{matrix} \right. \right] = H_{p,q}^{m,n} \left[z \left| \begin{matrix} (a_1 + \mu \alpha_1, \alpha_1), \ldots, (a_p + \mu \alpha_p, \alpha_p) \\ (b_1 + \mu \beta_1, \beta_1), \ldots, (b_q + \mu \beta_q, \beta_q) \end{matrix} \right. \right].$$

34. Show that

$$G_{2,2}^{2,2} \left[x \left| \begin{matrix} 0, \mu \\ 0, \mu \end{matrix} \right. \right] = \frac{\pi}{\sin(\mu \pi)} \frac{x^\mu - 1}{x - 1}.$$

imposing the necessary conditions for the parameter μ.

35. Let $\mu \neq \mathbb{N}$. Show that

$$e^x = \frac{\pi}{\sin(\pi \mu)} G_{1,2}^{1,0} \left[x \left| \begin{matrix} 1 - \mu \\ 0, 1 - \mu \end{matrix} \right. \right].$$

36. Let $\mathrm{erf}(\cdot)$ be the error function. Show that

$$\mathrm{erf}(\sqrt{x}) = \frac{1}{\sqrt{\pi}} G_{1,2}^{1,1} \left[x \left| \begin{matrix} 1 \\ 1/2, 0 \end{matrix} \right. \right].$$

37. With due restrictions on the parameters, show that

$$H_{1,2}^{1,1} \left[z \left| \begin{matrix} (1 - a, 1) \\ (0, 1), (1 - c, 1) \end{matrix} \right. \right] = \frac{\Gamma(a)}{\Gamma(c)} \, {}_1F_1(a; c; -z),$$

where ${}_1F_1(a; c; -z)$ is the confluent hypergeometric function.

38. Use the integral representation for the hypergeometric function and an adequate variable change to show that

$$_2F_1(a, b; c; x) = (1 - x)^{-a} \, {}_2F_1 \left(a, c - b; c; \frac{x}{x - 1} \right).$$

39. From the series and integral representations related to the hypergeometric function, show that

$$(1 - t)^{b-c}(1 - t + xt)^{-b} = \sum_{k=0}^{\infty} \frac{(c)_k}{k!} \, {}_2F_1(-k, b; c; x) \, t^k$$

with c different from an integer and $(c)_k$ the Pochhammer symbol.

40. Use the reflection formula to show that

$$\Gamma\left(\frac{1}{2} + x\right) \Gamma\left(\frac{1}{2} - x\right) = \frac{\pi}{\cos \pi x}$$

explaining their respective validity domain.

41. Let $\mu \in \mathbb{R}_+$ and $x \neq -1$. Show that

$$G_{1,1}^{1,1}\left[x \left| \begin{matrix} 1 - \mu \\ 0 \end{matrix} \right. \right] = \frac{\Gamma(\mu)}{(1 + x)^{\mu}},$$

with $G_{1,1}^{1,1}\left[\cdot \left| \begin{matrix} \cdot \\ \cdot \end{matrix} \right. \right]$ a Meijer's function.

42. Let $\mu \in \mathbb{R}_+$ and $x \neq 1$. Show that

$$G_{2,2}^{1,1}\left[x \left| \begin{matrix} 0, \, 1/2 \\ 0, \, 1/2 \end{matrix} \right. \right] = \frac{1}{1 - x},$$

with $G_{2,2}^{1,1}\left[\cdot \left| \begin{matrix} \cdot \\ \cdot \end{matrix} \right. \right]$ a Meijer's function.

References

1. Capelas de Oliveira, E.: Special Functions and Applications. (in Portuguese), 2nd edn. Livraria Editora da Física, São Paulo (2012)
2. Gorenflo, R., Kilbas, A.A., Mainardi, F., Rogosin, S.V.: Mittag-Leffler Functions, Related Topics and Applications. Springer, Heidelberg (2014)
3. Kilbas, A.A.: Fractional calculus on the generalized Wright functions. Fract. Calc. Appl. Anal. **8**, 113–126 (2005)
4. Boersma, J.: On a function which is a special case of Meijer's G-functions. Compositio Math. **15**, 34–63 (1962–1964)
5. Mathai, A.M.: A Handbook of Generalized Special Functions for Statistical and Physical Sciences. Oxford University Press, Oxford (1993)
6. Luke, Y.L.: The Special Functions and Their Applications. Academic Press, New York (1969)
7. Bateman, H., Erdélyi, A., Magnus, W., Oberhettinger, F., Tricomi, F.G.: Table of Integral Transform. McGraw-Hill, New York (1954)
8. Prudnikov, A.P., Brychkov, Y.A., Marichev, O.I.: Integral and Series (Direct Laplace Transforms). Gordon and Breach Science Publishers, New York (1992)

9. Prudnikov, A.P., Brychkov, Y.A., Marichev, O.I.: Integral and Series (Inverse Laplace Transforms). Gordon and Breach Science Publishers, New York (1992)
10. El-Shahed, M., Salem, A.: An extension of Wright functions and its properties. J. Math. **2015**, Article ID 950728 (2015)
11. Karp, D., Prilepkina, E.: Hypergeometric differential equation and new identities for the coefficients of Nørlund and Bühring. Symmetry Integr. Geom. Meth. Appl. (SIGMA) **12**, 52–75 (2016)
12. Fox, C.: The G and H functions as symmetrical Fourier kernels. Trans. Am. Math. Soc. **98**, 395–429 (1961)
13. Kilbas, A.A., Saigo, M.: H-Transform Theory and Applications. Chapman and Hall/CRC, Boca Raton (2004)

Chapter 3
Mittag-Leffler Functions

*Just as integer order calculus has several special functions
associated with it, fractional calculus also admits a vast class of
functions. In ordinary calculus, the solution of an ordinary,
linear differential equation with constant coefficients is given in
terms of exponential functions, whereas in fractional calculus
the analogue is the Mittag-Leffler function, which generalizes the
exponential function. The classical Mittag-Leffler function,
which contains a single parameter and is known as the queen of
the functions of fractional calculus, is for fractional calculus
just what the exponential function is for integer order calculus.*

In the previous chapter, we presented the classic hypergeometric functions that constitute the functions associated with the integer order calculus, in particular, a generalization of the factorial concept by the gamma function. In a similar way, we can understand why fractional calculus is an important tool for refining the description of many natural phenomena, particularly those with temporal dependence, where the called memory effect becomes present, from the understanding of form that the functions related to it generalize the functions relative to the integer order calculus.

The resolution of an ordinary, linear differential equation with constant coefficients has its solution given in terms of the exponential function. On the other hand, a fractional differential equation with constant coefficients has in many cases its solution given in terms of the called Mittag-Leffler function. In this sense, we can say that the Mittag-Leffler function is the fractional generalization of the exponential function.

In this way, in understanding the way the Mittag-Leffler function generalizes the exponential function we can, in a way, understand why, in some cases, a non-integer differential equation gives a more adequate description of a given phenomenon that

© Springer Nature Switzerland AG 2019
E. Capelas de Oliveira, *Solved Exercises in Fractional
Calculus*, Studies in Systems, Decision and Control 240,
https://doi.org/10.1007/978-3-030-20524-9_3

the respective integer order differential equation, obtained as a limiting case with respect to the order of the non-integer differential equation.

Just as the development of integer-order calculus depends intrinsically on the knowledge of the functions related to it, its characteristics and properties, the development of the fractional calculus is directly related to the knowledge of the functions associated with it. Then, before moving on to the study of applications, presented from fractional differential equations, that is, to the applications of the fractional calculus, we present the classical Mittag-Leffler function, as well as extensions of it involving more than one parameter, one of the most important and common functions related to non-integer order calculus. We present the definition, some possible generalizations, properties and relations with other special functions, in particular, the Mittag-Leffler function with three parameters is obtained as a particular case of the Fox's H-function.

The systematic study of the Mittag-Leffler function and its extensions is the main objective of this chapter, which is arranged as follows: after defining the classical Mittag-Leffler function and presenting a possible geometric interpretation, we introduced the Mittag-Leffler function with two parameters, also know as Wiman function, due to Agarwal, as well as the Mittag-Leffler function with three parameters, as proposed by Prabhakar. We present an important relation involving the Mittag-Leffler function with three parameters and the Mittag-Leffler function derivative with two parameters. We conclude the chapter by presenting the Wright and Mainardi functions, as well as possible variations and particular cases of such functions.

3.1 Mittag-Leffler Functions

The function introduced by Mittag-Leffler in 1903, containing one parameter, and which today bears its own name, can be considered a generalization of the exponential function, since it is reduced when the parameter is unitary. Although it has been defined for complex arguments, in this book we only discuss the case of a real variable. Here, we present and discuss only the Mittag-Leffler function introduced by itself and the Mittag-Leffler function with two and three parameters.

Definition 3.1.1 (*The Mittag-Leffler function*) We define the Mittag-Leffler function, also know as classical Mittag-Leffler function, denoted by $E_\alpha(x)$, dependent on a parameter α, with $\mathrm{Re}(\alpha) > 0$, from the power series

$$E_\alpha(x) := \sum_{k=0}^{\infty} \frac{x^k}{\Gamma(\alpha k + 1)}. \tag{3.1}$$

An alternative to this definition is

$$E_\alpha(x^\alpha) := \sum_{k=0}^{\infty} \frac{x^{\alpha k}}{\Gamma(\alpha k + 1)} \tag{3.2}$$

because, as we have already mentioned, both $\alpha = 1$, recover the exponential function and therefore can be considered as fractional generalizations of the exponential function [1].

Although there are several generalizations of the Mittag-Leffler function involving the number of parameters as the number of independent variables [2, 3], but in this book, we introduced only the Mittag-Leffler function with two parameters, due to Agarwal [4] and the Mittag-Leffler function with three parameters, due to Prabhakar [5], as well as the relation involving the integer-order derivative of the Mittag-Leffler function with two parameters and the Mittag-Leffler function with three parameters.

Definition 3.1.2 (*Mittag-Leffler function with two parameters*) Let $x \in \mathbb{C}$ and $\alpha, \beta \in \mathbb{C}$, with $\text{Re}(\alpha) > 0$ and $\text{Re}(\beta) > 0$ two parameters. We define the Mittag-Leffler function with two parameters from the power series [4]

$$E_{\alpha,\beta}(x) := \sum_{k=0} \frac{x^k}{\Gamma(\alpha k + \beta)}. \tag{3.3}$$

This Mittag-Leffler function with two parameters generalizes the function given by Eq. (3.1), since for $\beta = 1$ we have $E_{\alpha,1}(x) = E_\alpha(x)$.

Definition 3.1.3 (*Mittag-Leffler function with three-parameter*) Let $x \in \mathbb{C}$ and α, β, $\rho \in \mathbb{R}$, with $\text{Re}(\alpha) > 0$, $\text{Re}(\beta) > 0$ and $\text{Re}(\rho) > 0$ three parameters. We define the Mittag-Leffler function with three parameters from the power series [5]

$$E_{\alpha,\beta}^\rho(x) := \sum_{k=0} \frac{(\rho)_k}{\Gamma(\alpha k + \beta)} \frac{x^k}{k!}, \tag{3.4}$$

with $(\rho)_k$ the Pochhammer symbol.

Note that, in analogy to the Mittag-Leffler function with two parameters, this function generalizes the function given by Eq. (3.3), since for $\rho = 1$ we have $E_{\alpha,\beta}^1(x) = E_{\alpha,\beta}(x)$.

Let us, through a proposition, justify the convenience of working with the Mittag-Leffler function with three parameters, $E_{\alpha,\beta}^k(x)$, instead of the k-th derivative of the Mittag-Leffler function with two parameters.

Proposition 3.1.1 (Relationship between two Mittag-Leffler functions) *Let $x \in \mathbb{C}$, $\alpha, \beta \in \mathbb{C}$, with $\text{Re}(\alpha) > 0$ and $\text{Re}(\beta) > 0$ two parameters and $k = 0, 1, 2, \ldots$ It is valid a relation involving Mittag-Leffler functions with two and three parameters [6]*

$$\frac{d^k}{dx^k} E_{\alpha,\beta}(x) = k! \, E_{\alpha,\beta+\alpha k}^{k+1}(x).$$

We then introduce alternative forms involving a Mittag-Leffler function, that is, some functions that were introduced, associated with a specific situation, but are related to a particular Mittag-Leffler function, as we will show.

Definition 3.1.4 (*Miller-Ross function*) Let $x \in \mathbb{R}$. We define the Miller-Ross function, as introduced by the authors [7], denoted by $E_x(\nu, \alpha)$, by means of the series

$$E_x(\nu, \alpha) = x^\nu \sum_{k=0}^{\infty} \frac{(ax)^k}{\Gamma(\nu + k + 1)}$$

with $\mathrm{Re}(\nu) > -1$ and $a \in \mathbb{R}$.

Definition 3.1.5 (*Rabotnov function*) Let $x \in \mathbb{R}$. We define the Rabotnov function, as introduced by the author [8], denoted by $R_\alpha(\beta, x)$, by means of the series

$$R_\alpha(\beta, x) = x^\alpha \sum_{k=0}^{\infty} \frac{\beta^k x^{k(\alpha+1)}}{\Gamma((1 + \alpha)(k + 1))}$$

with $\mathrm{Re}(\alpha) > -1$ and $\beta \in \mathbb{R}$.

3.2 Wright and Mainardi Functions

The Wright function [9, 10] is also a particular case of Fox's H-function, which can be written in terms of the hypergeometric function with p terms in the numerator and q terms in the denominator, but in this book we are interested only in a specific Wright function, as well as in one particular case the called Mainardi functions. Further, we use the same notation as proposed by Mainardi [11].

Definition 3.2.1 (*Wright function*) Let $x \in \mathbb{R}$ and $\alpha, \beta \in \mathbb{C}$ with $\mathrm{Re}(\alpha > 0)$ and $\mathrm{Re}(\beta > 0)$. We define the Wright function, denoted by $W(x; \alpha, \beta)$, in terms of the power series

$$W(x; \alpha, \beta) = \sum_{k=0}^{\infty} \frac{1}{\Gamma(\alpha k + \beta)} \frac{x^k}{k!}. \tag{3.5}$$

Definition 3.2.2 (*Mainardi function*) Let $x \in \mathbb{R}$ and $\alpha \in \mathbb{C}$ with $\mathrm{Re}(\alpha > 0)$. We define the Mainardi function, denoted by $M(x; \alpha)$, in terms of the power series

$$W(-x; -\alpha, 1 - \alpha) \equiv M(x; \alpha) = \sum_{k=0}^{\infty} \frac{(-1)^k}{\Gamma(\alpha k + \beta)} \frac{x^k}{k!}. \tag{3.6}$$

It is important to note that, the classical Wright function is an entire function only if $\alpha > -1$ [12]. Also, this function with negative first parameter enters in the fundamental solution of the time fractional diffusion-wave equations [13, 14].

As we have already mentioned, the Mittag-Leffler functions play a prominent role in the fractional calculus. After introducing an appropriate tool to discuss partial differential equations, we will explain a relation involving the Wright function and the Mittag-Leffler function with two parameters, using the Laplace transform, in Chap. 4.

3.3 Exercises

We separate this section into four subsections. The first one contains only the statements of the exercises (Exercise list), the second account with a suggestion (Suggestions) for the respective solution, while the third contains the resolutions (Solutions) themselves. The fourth subsection presents exercises (Proposed exercises) to be solved by the students, most of them similar to those discussed in the text.

3.3.1 Exercise List

1. Show that $E_{1,2}(x) = 1 + x E_{1,3}(x)$.
2. Let $\alpha \in \mathbb{R}_+$ and $|x| < 1$. Since $E_\alpha(x)$ is the classical Mittag-Leffler function, discuss the limit case $\alpha \to 0$.
3. Show that $E_1(\pm x)$ is an exponential function.
4. Let $x \in \mathbb{R}$. Evaluate $x E_{2,2}(-x^2)$.
5. Let $0 < \alpha \leq 1, 0 < \beta \leq 1$ and $0 < \rho \leq 1$. Show that

$$H_{1,2}^{1,1}\left[-x \left| \begin{matrix} (1-\rho, 1) \\ (0, 1), (1-\beta, \alpha) \end{matrix} \right.\right] = \Gamma(\rho) E_{\alpha,\beta}^\rho(x)$$

where $E_{\alpha,\beta}^\rho(\cdot)$ is a Mittag-Leffler function with three parameters.
6. Using the previous result, show that for $\alpha = 1$, it is worth the relation

$$\Gamma(\beta) E_{1,\beta}^\rho(x) = {}_1F_1(\rho; \beta; x)$$

where ${}_1F_1(\rho; \beta; x)$ is a confluent hypergeometric function.
7. Let α and β be real positive numbers. Show that

$$G_{1,1}^{1,1}\left[x^\alpha \left| \begin{matrix} \beta/\alpha \\ \beta/\alpha \end{matrix} \right.\right] = \frac{x^\beta}{1+x^\alpha}.$$

8. Let $\alpha > 0$ and $x \in \mathbb{R}$. Show the called duplication formula for the classical Mittag-Leffler function

$$\frac{1}{2}\left[E_\alpha(\sqrt{x}) + E_\alpha(-\sqrt{x})\right] = E_{2\alpha}(x).$$

9. Let $\alpha > 0$ and $0 < \beta \leq 1$. Show that it is worth the following integral representation for the Mittag-Leffler function with two parameters

$$E_{\alpha,\beta+1}(x) = \frac{1}{\Gamma(\beta)} \int_0^1 (1-\xi)^{\beta-1} E_\alpha(x\xi^\alpha)\, d\xi.$$

10. Let $|x| < 1$ and $\alpha > 0$. Show the following equalities

$$\int_0^\infty e^{-t} t^{\beta-1} E_{\alpha,\beta}(xt^\alpha)\, dt = \int_0^\infty e^{-t} E_\alpha(xt^\alpha)\, dt = \frac{1}{1-x}.$$

11. Let $\alpha > 0$. Get the Mellin-Barnes integral representation for the classical Mittag-Leffler function,

$$E_\alpha(x) = \frac{1}{2\pi i} \int_{\mathcal{L}} \frac{\Gamma(s)\Gamma(1-s)}{\Gamma(1-\alpha s)} (-x)^{-s}\, ds.$$

The contour $\mathcal{L}$, as in Fig. 3.1, is a straight line, parallel to the vertical axis, starting at $c - i\infty$ and ending at $c + i\infty$, such that the constant $0 < c < 1$, thus leaving all poles $s = 0, -1, -2, \ldots$ of $\Gamma(s)$ to the left and all poles $s = 1, 2, 3, \ldots$ of $\Gamma(1-s)$ to the right. So, only the former contribute to the sum of the residues.

12. Let $\alpha > 0$. Show that

$$1 + \frac{1}{\Gamma(\alpha)} \int_0^x \frac{E_\alpha(\xi^\alpha)}{(x-\xi)^{1-\alpha}}\, d\xi = E_\alpha(x^\alpha).$$

13. Let $\alpha, \beta, \gamma > 1$ and $x \neq y$. Show that

$$\int_0^1 t^{\gamma-1} E_{\alpha,\gamma}(xt^\alpha)(1-t)^{\beta-1} E_{\alpha,\beta}(y(1-t)^\alpha)\, dt = \frac{x E_{\alpha,\gamma+\beta}(x) - y E_{\alpha,\gamma+\beta}(y)}{x-y}.$$

This expression is similar to the well-known Christofell-Darboux formula [15] associated with orthogonal polynomials.

14. Use the previous one to discuss the case $y = x$.

15. Let $\alpha > 0$, $\beta > 0$, $\text{Re}(s) > 0$ and $s > |a|^{\frac{1}{\beta}}$. Show that

$$\int_0^\infty t^{\beta-1} e^{-st} E_{\alpha,\beta}^\rho(at^\alpha)\, dt = s^{-\beta}(1-as^{-\alpha})^{-\rho}.$$

This integral can be interpreted as the Laplace transform (Chap. 4) of the function $t^{\beta-1} E_{\alpha,\beta}^\rho(at^\alpha)$, being s the parameter of the Laplace transform.

16. Let $\alpha > 0$ and $\beta > 0$. Show that

$$\left(\frac{d}{dx}\right)^k E_{\alpha,\beta}^{\rho}(x) = (\rho)_k \, E_{\alpha,\beta+\alpha k}^{\rho+k}(x),$$

being $E_{\alpha,\beta}^{\rho}(\cdot)$ Mittag-Leffler functions with three parameters and $(\rho)_k$ the Pochhammer symbol.

17. Let $z \in \mathbb{C}$ and $0 < \alpha < 1$. Show that

$$E_\alpha(z) = \frac{1}{2\pi i} \int_{C_\gamma} \frac{e^\xi \xi^{\alpha-1}}{\xi^\alpha - z} \, d\xi$$

where C_γ is the contour given as in Eq. (2.15).

18. Let $k \in \mathbb{N}$ and $E_k(\cdot)$ the classical Mittag-Leffler function. Show that

$$\left(\frac{d}{dx}\right)^k E_k(x^k) = E_k(x^k).$$

19. Let $\alpha > 0$, $\beta > 0$, $\gamma > 0$, $x > 0$ and $a \in \mathbb{R}$. Show that

$$\int_0^x t^{\beta-1} E_{\alpha,\beta}^{\gamma}(at^\alpha) \, dt = x^\beta E_{\alpha,\beta+1}^{\gamma}(at^\alpha),$$

where $E_{\alpha,\beta}^{\gamma}(\cdot)$ is the Mittag-Leffler function with three parameters.

20. With the data from the previous exercise, discuss the case $a = -1$, $\alpha = 2$ and $\beta = 1 = \gamma$.

21. Express the derivative of order one of $E_\alpha(x^\alpha)$, with $\alpha > 0$ and $x > 0$, as a Mittag-Leffler function. Discuss the case $\alpha = 1$.

22. Let $\alpha > 0$. Show that it's worth the

$$E_\alpha(-x) = E_{2\alpha}(x^2) - x E_{2\alpha,\alpha+1}(x^2)$$

for the classical Mittag-Leffler function.

23. Let $x > 0$ and $\alpha, \beta, \gamma > 0$. Show that

$$\frac{1}{\Gamma(\gamma)} \int_0^x (x-\xi)^{\gamma-1} \xi^{\beta-1} E_{\alpha,\beta}(\xi^\alpha) \, d\xi = x^{\beta+\gamma-1} E_{\alpha,\beta+\gamma}(x^\alpha).$$

24. Let $\alpha > 0$. Show that

$$\frac{1}{\Gamma(\alpha)} \int_0^x (x-\xi)^{\alpha-1} E_{2\alpha}(\xi^{2\alpha}) \, d\xi = E_\alpha(x^\alpha) - E_{2\alpha}(x^{2\alpha}).$$

25. Let $\alpha > 0$, $\beta > 0$ and $x \neq 0$. Show the extension to negative values of the first parameter in the Mittag-Leffler function with two parameters

$$E_{-\alpha,\beta}(x) = \frac{1}{\Gamma(\beta)} - E_{\alpha,\beta}\left(\frac{1}{x}\right) \cdot \cdot$$

26. Let $x \neq 0$. Show that

$$E_{-2,1}\left(-\frac{1}{x^2}\right) = 1 - \cos x \cdot$$

27. Let $\beta > \alpha > 0$. Show that

$$E_{\alpha,\beta}(-x) = \frac{1}{\alpha\Gamma(\beta - \alpha)} \int_0^1 \left(1 - \xi^{\frac{1}{\alpha}}\right)^{\beta - \alpha - 1} E_{\alpha,\alpha}(-x\xi)\,d\xi. \qquad (3.7)$$

28. Let $\rho > 1$, $\beta > \alpha > 0$ and $x \in \mathbb{R}$. Show that the relation involving the Mittag-Leffler function with three parameters is valid

$$E_{\alpha,\beta-\alpha}^{\rho}(x) - E_{\alpha,\beta-\alpha}^{\rho-1}(x) = x E_{\alpha,\beta}^{\rho}(x).$$

29. Let $\alpha, \beta, \rho \in \mathbb{R}_+$. Show that

$$\int_0^x \xi^{\beta-1} E_{\alpha,\beta}^{\rho}(\lambda\xi^{\alpha})\,d\xi = x^{\beta} E_{\alpha,\beta+1}^{\rho}(\lambda x^{\alpha})$$

with λ a positive parameter.

30. Let us denote the classical Laguerre polynomials by $L_k(x)$. Show the relation

$$E_{1,1}^{-k}(x) = L_k(x)$$

with $k = 0, 1, 2, \ldots$

31. Let λ a positive constant. Show that

$$\frac{d^k}{dx^k}\left[x^{n-1} E_{k,n}(\lambda x^k)\right] = \lambda\, x^{n-1} E_{k,n}(\lambda x^k)$$

with $1 \leq n \leq k$.

32. Let $\alpha, \beta, \rho \in \mathbb{C}$, all with positive real parts. Show that

$$\left(x\frac{d}{dx} + \rho\right) E_{\alpha,\beta}^{\rho}(x) = \rho\, E_{\alpha,\beta}^{\rho+1}(x),$$

a relation involving the derivative of a Mittag-Leffler function with three parameters.

33. The generalized Pochhammer symbol, denoted by $(\rho)_{qk}$, is defined by

$$(\rho)_{qk} =: \frac{\Gamma(\rho + qk)}{\Gamma(\rho)}$$

being $\rho \in \mathbb{C}$ and $q \in (0, 1) \cup \mathbb{N}$. Show that

$$\frac{1}{(\delta)_{pk}} - \frac{1}{(\delta - 1)_{pk}} = \frac{\Gamma(\delta)\frac{pk}{1-\delta}}{\Gamma(\delta + pk)},$$

with $\delta \in \mathbb{C}$, $\delta \neq 1$ and $p \in (0, 1) \cup \mathbb{N}$.

34. We define the Mittag-Leffler function with six parameters [16] by means of the series

$$E_{\alpha,\beta,p}^{\rho,\delta,q}(x) = \sum_{k=0}^{\infty} \frac{(\rho)_{kq}}{\Gamma(\alpha k + \beta)(\delta)_{kp}} x^k$$

being the parameters such that $\alpha, \beta, \delta, \rho \in \mathbb{C}$, $p, q > 0$, $\mathrm{Re}(\alpha) > 0$, $\mathrm{Re}(\beta) > 0$, $\mathrm{Re}(\rho) > 0$, $\mathrm{Re}(\delta) > 0$ and $\mathrm{Re}(\alpha) + p \geq q$. Let $\delta \neq 1$. Show that

$$E_{\alpha,\beta,p}^{\rho,\delta,q}(x) - E_{\alpha,\beta,p}^{\rho,\delta-1,q}(x) = \frac{xp}{1-\delta} \frac{d}{dx} E_{\alpha,\beta,p}^{\rho,\delta,q}(x).$$

35. Use the Mittag-Leffler function with four parameters, according to the Proposed exercise (30) to show that the relation involving the n-th derivative is valid

$$\frac{d^n}{dx^n} E_{\alpha,\beta}^{\rho,q}(x) = (\rho)_{qn} E_{\alpha,\beta+n\alpha}^{\rho+qn,q}(x),$$

with $n = 0, 1, 2, \ldots$

36. Consider the Mittag-Leffler type function as introduced by Miller-Ross. Show that it's worth the

$$E_x(\nu, \alpha) = \alpha E_x(1 + \nu, \alpha) + \frac{x^\nu}{\Gamma(\nu + 1)}$$

with $\alpha, \nu \in \mathbb{R}$ and $x \in \mathbb{R}$. By means of this result, show that

$$E_x(\nu, 0) = \frac{x^\nu}{\Gamma(\nu + 1)}.$$

37. Consider the function defined by Rabotnov, denoted by $R_\alpha(\beta, x)$. Since $E_\mu(\cdot)$ is the classical Mittag-Leffler function, show that the relation

$$\frac{1}{\beta} \frac{d}{dx} E_{\alpha+1}(\beta x^{\alpha+1}) = R_\alpha(\beta, x).$$

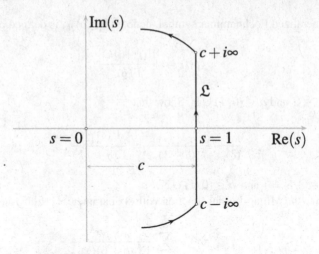

Fig. 3.1 $\mathcal{L}$ is a straight line, parallel to the vertical axis

38. The function $f(x, t) = e^{-x^2/4t}/\sqrt{\pi t}$ is known as the fundamental solution of the standard diffusion equation. Being $E_\mu(\cdot)$ the classical Mittag-Leffler function, show that

$$\frac{1}{\sqrt{\pi t}} \int_0^\infty e^{-x^2/4t} E_{2\alpha}(-x^{2\alpha}) \, dx = E_\alpha(-t^\alpha)$$

    known as an integral representation for the classical Mittag-Leffler function.

39. Let $x \in \mathbb{R}$ and $0 \le \operatorname{Re}(\alpha) \le 1$. Show that

$$E_\alpha(-x) = \frac{2x}{\pi} \int_0^\infty \frac{E_{2\alpha}(-t^2)}{x^2 + t^2} \, dt$$

    being $E_\alpha(\cdot)$ the classical Mittag-Leffler function.

40. Consider the Mittag-Leffler function with four parameters [17]. Assuming we have parameters such that $\alpha = k \in \mathbb{N}$ and $q \in \mathbb{N}$, write this function

$$E_{k,\beta}^{\gamma,q}(z) = \sum_{n=0}^\infty \frac{(\gamma)_{qn}}{\Gamma(kn + \beta)} \frac{z^n}{n!}$$

    in terms of a generalized hypergeometric function, according to Eq. (2.5).

3.3.2 Suggestions

1. Use the definition of the Mittag-Leffler function with two parameters.
2. Direct from the definition, considering $\alpha \to 0$. Compare with the geometric series.
3. Replace $\alpha = 1$ in the definition, the relation between the gamma function and the factorial and compare with the exponential series.
4. Replace $\alpha = 2 = \beta$ in the definition of the Mittag-Leffler function with two parameters and compare with the sine series expansion.
5. Write the respective integral representation for Fox's H-function, use the residue theorem, the definition of Pochhammer symbol, and compare with the series representation of the Mittag-Leffler function with three parameter.
6. Replace $\alpha = 1$ and use the Exercise (37) in Chap. 2.
7. Write an integral representation, use the residue theorem and the result of the sum of a geometric progression with infinite terms.
8. Use the definition and rearrange the sum of the series.
9. Use the definition of the classical Mittag-Leffler function as well as the definition of the beta function.
10. For the first equality, use the definition of the Mittag-Leffler function with two parameters, the definition of gamma function and take into account the sum of the infinite terms of a geometric progression. For the second equality, it is sufficient to consider the case $\beta = 1$.
11. Use the residue theorem and the definition of the classical Mittag-Leffler function.
12. Use the definition of the classical Mittag-Leffler function, an adequate change of variable and the relationship between beta and gamma functions.
13. Use the definition of the Mittag-Leffler function with two parameters, change the order of integration with the summations, and use the definition of the beta function as well as its relation to the gamma function.
14. Use l'Hôpital rule to raise indetermination.
15. Replace the Mittag-Leffler function with three-parameters by the series, change the order of integration, use the definition of gamma function to calculate the integral, and use the generalized geometric series.
16. Replace the series representation of the Mittag-Leffler function with three parameters, compute the derivatives, and manipulate a relation involving the Pochhammer symbol.
17. Expand the denominator in a power series, integrate term by term, and use the integral representation for the inverse of the gamma function.
18. Calculate the derivative by taking into account the index since the first term in the series of the classical Mittag-Leffler function is a constant and the derivative of any integer order of a constant is zero.
19. Use the series representation for the Mittag-Leffler function with three parameters, change the order of integration, take a convenient change of variable and rearranging to obtain the desired result.

20. Replace the values so that the integral is equal to $\sin x$.
21. Derive, rearrange the indices and use the relation $\Gamma(z+1) = z\Gamma(z)$. The analysis for $\alpha = 1$ refers to the exponential.
22. Use power series expansion explicitly and rearrange the terms.
23. Enter the series representation for the Mittag-Leffler function with two parameters and change the order of integration with the summation. Use a variable change to calculate the remaining integral expressing it in terms of gamma functions.
24. Consider, from the previous, $\alpha \to 2\alpha$, $\beta = 1$ and $\gamma \to \alpha$. Add and subtract parcels containing only even powers and rearranging.
25. Use partial fractions of the Proposed exercise (23).
26. Direct from the previous for $\alpha = 2$ and $x \to -1/x^2$.
27. Use the power series representation for the Mittag-Leffler function with two parameters and the integral representation for the beta function.
28. Use the power series representation for the Mittag-Leffler functions with three parameters, manipulate the Pochhammer symbols, and take a change in the index of sum.
29. Enter the power series representation for the Mittag-Leffler function with three parameters and switch with the integral. Calculate the resulting integral and identify the result with a Mittag-Leffler function with three parameters.
30. Use the power series representation for the Mittag-Leffler function with three parameters and use the reflection formula involving the gamma function. The classical Laguerre polynomials are given by the following power series

$$L_k(x) = \sum_{m=0}^{k}(-1)^m \binom{k}{k-m}\frac{x^m}{m!}.$$

31. Enter the series representation of the Mittag-Leffler function with two parameters and change the order of the derivative of order k with the summation.
32. Use the definition and take a suitable change of index. Manipulate the Pochhammer symbols so that you can add the two series that emerge on the first member and get that on the second member.
33. From the first member, use the relation $\Gamma(z+1) = z\Gamma(z)$ and the definition of the generalized Pochhammer symbol.
34. Use the Exercise (33) and rearrange the sums.
35. Use the relation $(\rho)_{kq+nq} = (\rho)_{nq}(\rho+n)_{kq}$ involving the Pochhammer symbol and take an adequate change of index.
36. Use the definition and manipulate the sum indices. For the second part, just consider the parameter $\alpha = 0$.
37. Write the classical Mittag-Leffler function, derive it and rearrange the parameters.
38. Introduce the series representation for the classical Mittag-Leffler function, exchange the sum with the integral, and calculate the integral. Use the Legendre duplication formula, involving the gamma function and simplify.

39. Use the duplication formula for the classical Mittag-Leffler function and the residue theorem to evaluate an integral in the complex plane.
40. Use the definition of the generalized Pochhammer symbol and the definition of the generalized hypergeometric function, according to Eq. (2.5).

3.3.3 Solutions

1. We write the Mittag-Leffler function with two parameters, in the first member, in terms of a power series,

$$E_{1,2}(x) = \sum_{k=0}^{\infty} \frac{x^k}{\Gamma(k+2)}.$$

Separating the term from $k = 0$ we can write

$$E_{1,2}(x) = 1 + \sum_{k=1}^{\infty} \frac{x^k}{\Gamma(k+2)},$$

which, with the change of index $k \to k+1$ provides

$$E_{1,2}(x) = 1 + \sum_{k=1}^{\infty} \frac{x^{k+1}}{\Gamma(k+3)}.$$

Using the definition again, we obtain

$$E_{1,2}(x) = 1 + E_{1,3}(x)$$

which is the desired result   ◇

2. From the definition of the classical Mittag-Leffler function

$$E_{\alpha}(x) = \sum_{k=0}^{\infty} \frac{x^k}{\Gamma(\alpha k + 1)}$$

taking the limit $\alpha \to 0$ and denoting by $E_0(x)$ we can write

$$E_0(x) = \sum_{k=0}^{\infty} x^k = 1 + x + x^2 + x^3 + \cdots.$$

Using the condition $|x| < 1$, this is a geometric series with first term equal to unity and ratio x, whose sum provides

$$E_0(x) = \sum_{k=0}^{\infty} x^k = \frac{1}{1-x}$$

which is the desired result   ◇

3. Substituting $\alpha = 1$ into the expression for the classical Mittag-Leffler function and using the fact that $\Gamma(k+1) = k!$, we obtain

$$E_1(\pm x) = \sum_{k=0}^{\infty} \frac{(\pm x)^k}{\Gamma(k+1)}$$

or by using exponential expansion

$$E_1(\pm x) = \sum_{k=0}^{\infty} \frac{(\pm x)^k}{k!} = e^{\pm x}$$

which is the desired result   ◇

4. Using the expression for the Mittag-Leffler function with two parameters we have

$$x E_{2,2}(-x^2) = x \sum_{k=0}^{\infty} \frac{(-x^2)^k}{\Gamma(2k+2)}.$$

Using the relation $\Gamma(2k+2) = (2k+1)!$ we can write

$$x E_{2,2}(-x^2) = \sum_{k=0}^{\infty} \frac{(-1)^k x^{2k+1}}{(2k+1)!}$$

where, compared with the series for the sine function, it follows

$$x E_{2,2}(-x^2) = \sin x$$

which is the desired result   ◇

5. We begin by identifying the indices, $m = n = p = 1$ and $q = 2$. Comparing with the product of gamma functions, only those that are not 'empty', we have

$$\prod_{j=1}^{1} \Gamma(b_j - \beta_j s), \quad \prod_{j=1}^{1} \Gamma(1 - a_j + \alpha_j s), \quad \prod_{j=2}^{2} \Gamma(1 - b_j + \beta_j s).$$

We have, directly from Fox's H-function, $a_1 = 1 - \rho$, $\alpha_1 = 1$, $b_1 = 0$, $\beta_1 = 1$, $b_2 = 1 - \beta$ and $\beta_2 = \alpha$ from which follows, by definition, the integral representation

$$\frac{1}{2\pi i} \int_{\mathcal{L}} \frac{\Gamma(-s)\Gamma(\rho+s)}{\Gamma(\beta+\alpha s)} (-x)^s \, ds \,,$$

where the contour $\mathfrak{L}$ separates the poles of $\Gamma(-s)$, leaving them to the left of the vertical straight line $\mathrm{Re}(s) = b > 0$, of those of $\Gamma(\rho + s)$, leaving them to the right of the straight line $\mathrm{Re}(s) = b > 0$, as in Fig. 3.2.

Since the poles of the function $\Gamma(-s)$ are at the points $s = k$ with $k = 0, 1, 2, \ldots$ and using the residue theorem we can write

$$2\pi i \sum_{k=0}^{\infty} \operatorname*{Res}_{s=k} \Omega(s) = 2\pi i \sum_{k=0}^{\infty} \lim_{s \to k} (-s + k)\Omega(s)$$

where $\Omega(s)$ is the integrand of the contour integral.

In order to calculate the limit, we write the product $(-s + k)\Gamma(-s)$ in the form of the quotient

$$\frac{(-s+k)(-s+k-1)\cdots(-s)\Gamma(-s)}{(-s+k-1)(-s+k-2)\cdots(-s)} = \frac{\Gamma(-s+k+1)}{(-s+k-1)(-s+k-2)\cdots(-s)}$$

where the fundamental relation $\Gamma(a + 1) = a\Gamma(a)$ was used. Returning in the initial expression, we have

$$H_{1,2}^{1,1}\left[-x \,\middle|\, \begin{matrix}(1-\rho, 1)\\(0, 1), (1-\beta, \alpha)\end{matrix}\right] = \sum_{k=0}^{\infty} \lim_{s \to k} \frac{\Gamma(-s+k+1)}{(-s+k-1)(-s+k-2)\cdots(-s)} \frac{\Gamma(\rho+s)}{\Gamma(\beta+\alpha s)}(-x)^s$$

where it follows, after the calculation of the limit,

$$H_{1,2}^{1,1}\left[-x \,\middle|\, \begin{matrix}(1-\rho, 1)\\(0, 1), (1-\beta, \alpha)\end{matrix}\right] = \sum_{k=0}^{\infty} \frac{(-1)^k}{k!} \frac{\Gamma(\rho+k)}{\Gamma(\beta+\alpha k)}(-x)^k$$

which can be written using the Pochhammer symbol in the form

$$H_{1,2}^{1,1}\left[-x \,\middle|\, \begin{matrix}(1-\rho, 1)\\(0, 1), (1-\beta, \alpha)\end{matrix}\right] = \Gamma(\rho) \sum_{k=0}^{\infty} \frac{(\rho)_k}{\Gamma(\beta+\alpha k)} \frac{x^k}{k!}.$$

Comparing with Eq. (3.4), we obtain

$$H_{1,2}^{1,1}\left[-x \,\middle|\, \begin{matrix}(1-\rho, 1)\\(0, 1), (1-\beta, \alpha)\end{matrix}\right] = \Gamma(\rho) E_{\alpha,\beta}^{\rho}(x),$$

which is the desired result $\diamond$

6. Taking $\alpha = 1$ in the previous result, we have

$$H_{1,2}^{1,1}\left[-x \,\middle|\, \begin{matrix}(1-\rho, 1)\\(0, 1), (1-\beta, 1)\end{matrix}\right] = \Gamma(\rho) E_{1,\beta}^{\rho}(x) \tag{3.8}$$

which is the result of Proposed exercise (37) in Chap. 2. Let us write in terms of the Meijer G-function, since all α_i and β_i with $i = 1, 2, \dots$ are unitary,

$$G^{1,1}_{1,2}\left[-x\left|\begin{matrix}1-\rho\\0, 1-\beta\end{matrix}\right.\right]$$

and use the procedure of the above, that is, via the residue theorem.
Identifying the indices and parameters, we obtain the following integral representation

$$G^{1,1}_{1,2}\left[-x\left|\begin{matrix}1-\rho\\0, 1-\beta\end{matrix}\right.\right] = \frac{1}{2\pi i}\int_{\mathfrak{L}}\frac{\Gamma(-s)\Gamma(\rho+s)}{\Gamma(\beta+s)}(-x)^s\, ds$$

where the contour is analogous to the previous one. Using the residue theorem, procedure analogous to the previous one, we get

$$G^{1,1}_{1,2}\left[-x\left|\begin{matrix}1-\rho\\0, 1-\beta\end{matrix}\right.\right] = \sum_{k=0}^{\infty}\frac{\Gamma(\rho+k)}{\Gamma(\beta+k)}\frac{x^k}{k!}$$

which, using the definition of Pochhammer symbol, we can write

$$t^{\beta-1}e^{-st}E^{\rho}_{\alpha,\beta}(at^{\alpha})G^{1,1}_{1,2}\left[-x\left|\begin{matrix}1-\rho\\0, 1-\beta\end{matrix}\right.\right] = \frac{\Gamma(\rho)}{\Gamma(\beta)}\sum_{k=0}^{\infty}\frac{(\rho)_k}{(\beta)_k}\frac{x^k}{k!}$$

or, in terms of the confluent hypergeometric function

$$G^{1,1}_{1,2}\left[-x\left|\begin{matrix}1-\rho\\0, 1-\beta\end{matrix}\right.\right] = \frac{\Gamma(\rho)}{\Gamma(\beta)}\,_1F_1(\rho; \beta; x)\cdot$$

Comparing with Eq. (3.8) and simplifying, it follows

$$\Gamma(\beta)E^{\rho}_{1,\beta}(x) = \,_1F_1(\rho; \beta; x) \tag{3.9}$$

which is the desired result   ◇

7. Identifying with the definition we have, for the indices $m = n = p = q = 1$ while for the parameters $a_1 = b_1 = \beta/\alpha$. The integral representation is given by

$$G^{1,1}_{1,1}\left[-x\left|\begin{matrix}\beta/\alpha\\\beta/\alpha\end{matrix}\right.\right] = \frac{1}{2\pi i}\int_{\mathfrak{L}}\Gamma\left(\frac{\beta}{\alpha}-s\right)\Gamma\left(1-\frac{\beta}{\alpha}+s\right)x^{\alpha s}\, ds,$$

where the contour $\mathfrak{L}$ separates the poles from the two gamma functions, so using the residue theorem we can write

$$G_{1,1}^{1,1}\left[-x \left| \begin{matrix} \beta/\alpha \\ \beta/\alpha \end{matrix} \right.\right] = \sum_{k=0}^{\infty} \operatorname*{Res}_{s=k+\beta/\alpha} \Gamma\left(\frac{\beta}{\alpha}-s\right)\Gamma\left(1-\frac{\beta}{\alpha}+s\right) x^{\alpha s}$$

$$= \sum_{k=0}^{\infty} \lim_{k\to k+\beta/\alpha} \left(-s+\frac{\beta}{\alpha}\right)\Gamma\left(\frac{\beta}{\alpha}-s\right)\Gamma\left(1-\frac{\beta}{\alpha}+s\right) x^{\alpha s}$$

$$= \sum_{k=0}^{\infty} \frac{(-1)^k \Gamma(k+1)}{k!} x^{\alpha k+\beta} = x^{\beta}\sum_{k=0}^{\infty}(-x^{\alpha})^k.$$

By imposing the condition $|-x^{\alpha}| < 1$ we have the sum of the infinite terms of a geometric progression with the first term is equal to the unit and ratio equal to $-x^{\alpha}$ from where it follows

$$G_{1,1}^{1,1}\left[-x \left| \begin{matrix} \beta/\alpha \\ \beta/\alpha \end{matrix} \right.\right] = \frac{x^{\beta}}{1+x^{\alpha}}$$

which is the desired result ◇

8. Using the definition we can write

$$E_{\alpha}(\sqrt{x}) + E_{\alpha}(-\sqrt{x}) = \sum_{k=0}^{\infty} \frac{z^{k/2}}{\Gamma(\alpha k+1)} + \sum_{k=0}^{\infty}(-1)^k \frac{z^{k/2}}{\Gamma(\alpha k+1)}.$$

From these sums, it is clear that the odd exponent terms are canceled while the exponent pairs are added together, so we obtain

$$E_{\alpha}(\sqrt{x}) + E_{\alpha}(-\sqrt{x}) = 2\sum_{k=0}^{\infty} \frac{x^k}{\Gamma(2\alpha k+1)}$$

which, from the definition of the Mittag-Leffler function with two parameters, allows to write

$$\frac{1}{2}\left[E_{\alpha}(\sqrt{x}) + E_{\alpha}(-\sqrt{x})\right] = E_{2\alpha}(x),$$

which is the desired result ◇

9. We start from the second member II_M. By introducing the definition of the classical Mittag-Leffler function into the integral, we obtain

$$\mathrm{II}_M = \frac{1}{\Gamma(\beta)}\int_0^1 (1-\xi)^{\beta-1}\sum_{k=0}^{\infty}\frac{(x\xi^{\alpha})^k}{\Gamma(\alpha k+1)}\,d\xi$$

which, by changing the order of the integral with the summation, allows us to write

$$\mathrm{II}_M = \frac{1}{\Gamma(\beta)}\sum_{k=0}^{\infty}\frac{x^k}{\Gamma(\alpha k+1)}\int_0^1 (1-\xi)^{\beta-1}\xi^{\alpha k}\,d\xi.$$

Using the definition of the beta function, we have

$$\text{II}_M = \frac{1}{\Gamma(\beta)} \sum_{k=0}^{\infty} \frac{x^k}{\Gamma(\alpha k + 1)} B(\beta, \alpha k + 1) \cdot$$

From the relation between the gamma and beta functions and rearranging, we obtain, using the definition of the beta function,

$$\text{II}_M = \sum_{k=0}^{\infty} \frac{x^k}{\Gamma(\alpha k + \beta + 1)},$$

that is, a Mittag-Leffler function with two parameters. So, matching the first member, we have

$$\frac{1}{\Gamma(\beta)} \int_0^1 (1 - \xi)^{\beta-1} \sum_{k=0}^{\infty} \frac{(x\xi^\alpha)^k}{\Gamma(\alpha k + 1)} \, d\xi = E_{\alpha,\beta+1}(x)$$

which is the desired result   ◇

10. Let us call this double equality of Λ. For the first equality we introduce the definition of the Mittag-Leffler function with two parameters and we change the order of the integration with the summation, so

$$\Lambda = \sum_{k=0}^{\infty} \frac{x^k}{\Gamma(k\alpha + \beta)} \int_0^\infty e^{-t} t^{\alpha k + \beta - 1} \, dt \cdot$$

The remaining integral is nothing more than the definition of the gamma function, from which it follows, simplifying

$$\Lambda = \sum_{k=0}^{\infty} x^k$$

whereas, with the $|x| < 1$ results in the sum of the infinite terms of a geometric progression, so we obtain

$$\int_0^\infty e^{-t} t^{\beta-1} E_{\alpha,\beta}(xt^\alpha) \, dt = \frac{1}{1-x} \cdot$$

Since this equality is independent of the parameters α and β, we can consider $\beta = 1$ in the first integral which results in the second, also independent of α from which follows the second equality, which concludes the exercise   ◇

11. Using the residue theorem to calculate the integral and knowing that we only have contributions from the poles that lie to the left of the straight line $\text{Re}(s) = c > 0$, we obtain

$$\frac{1}{2\pi i}\int_{\mathcal{L}}\frac{\Gamma(s)\Gamma(1-s)}{\Gamma(1-\alpha s)}(-x)^{-s}\,ds = \sum_{k=0}^{\infty}\lim_{s\to -k}\frac{(s+k)\Gamma(s)\Gamma(1-s)}{\Gamma(1-\alpha s)}(-x)^{-s}$$

which, by calculating the limit and simplifying, allows to write

$$\frac{1}{2\pi i}\int_{\mathcal{L}}\frac{\Gamma(s)\Gamma(1-s)}{\Gamma(1-\alpha s)}(-x)^{-s}\,ds = \sum_{k=0}^{\infty}\frac{x^k}{\Gamma(\alpha k+1)}$$

which is exactly the classical Mittag-Leffler function, so

$$E_{\alpha}(x) = \frac{1}{2\pi i}\int_{\mathcal{L}}\frac{\Gamma(s)\Gamma(1-s)}{\Gamma(1-\alpha s)}(-x)^{-s}\,ds$$

which is the desired result   ◇

12. Denoting by I_M the first member, introducing the definition of the classical Mittag-Leffler function and permuting the summation with the integral, we obtain

$$I_M = 1 + \frac{1}{\Gamma(\alpha)}\sum_{k=0}^{\infty}\frac{1}{\Gamma(\alpha k+1)}\int_0^x (x-\xi)^{\alpha-1}\xi^{\alpha k}\,d\xi.$$

By introducing the change of variable $\xi = xt$ and rearranging, we can write

$$I_M = 1 + \frac{1}{\Gamma(\alpha)}\sum_{k=0}^{\infty}\frac{x^{\alpha k+\alpha}}{\Gamma(\alpha k+1)}\int_0^1 (1-t)^{\alpha-1}t^{\alpha k}\,dt.$$

The remaining integral is nothing more than the definition of the beta function which, written in terms of the gamma function and already simplifying, provides

$$I_M = 1 + \sum_{k=0}^{\infty}\frac{x^{\alpha k+\alpha}}{\Gamma(\alpha k+\alpha+1)}.$$

Changing the index $k \to k-1$ we obtain

$$I_M = 1 + \sum_{k=1}^{\infty}\frac{x^{\alpha k}}{\Gamma(\alpha k+1)},$$

which, by reincorporating the term $k = 0$, finally provides

$$I_M = \sum_{k=0}^{\infty}\frac{x^{\alpha k}}{\Gamma(\alpha k+1)}$$

which is exactly the classical Mittag-Leffler function, then

$$1 + \frac{1}{\Gamma(\alpha)} \int_0^x \frac{E_\alpha(\xi^\alpha)}{(x - \xi)^{1-\alpha}} d\xi = E_\alpha(x^\alpha).$$

which is the desired result   ◇

13. Denote by I_M the first member. By introducing the definition of the Mittag-Leffler function with two parameters and changing the order with the integration we obtain

$$I_M = \int_0^1 t^{\gamma-1} E_{\alpha,\gamma}(xt^\alpha)(1 - t)^{\beta-1} E_{\alpha,\beta}(y(1 - t)^\alpha) \, dt$$

$$= \sum_{n=0}^\infty \sum_{m=0}^\infty \frac{x^n y^m}{\Gamma(\alpha n + \gamma)\Gamma(\alpha m + \beta)} \int_0^1 t^{\alpha n+\gamma-1}(1 - t)^{\alpha m+\beta-1} \, dt.$$

The remaining integral is nothing more than the definition of the beta function which, using the relation with the gamma function and simplifying, allows to write

$$I_M = \sum_{n=0}^\infty \sum_{m=0}^\infty \frac{x^n y^m}{\Gamma(\alpha n + \alpha m + \gamma + \beta)}.$$

Entering the index change $k = m + n$ from where $m = k - n$, and rearranging, we obtain

$$I_M = \sum_{n=0}^\infty \sum_{k=n}^\infty \frac{x^n y^{k-n}}{\Gamma(\alpha k + \gamma + \beta)},$$

or in the following form

$$I_M = \sum_{k=0}^\infty \frac{y^k}{\Gamma(\alpha k + \gamma + \beta)} \sum_{n=0}^k \left(\frac{x}{y}\right)^n.$$

Adding the variable n, a geometric series with finite number of terms, we can write

$$I_M = \frac{1}{x - y} \sum_{k=0}^\infty \frac{x^{k+1} - y^{k+1}}{\Gamma(\alpha k + \gamma + \beta)}.$$

Finally, using the definition of the Mittag-Leffler function with two parameters, we have

$$\int_0^1 t^{\gamma-1} E_{\alpha,\gamma}(xt^\alpha)(1 - t)^{\beta-1} E_{\alpha,\beta}(y(1 - t)^\alpha) \, dt = \frac{x E_{\alpha,\gamma+\beta}(x) - y E_{\alpha,\gamma+\beta}(y)}{x - y}$$

which is the desired result   ◇

14. Note that for $y = x$ we have an indeterminacy on the second member. To raise it, we use the l'Hôpital rule from which we can write

$$\int_0^1 t^{\gamma-1} E_{\alpha,\gamma}(xt^\alpha)(1-t)^{\beta-1} E_{\alpha,\beta}(x(1-t)^\alpha)\,dt = E_{\alpha,\gamma+\beta}(x) + x E'_{\alpha,\gamma+\beta}(x).$$

Using the relation involving the functions of Mittag-Leffler and its derivative

$$\frac{d}{dx} E_{\alpha,\beta}(x) = x\frac{d}{dx} E_{\alpha,\alpha+\beta}(x) + E_{\alpha,\alpha+\beta}(x)$$

which can be rewritten, simplifying, in the form

$$\int_0^1 t^{\gamma-1} E_{\alpha,\gamma}(xt^\alpha)(1-t)^{\beta-1} E_{\alpha,\beta}(x(1-t)^\alpha)\,dt = E'_{\alpha,\gamma+\beta-\alpha}(x),$$

which is the desired result   ◇

15. We introduce the notation I_M to denote the first member. Substituting the Mittag-Leffler function with three parameters and permuting the order of integration, we can write

$$I_M = \sum_{k=0}^\infty \frac{(\rho)_k}{\Gamma(\alpha k+\beta)}\frac{a^k}{k!}\int_0^\infty t^{\alpha k+\beta-1} e^{-st}\,dt.$$

In order to calculate this integral, interpreted as the Laplace transform of a power function, we begin with the gamma function, in terms of an integral

$$\Gamma(x) = \int_0^\infty t^{x-1} e^{-t}\,dt$$

with $x > 0$. We change the variable $t \to st$ with $\operatorname{Re}(s) > 0$ from where it follows

$$\Gamma(x) = s^x \int_0^\infty t^{x-1} e^{-st}\,dt.$$

Entering $x = \alpha k + \beta$ and rearranging, we have

$$\int_0^\infty t^{\alpha k+\beta-1} e^{-st}\,dt = \frac{\Gamma(\alpha k+\beta)}{s^{\alpha k+\beta}}.$$

Returning with this result in the expression for I_M, we have, simplifying

$$I_M = \frac{1}{s^\beta}\sum_{k=0}^\infty \frac{(\rho)_k}{k!}\left(\frac{a}{s^\alpha}\right)^k.$$

Using the generalized geometric series,

$$\sum_{k=0}^{\infty} \frac{(\rho)_k}{k!} x^k = \left(\frac{1}{1-x}\right)^{\rho} \tag{3.10}$$

being $|x| < 1$, we can write, already rearranging

$$\int_0^{\infty} t^{\beta-1} e^{-st} E_{\alpha,\beta}^{\rho}(at^{\alpha}) \, dt = s^{-\beta}(1 - as^{-\alpha})^{-\rho},$$

which is the desired result   ◇

16. We introduce the notation I_M to denote the first member. Substituting the Mittag-Leffler function with three parameteres in terms of a series and rearranging, we obtain

$$I_M = \sum_{n=0}^{\infty} \frac{(\rho)_n}{\Gamma(\alpha n + \beta) n!} \left(\frac{d}{dx}\right)^k x^n.$$

Using the relation

$$\left(\frac{d}{dx}\right)^k x^n = \frac{n!}{(n-k)!} x^{n-k}, \qquad n \geq k$$

which leads to a change in the lower index, we have,

$$I_M = \sum_{n=k}^{\infty} \frac{(\rho)_n}{\Gamma(\alpha n + \beta)} \frac{x^{n-k}}{(n-k)!}.$$

By manipulating the Pochhammer symbol, we can show $(\rho)k + m = (\rho)_k(\rho + k)_m$, from where it follows, identifying with the Mittag-Leffler function with three parameters

$$\left(\frac{d}{dx}\right)^k E_{\alpha,\beta}^{\rho}(x) = (\rho)_k E_{\alpha,\beta+\alpha k}^{\rho+k}(x),$$

which is the desired result   ◇

17. Expanding $(\xi^{\alpha} - z)^{-1}$ in a power series

$$\frac{1}{\xi^{\alpha} - z} = \frac{1}{\xi^{\alpha}} \sum_{k=0}^{\infty} \left(\frac{z}{\xi^{\alpha}}\right)^k$$

valid for $|\xi| > |z|^{1/\alpha}$, follows, for the integral, rearranging

$$\frac{1}{2\pi i} \int_{C_{\gamma}} \frac{\xi^{\alpha-1} e^{\xi}}{\xi^{\alpha} - z} \, d\xi = \sum_{k=0}^{\infty} z^k \frac{1}{2\pi i} \int_{C_{\gamma}} \xi^{-\alpha k - 1} e^{\xi} \, d\xi.$$

In the second member we have the integral representation for the gamma function with argument $\mu = \alpha k + 1$, so we obtain, using the definition of Mittag-Leffler function

$$\frac{1}{2\pi i} \int_{C_\gamma} \frac{\xi^{\alpha-1} e^\xi}{\xi^\alpha - z} \, d\xi = \sum_{k=0}^\infty \frac{z^k}{\Gamma(\alpha k + 1)} = E_\alpha(z)$$

which is the desired result   ◇

18. We introduce the notation

$$I_M = \left(\frac{d}{dx}\right)^k E_k(x^k).$$

From the series for the classical Mittag-Leffler function, changing the order of the derivative with the summation and taking into account that the index starts at one, since for any k the derivative of a constant is zero, we can write

$$I_M = \sum_{\ell=1}^\infty \frac{1}{\Gamma(k\ell + 1)} \left(\frac{d}{dx}\right)^k (x^{k\ell}).$$

Deriving, in relation to k, we have

$$I_M = \sum_{\ell=1}^\infty \frac{1}{\Gamma(k\ell + 1)} \frac{\Gamma(k\ell + 1)}{\Gamma(k\ell - k + 1)} (x^{k\ell-k})$$

that by introducing the change $\ell \to \ell + 1$ and simplifying, it provides

$$I_M = \sum_{\ell=0}^\infty \frac{x^{k\ell}}{\Gamma(k\ell + 1)}.$$

Identifying with the series representation for the classical Mittag-Leffler function, we can write

$$\left(\frac{d}{dx}\right)^k E_k(x^k) = E_k(x^k)$$

which is the desired result   ◇

19. We introduce the notation

$$I_M = \int_0^x t^{\beta-1} E_{\alpha,\beta}^\gamma(at^\alpha) \, dt.$$

Using the series representation for the Mittag-Leffler function with three parameters, changing the order of the integral and rearranging, we can write

I

$$\mathsf{I_M} = \sum_{k=0}^{\infty} \frac{(\gamma)_k}{\Gamma(\alpha k + \beta)} \frac{a^k}{k!} \int_0^x t^{\alpha k + \beta - 1} \, dt \cdot$$

Integrating and using the property of the gamma function $\Gamma(z+1) = z\Gamma(z)$, we have

$$\mathsf{I_M} = \sum_{k=0}^{\infty} \frac{(\gamma)_k}{\Gamma(\alpha k + \beta + 1)} \frac{(ax^{\alpha})^k}{k!} x^{\beta}$$

from where, identifying with the representation in series for the Mittag-Leffler function with three parameters, allows to write

$$\int_0^x t^{\beta - 1} E_{\alpha, \beta}^{\gamma}(at^{\alpha}) \, dt = x^{\beta} E_{\alpha, \beta + 1}^{\gamma}(at^{\alpha}),$$

which is the desired result   ◇

20. Substituting $a = -1$, $\alpha = 2$ and $\beta = 1 = \gamma$ we get for the first member

$$\mathsf{I_M} = \int_0^x E_2(-t^{\alpha}) \, dt$$

where we use the identity $E_{2,1}^1(\cdot) = E_2(\cdot)$. Using the series representation for the classical Mittag-Leffler function with $\alpha = 2$, we obtain exactly the cosine series, that is,

$$E_2(-t^2) = \cos t$$

which, replacing in the previous one, after integration, provides

$$\mathsf{I_M} = \sin x \cdot$$

Using the result of the Exercise (4) we obtain

$$\int_0^x E_2(-t^{\alpha}) \, dt = x E_{2,2}(-x^2)$$

which is the desired result   ◇

21. Using the power series for the Mittag-Leffler function with two parameters and deriving, in relation to x, both members, we can write

$$\frac{d}{dx} E_{\alpha}(x^{\alpha}) = \frac{d}{dx} \sum_{k=0}^{\infty} \frac{x^{\alpha k}}{\Gamma(\alpha k + 1)}$$

or by explaining the derivative and changing the index, we have

$$\frac{d}{dx}E_\alpha(x^\alpha) = \sum_{k=1} \alpha k \frac{x^{\alpha k-1}}{\Gamma(\alpha k + 1)}.$$

Using the relation $\Gamma(z+1) = z\Gamma(z)$ and simplifying, we obtain

$$\frac{d}{dx}E_\alpha(x^\alpha) = \sum_{k=1} \frac{x^{\alpha k-1}}{\Gamma(\alpha k)}.$$

Changing the index $k \to k+1$ and rearranging, we can write

$$\frac{d}{dx}E_\alpha(x^\alpha) = x^{\alpha-1} \sum_{k=0} \frac{x^{\alpha k}}{\Gamma(\alpha k + \alpha)}$$

which, identifying with the Mittag-Leffler function with two parameters, provides

$$\frac{d}{dx}E_\alpha(x^\alpha) = x^{\alpha-1} E_{\alpha,\alpha}(x^\alpha).$$

For $\alpha = 1$ we have an identity involving exponentials, since the derivative of the exponential function $\exp(x)$, is itself ◇

22. Using the power series expressions for the Mittag-Leffler function with one- and two-parameters, in the second member, we have

$$\Omega_0 \equiv E_{2\alpha}(x^2) - xE_{2\alpha,\alpha+1}(x^2) = \sum_{k=0}^{\infty} \frac{(x^2)^k}{\Gamma(2\alpha k + 1)} - x\sum_{k=0}^{\infty} \frac{(x^2)^k}{\Gamma(2\alpha k\alpha + 1)}.$$

Explaining the two sums, we obtain

$$\Omega_0 = \left\{ 1 + \frac{x^2}{\Gamma(2\alpha+1)} + \frac{x^4}{\Gamma(4\alpha+1)} + \frac{x^6}{\Gamma(6\alpha+1)} + \cdots \right\}$$

$$- \left\{ \frac{x}{\Gamma(\alpha+1)} + \frac{x^3}{\Gamma(3\alpha+1)} + \frac{x^5}{\Gamma(5\alpha+1)} + \cdots \right\}$$

rearranging provides

$$\Omega_0 = \left\{ 1 + \frac{(-x)}{\Gamma(\alpha+1)} + \frac{(-x)^2}{\Gamma(2\alpha+1)} + \frac{(-x)^3}{\Gamma(3\alpha+1)} + \frac{(-x)^4}{\Gamma(4\alpha+1)} + \frac{(-x)^5}{\Gamma(5\alpha+1)} \cdots \right\}.$$

Identifying with the series for the classical Mittag-Leffler function, we have

$$\Omega_0 = \sum_{k=0}^{\infty} \frac{(-x)^k}{\Gamma(k\alpha+1)}$$

of which follows

$$E_{2\alpha}(x^2) - x E_{2\alpha,\alpha+1}(x^2) = E_\alpha(-x)$$

which is the desired result   ◇

23. First, we denote the first member by I_M. Introducing the representation of power series for the Mittag-Leffler function with two parameters and changing the order of the integral with the summation, we can write

$$\mathsf{I}_M \equiv \frac{1}{\Gamma(\gamma)} \sum_{k=0}^{\infty} \frac{1}{\Gamma(\alpha k + \beta)} \int_0^x (x - \xi)^{\gamma-1} \xi^{\beta-1} \xi^{\alpha k} \, d\xi.$$

Introducing in the previous a change of variable $\xi = xt$ and rearranging, we obtain

$$\mathsf{I}_M = \frac{1}{\Gamma(\gamma)} \sum_{k=0}^{\infty} \frac{x^{\gamma+\beta+\alpha k-1}}{\Gamma(\alpha k + \beta)} \int_0^1 (1 - t)^{\gamma-1} t^{\alpha k+\beta-1} \, dt.$$

Identifying the remaining integral with a beta function and using the relation between the beta function and the gamma function, we can write

$$\mathsf{I}_M = \frac{1}{\Gamma(\gamma)} \sum_{k=0}^{\infty} \frac{x^{\gamma+\beta+\alpha k-1}}{\Gamma(\alpha k + \beta)} \frac{\Gamma(\gamma)\Gamma(\alpha k + \beta)}{\Gamma(\alpha k + \beta + \gamma)}$$

which, by simplifying, provides, already coming back with the expression of the first member

$$\frac{1}{\Gamma(\gamma)} \int_0^x (x - \xi)^{\gamma-1} \xi^{\beta-1} E_{\alpha,\beta}(\xi^\alpha) \, d\xi = x^{\beta+\gamma-1} E_{\alpha,\beta+\gamma}(x^\alpha)$$

which is the desired result   ◇

24. Considering the previous one with $\alpha \to 2\alpha$, $\beta = 1$ and $\gamma \to \alpha$, we obtain

$$\frac{1}{\Gamma(\alpha)} \int_0^x (x - \xi)^{\alpha-1} E_{2\alpha}(\xi^{2\alpha}) \, d\xi = x^\alpha E_{2\alpha,\alpha+1}(x^{2\alpha})$$

where we have used the relation $E_{\alpha,1}(\cdot) = E_\alpha(\cdot)$.

From the power series representation for the Mittag-Leffler function with two parameters, summing and subtracting the parcel with the even and rearranging terms, we can write for the second member

$$x^\alpha E_{2\alpha,\alpha+1}(x^{2\alpha}) = \sum_{k=0}^{\infty} \frac{x^{(2k+1)\alpha}}{\Gamma[(2k + 1)\alpha + 1]} + \sum_{k=0}^{\infty} \frac{x^{(2k)\alpha}}{\Gamma[(2k)\alpha + 1]} - \sum_{k=0}^{\infty} \frac{x^{(2k)\alpha}}{\Gamma[(2k)\alpha + 1]}.$$

The sum of the first two series in the second member allows to write

$$x^\alpha E_{2\alpha,\alpha+1}(x^{2\alpha}) = \sum_{k=0}^{\infty} \frac{x^{k\alpha}}{\Gamma(k\alpha+1)} - \sum_{k=0}^{\infty} \frac{x^{2\alpha k}}{\Gamma[(2k)\alpha+1]}$$

which, returning with the notation for the classical Mittag-Leffler function, provides

$$x^\alpha E_{2\alpha,\alpha+1}(x^{2\alpha}) = E_\alpha(x^\alpha) - E_{2\alpha}(x^{2\alpha})$$

which is the desired result   ◇

25. We begin by explicitly in partial fractions the integrating of the Proposed exercise (23), thus omitting the exponential, we can write

$$\frac{\xi^{\alpha-\beta}}{\xi^\alpha - x} = \frac{1}{\xi^\beta - x\,\xi^{-\alpha+\beta}} = \frac{1}{\xi^\beta} \; \frac{1}{\xi^\beta - \frac{1}{x}\xi^{\alpha+\beta}}.$$

Using the previous one, we can write for the Mittag-Leffler function of two parameters, according to the Proposed exercise (23)

$$E_{\alpha,\beta}(x) = \frac{1}{2\pi i} \int_C \frac{e^\xi}{\xi^\beta} \, d\xi - \frac{1}{2\pi i} \int_C \frac{e^\xi \, \xi^{-\alpha-\beta}}{\xi^{-\alpha} - \frac{1}{x}} \, d\xi.$$

Considering the change of variable $\alpha \to -\alpha$ we obtain

$$E_{-\alpha,\beta}(x) = \frac{1}{2\pi i} \int_C \frac{e^\xi}{\xi^\beta} \, d\xi - \frac{1}{2\pi i} \int_C \frac{e^\xi \, \xi^{\alpha-\beta}}{\xi^\alpha - \frac{1}{x}} \, d\xi.$$

The second portion in the second member of the preceding one is nothing more than the Mittag-Leffler function with two-parameter, with $1/x$ argument, and the first portion is the integral representation of the inverse of the gamma function, as given in Definition 2.1.3, then, we obtain

$$E_{-\alpha,\beta}(x) = \frac{1}{\Gamma(\beta)} - E_{\alpha,\beta}\left(\frac{1}{x}\right)$$

which is the desired result   ◇

·26. Using the result of the precedent, with $\alpha = 2$, $\beta = 1$ and $x \to -1/x^2$, we can write

$$E_{-2,1}\left(-\frac{1}{x^2}\right) = \frac{1}{\Gamma(1)} - E_{2,1}\left(-x^2\right).$$

Using known results, this is $\Gamma(1) = 1$ and $E_{2,1}(\cdot) = E_2(\cdot)$, we obtain

$$E_{-2}\left(-\frac{1}{x^2}\right) = 1 - E_2\left(-x^2\right)$$

or, in the following form, because $E_2(-x^2) = \cos x$ is a known result

$$E_{-2}\left(-\frac{1}{x^2}\right) = 1 - \cos x$$

which is the desired result   ◇

27. Let us consider, first, the integral, denoted by J. By introducing the power series representation for the classical Mittag-Leffler function and changing the order of the summation with the integral, we can write

$$J \equiv \sum_{k=0}^{\infty} \frac{(-x)^k}{\Gamma(\alpha k + \alpha)} \int_0^1 \left(1 - \xi^{\frac{1}{\alpha}}\right)^{\beta-\alpha-1} \xi^k \, d\xi.$$

Considering the variable change $\xi^{\frac{1}{\alpha}} = u$, we obtain

$$J = \sum_{k=0}^{\infty} \frac{\alpha(-x)^k}{\Gamma(\alpha k + \alpha)} \int_0^1 (1 - u)^{\beta-\alpha-1} u^{\alpha k + \alpha - 1} \, du.$$

The remaining integral is nothing more than the integral representation for the beta function which, expressed in terms of the gamma function, allows to write, already simplifying

$$J = \alpha\Gamma(\beta - \alpha) \sum_{k=0}^{\infty} \frac{(-x)^k}{\Gamma(\alpha k + \alpha)} = \alpha\Gamma(\beta - \alpha) \, E_{\alpha,\beta}(-x)$$

which, substituted in Eq. (3.7) and simplified, provides

$$E_{\alpha,\beta}(-x) = \frac{1}{\alpha\Gamma(\beta - \alpha)} \int_0^1 \left(1 - \xi^{\frac{1}{\alpha}}\right)^{\beta-\alpha-1} E_{\alpha,\alpha}(-x\xi) \, d\xi$$

which is the desired result   ◇

28. Consider the first member, denoted by I_M. Replacing the power series for the two Mittag-Leffler functions with three parameters, we have

$$I_M = \sum_{k=0}^{\infty} \frac{(\rho)_k}{\Gamma(\alpha k + \beta - \alpha)} \frac{x^k}{k!} - \sum_{k=0}^{\infty} \frac{(\rho - 1)_k}{\Gamma(\alpha k + \beta - \alpha)} \frac{x^k}{k!}.$$

Rearranging in a single summation and using the definition of the Pochhammer symbol, we can write

$$I_M = \sum_{k=0}^{\infty} \frac{x^k}{\Gamma(\alpha k + \beta - \alpha)k!} \left\{ \frac{\Gamma(\rho + k)}{\Gamma(\rho)} - \frac{\Gamma(\rho + k - 1)}{\Gamma(\rho - 1)} \right\}.$$

Using the relation $\Gamma(z + 1) = z\Gamma(z)$ and simplifying, we obtain

$$I_M = \sum_{k=0}^{\infty} \frac{x^k}{\Gamma(\alpha k + \beta - \alpha)k!} \frac{k\Gamma(\rho + k - 1)}{\Gamma(\rho)},$$

that, with the change $k \to k + 1$, leads us to

$$I_M = x \sum_{k=0}^{\infty} \frac{x^k}{\Gamma(\alpha k + \beta)k!} \cdot \frac{\Gamma(\rho + k)}{\Gamma(\rho)}$$

and now coming back with the Pochhammer symbol, we have

$$I_M = x \sum_{k=0}^{\infty} \frac{(\rho)_k}{\Gamma(\alpha k + \beta)k!} \frac{x^k}{k!}$$

or in the following form

$$E_{\alpha,\beta-\alpha}^{\rho}(x) - E_{\alpha,\beta-\alpha}^{\rho-1}(x) = xE_{\alpha,\beta}^{\rho}(x)$$

which is the desired result ◇

29. Denote by I_M the first member. We introduce the series representation for the Mittag-Leffler function with three parameters and change the order of the sum with the integral, we can write

$$I_M \equiv \int_0^x \xi^{\beta-1} E_{\alpha,\beta}^{\rho}(\lambda\xi^{\alpha}) \, d\xi = \sum_{k=0}^{\infty} \frac{(\rho)_k}{\Gamma(\alpha k + \beta)} \frac{\lambda^k}{k!} \int_0^x \xi^{\beta-1} \xi^{\alpha k} d\xi.$$

By introducing a change of variable $\xi = xt$, evaluating integration and rearranging, we have

$$I_M = \sum_{k=0}^{\infty} \frac{(\rho)_k}{\Gamma(\alpha k + \beta)} \frac{(\lambda x^{\alpha})^k}{k!} x^{\beta} \frac{t^{\alpha k + \beta}}{\alpha k + \beta}\bigg|_{t=0}^{t=1}$$

or, in the following way

$$I_M = x^{\beta} \sum_{k=0}^{\infty} \frac{(\rho)_k}{\Gamma(\alpha k + \beta + 1)} \frac{(\lambda x^{\alpha})^k}{k!}$$

which, compared to the second member, allows writing

$$\int_0^x \xi^{\beta-1} E_{\alpha,\beta}^{\rho}(\lambda\xi^{\alpha}) \, d\xi = x^{\beta} E_{\alpha,\beta+1}^{\rho}(\lambda x^{\alpha})$$

which is the desired result   ◊

30. Using the power series representation for the Mittag-Leffler function with three parameters, we have

$$E_{1,1}^{-k}(x) = \sum_{n=0}^{k} \frac{(-k)_n}{\Gamma(n+1)} \frac{x^n}{n!}$$

which, by explaining the Pochhammer symbols, allows us to write

$$E_{1,1}^{-k}(x) = \sum_{n=0}^{k} \frac{\Gamma(n-k)}{\Gamma(-k)} \frac{x^n}{n!n!}.$$

Note that the index runs from zero to k, the degree of the polynomial, since it can be noted that this index is at most k, as seen by binomial coefficient. Using the reflection formula, we can write for the quotient of the two gamma functions,

$$\frac{\Gamma(n-k)}{\Gamma(-k)} = \frac{\pi}{\sin \pi(n-k)\Gamma(1+k-n)} \frac{\sin(-k)\Gamma(n+1)}{\pi} = (-1)^n \frac{k!}{\Gamma(1+k-n)}.$$

Returning with this result in the expression for the particular Mittag-Leffler function with three parameters, we obtain

$$E_{1,1}^{-k}(x) = k! \sum_{n=0}^{k} \frac{(-x)^n}{(k-n)!n!n!}$$

which can be written as follows

$$E_{1,1}^{-k}(x) = \sum_{n=0}^{k} (-1)^n \binom{k}{k-n} \frac{x^n}{n!}.$$

Comparing this result with the expression for the Laguerre polynomials, it follows

$$E_{1,1}^{-k}(x) = L_k(x)$$

which is the desired result   ◊

31. Denote by I_M the first member. Replacing the power series representation for the Mittag-Leffler function with two parameters and permuting the order of the derivative with the summation, we have

$$I_M \equiv \frac{d^k}{dx^k} \left[x^{n-1} E_{k,n}(\lambda x^k) \right] = \sum_{\ell=0}^{\infty} \frac{\lambda^\ell}{\Gamma(k\ell+n)} \frac{d^k}{dx^k} \left(x^{k\ell+n-1} \right).$$

By doing the derivative, we can write

$$I_M = \sum_{\ell=0}^{\infty} \frac{\lambda^\ell}{\Gamma(k\ell+n)} \frac{\Gamma(k\ell+n)}{\Gamma(k\ell+n-k)} x^{k\ell+n-1-k}$$

which simplifying provides

$$I_M = x^{n-1} \sum_{\ell=1}^{\infty} \frac{\lambda^\ell}{\Gamma(k\ell+n-k)} x^{k\ell-k}.$$

Note that the lower index had to be changed, since the exponent $k\ell - k$ does not allow $\ell = 0$. Entering the index change $\ell \to \ell+1$ and rearranging, we get

$$I_M = \lambda x^{n-1} \sum_{\ell=0}^{\infty} \frac{(\lambda x^k)^\ell}{\Gamma(k\ell+n)}$$

that identifying with the second member allows writing

$$\frac{d^k}{dx^k} \left[x^{n-1} E_{k,n}(\lambda x^k) \right] = \lambda x^{n-1} E_{k,n}(\lambda x^k)$$

which is the desired result   ◇

32. Consider the first member, denoting it by I_M

$$I_M \equiv \left(x\frac{d}{dx} + \rho \right) E_{\alpha,\beta}^{\rho}(x) = x \frac{d}{dx} E_{\alpha,\beta}^{\rho}(x) + \rho\, E_{\alpha,\beta}^{\rho}(x).$$

By introducing the series representation for the Mittag-Leffler functions with three parameters and calculating the derivative of the first parcel, we get

$$I_M = \sum_{k=0}^{\infty} \frac{(\rho)_k k\, x^k}{\Gamma(\alpha k + \beta)k!} + \rho \sum_{k=0}^{\infty} \frac{(\rho)_k\, x^k}{\Gamma(\alpha k + \beta)k!}$$

or rearranging in the following manner

$$I_M = \sum_{k=0}^{\infty} \frac{(\rho)_k\, x^k}{\Gamma(\alpha k + \beta)k!}(k+\rho).$$

Using the relation involving Pochhammer symbols,

$$(\rho)_k = \frac{\rho}{\rho+k}(\rho+1)_k\,,$$

replacing in the previous one and simplifying, we can write

$$I_M = \rho \sum_{k=0}^{\infty} \frac{(\rho + 1)_k \, x^k}{\Gamma(\alpha k + \beta) k!}.$$

Finally, returning with the series representation for the Mittag-Leffler function with three parameters, we have

$$\left(x \frac{d}{dx} + \rho \right) E_{\alpha, \beta}^{\rho}(x) = \rho \, E_{\alpha, \beta}^{\rho + 1}(x),$$

which is the desired result   ◇

33. Entering the notation I_M for the first member and using the definition for the generalized Pochhammer symbol, we obtain

$$I_M \equiv \frac{1}{(\delta)_{pk}} - \frac{1}{(\delta - 1)_{pk}} = \frac{\Gamma(\delta)}{\Gamma(\delta + pk)} - \frac{\Gamma(\delta - 1)}{\Gamma(\delta - 1 + pk)}.$$

Using the relation $\Gamma(z + 1) = z\Gamma(z)$, in the second parcel, we can write

$$I_M = \frac{\Gamma(\delta)}{\Gamma(\delta + pk)} - \frac{\frac{\Gamma(\delta)}{\delta - 1}(\delta - 1 + pk)}{\Gamma(\delta + pk)}.$$

Bringing in evidence the first parcel, rearranging and simplifying, we have

$$\frac{1}{(\delta)_{pk}} - \frac{1}{(\delta - 1)_{pk}} = \frac{\Gamma(\delta)\frac{pk}{1 - \delta}}{\Gamma(\delta + pk)},$$

which is the desired result   ◇

34. By introducing the notation I_M for the first member and using the definition of the Mittag-Leffler function with six parameters, through the power series, we have

$$I_M \equiv E_{\alpha, \beta, p}^{\rho, \delta, q}(x) - E_{\alpha, \beta, p}^{\rho, \delta - 1, q}(x) = \sum_{k=0}^{\infty} \frac{(\rho)_{kq} \, x^k}{\Gamma(\alpha k + \beta)(\delta)_{kp}} - \sum_{k=0}^{\infty} \frac{(\rho)_{kq} \, x^k}{\Gamma(\alpha k + \beta)(\delta - 1)_{kp}}.$$

Rewriting the previous expression in the form

$$I_M = \sum_{k=0}^{\infty} \frac{(\rho)_{kq} \, x^k}{\Gamma(\alpha k + \beta)} \left(\frac{1}{(\delta)_{kp}} - \frac{1}{(\delta - 1)_{kp}} \right)$$

and using the result of the Exercise (33) we can write

$$I_M = \sum_{k=0}^{\infty} \frac{(\rho)_{kq} \, x^k}{\Gamma(\alpha k + \beta)} \left(\frac{\Gamma(\delta)\frac{pk}{1 - \delta}}{\Gamma(\delta + kp)} \right).$$

Manipulating the Pochhammer symbol and simplifying it, we obtain

$$I_M = \frac{xp}{1-\delta} \sum_{k=0}^{\infty} \frac{(\rho)_{kq}\, k\, x^{k-1}}{\Gamma(\alpha k + \beta)} \frac{1}{(\delta)_{kp}}$$

which, by simulating a derivative, allows to write, already returning with the first member

$$E_{\alpha,\beta,p}^{\rho,\delta,q}(x) - E_{\alpha,\beta,p}^{\rho,\delta-1,q}(x) = \frac{xp}{1-\delta} \frac{d}{dx} E_{\alpha,\beta,p}^{\rho,\delta,q}(x)$$

which is the desired result ◇

35. We begin by observing that

$$
\begin{aligned}
(\rho)_{kq+nq} &= \rho(\rho+1)\cdots(\rho+kq+nq-1) \\
&= \underbrace{\rho(\rho+1)\cdots(\rho+nq-1)}_{(\rho)_{nq}}\, \underbrace{(\rho+nq)(\rho+nq+1)\cdots(\rho+nq+kq-1)}_{(\rho+n)_{kq}} \cdot
\end{aligned}
$$

Deriving n times, we can write to the first member

$$\frac{d^n}{dx^n} E_{\alpha,\beta}^{\rho,q}(x) = \sum_{k=0}^{\infty} \frac{(\rho)_{kq}\, k(k-1)\cdots(k-n+1)}{\Gamma(\alpha k + \beta)} \frac{x^{k-n}}{k!}.$$

Using the definition of binomial and simplifying, we obtain

$$\frac{d^n}{dx^n} E_{\alpha,\beta}^{\rho,q}(x) = \sum_{k=0}^{\infty} \frac{(\rho)_{kq}}{\Gamma(\alpha k + \beta)} \frac{x^{k-n}}{(k-n)!}.$$

Introducing the index change $k \to k+n$ we have

$$\frac{d^n}{dx^n} E_{\alpha,\beta}^{\rho,q}(x) = \sum_{k=0}^{\infty} \frac{(\rho)_{kq+nq}}{\Gamma(\alpha k + \alpha n + \beta)} \frac{x^k}{k!}$$

where we already came back with the index starting at zero. Using the relation involving the Pochhammer symbols, we can write

$$\frac{d^n}{dx^n} E_{\alpha,\beta}^{\rho,q}(x) = \sum_{k=0}^{\infty} \frac{(\rho)_{nq}(\rho+nq)_{kq}}{\Gamma(\alpha k + \alpha n + \beta)} \frac{x^k}{k!}$$

or in the following manner

$$\frac{d^n}{dx^n} E_{\alpha,\beta}^{\rho,q}(x) = (\rho)_{nq} \sum_{k=0}^{\infty} \frac{(\rho+nq)_{kq}}{\Gamma(\alpha k + \alpha n + \beta)} \frac{x^k}{k!}$$

which, using the series representation of the Mittag-Leffler function with four parameters, allows us to write

$$\frac{d^n}{dx^n} E_{\alpha,\beta}^{\rho,q}(x) = (\rho)_{qn} E_{\alpha,\beta+n\alpha}^{\rho+qn,q}(x),$$

which is the desired result   ◇

36. We denote the second member by $\|_M$ and explicitly write the function in terms of the power series representation, so

$$\|_M \equiv \alpha E_x(1+\nu,\alpha) + \frac{x^\nu}{\Gamma(\nu+1)} = \alpha x^{\nu+1} \sum_{k=0}^{\infty} \frac{(\alpha x)^k}{\Gamma(\nu+k+2)} + \frac{x^\nu}{\Gamma(\nu+1)}$$

with $\mathrm{Re}(\nu) > -1$ and $\alpha \in \mathbb{R}$. Introducing an index change $k \to k-1$, we have

$$\|_M = \alpha x^{\nu+1} \sum_{k=1}^{\infty} \frac{(\alpha x)^{k-1}}{\Gamma(\nu+k+1)} + \frac{x^\nu}{\Gamma(\nu+1)}$$

or, rearranging, in the following form

$$\|_M = x^\nu \sum_{k=1}^{\infty} \frac{(\alpha x)^k}{\Gamma(\nu+k+1)} + \frac{x^\nu}{\Gamma(\nu+1)}.$$

Reincorporating the term $k = 0$ we can write

$$\|_M = x^\nu \sum_{k=0}^{\infty} \frac{(\alpha x)^k}{\Gamma(\nu+k+1)}$$

from which exactly follows the first member, that is,

$$\alpha E_x(1+\nu,\alpha) + \frac{x^\nu}{\Gamma(\nu+1)} = E_x(\nu,\alpha).$$

On the other hand, considering $\alpha = 0$, we obtain

$$E_x(\nu,0) = \frac{x^\nu}{\Gamma(\nu+1)}$$

which is the desired result   ◇

37. Let us denote the first member by I_M and introduce the power series representation for the classical Mittag-Leffler function, then

$$I_M \equiv \frac{1}{\beta}\frac{d}{dx}E_{\alpha+1}(\beta x^{\alpha+1}) = \frac{1}{\beta}\frac{d}{dx}\sum_{k=0}^{\infty}\frac{(\beta x^{\alpha+1})^k}{\Gamma((\alpha+1)k+1)}.$$

Evaluate the derivative in relation to x and changing the lower index, we have

$$I_M = \frac{1}{\beta}\sum_{k=1}^{\infty}\frac{(\alpha+1)k\,\beta^k x^{(\alpha+1)k-1}}{\Gamma((\alpha+1)k+1)}.$$

Introducing the change $k \to k+1$, simplifying and rearranging, we obtain

$$I_M = \sum_{k=0}^{\infty}\frac{\beta^k x^{(\alpha+1)k+\alpha}}{\Gamma((\alpha+1)(k+1))}$$

that, returning with the first member, provides

$$\frac{1}{\beta}\frac{d}{dx}E_{\alpha+1}(\beta x^{\alpha+1}) = x^{\alpha}\sum_{k=0}^{\infty}\frac{\beta^k x^{(\alpha+1)k}}{\Gamma((\alpha+1)(k+1))} = R_{\alpha}(\beta, x)$$

which is the desired result ◇

38. Let us denote the first member by I_M. Introducing the series representation for the classical Mittag-Leffler function and permuting the sum with the integral, we get

$$I_M \equiv \frac{1}{\sqrt{\pi t}}\int_0^{\infty}e^{-x^2/4t}E_{2\alpha}(-x^{2\alpha})\,dx = \frac{1}{\sqrt{\pi t}}\sum_{k=0}^{\infty}\frac{(-1)^k}{\Gamma(2\alpha k+1)}\int_0^{\infty}e^{-x^2/4t}x^{2\alpha k}\,dx.$$

In order to solve the integral, we introduce the change of variable $x^2 = 4t\,z$ from where we can write, already simplifying

$$I_M = \frac{1}{\sqrt{\pi}}\sum_{k=0}^{\infty}\frac{(-1)^k(4t)^{\alpha k}}{\Gamma(2\alpha k+1)}\int_0^{\infty}e^{-z}z^{\alpha k-1/2}\,dz.$$

The remaining integral is exactly the gamma function, so

$$I_M = \frac{1}{\sqrt{\pi}}\sum_{k=0}^{\infty}\frac{(-1)^k(4t)^{\alpha k}}{\Gamma(2\alpha k+1)}\Gamma(\alpha k+1/2).$$

For simplicity, let's make use of the Legendre duplication formula

$$\sqrt{\pi}\Gamma(2y) = 2^{2y-1}\Gamma(y)\Gamma(y+1/2)$$

with $y = \alpha k + 1/2$, of which follows

$$I_M = \sum_{k=0}^{\infty} \frac{(-1)^k 2^{2\alpha k} t^{\alpha k}}{2^{2\alpha k} \Gamma(\alpha k + 1) \Gamma(\alpha k + 1/2)} \Gamma(\alpha k + 1/2).$$

By simplifying and reincorporating the first member, we can write

$$\frac{1}{\sqrt{\pi t}} \int_0^{\infty} e^{-x^2/4t} E_{2\alpha}(-x^{2\alpha}) \, dx = \sum_{k=0}^{\infty} \frac{(-t^{\alpha})^k}{\Gamma(\alpha k + 1)} = E_{\alpha}(-t^{\alpha})$$

which is the desired result   ◇

39. We denote the second member by II_M, using the duplication formula for the classical Mittag-Leffler function and rearranging, we can write

$$II_M \equiv \frac{2x}{\pi} \int_0^{\infty} \frac{E_{2\alpha}(-t^2)}{x^2 + t^2} \, dt = \frac{x}{\pi} \int_0^{\infty} \frac{E_{\alpha}(it) + E_{\alpha}(-it)}{x^2 + t^2} \, dt.$$

Separating the integral into two others, considering a change of variable $t \to -t$ and returning to write as a single integral, with the different ends, we obtain

$$II_M = \frac{x}{\pi} \int_{-\infty}^{\infty} \frac{E_{\alpha}(it)}{x^2 + t^2} \, dt.$$

To evaluate the integral, we first introduce the series representation for the classical Mittag-Leffler function, change the order of integration with the sum, then

$$II_M = \frac{x}{\pi} \sum_{k=0}^{\infty} \frac{i^k}{\Gamma(\alpha k + 1)} \int_{-\infty}^{\infty} \frac{t^k}{x^2 + t^2} \, dt.$$

The remaining integral is evaluated through the complex plane, with the residue theorem. Consider an integration contour, as in Fig. 3.3, in the complex plane, composed by a semicircle, in the upper half-plane, centered at the origin and radius $R > 0$ and a straight line segment of $-R$ to R, closed and counterclockwise oriented. Consider the function $f(z) = z^k/(z^2 + x^2)$ where $z = t + iy$ with $t, y \in \mathbb{R}$.

The poles of this function lie at the points $z_1 = ix$ and $z_2 = -ix$ where the only one that is within the contour is z_1, that is, it is the only one that contributes with the residue. Then, through the integration, we can write, only for the integral

$$\oint_{\Gamma} \frac{z^k}{z^2 + x^2} \, dz = \int_{\Gamma_R} \frac{z^k}{z^2 + x^2} \, dz + \int_{-R}^{+R} \frac{t^k}{t^2 + x^2} \, dt,$$

where Γ_R denotes the integration in the semicircle.

Taking the limit $R \to \infty$, using the Jordan lemma and the residue theorem, we have

$$\oint_\Gamma \frac{z^k}{z^2 + x^2}\, dz = \int_{-\infty}^{+\infty} \frac{t^k}{t^2 + x^2}\, dt = 2\pi i \lim_{z \to ix} (z - ix) \frac{z^k}{z^2 + x^2}$$

or only for the real integral, simplifying, in the following manner

$$\int_{-\infty}^{+\infty} \frac{t^k}{t^2 + x^2}\, dt = \frac{\pi}{x}(ix)^k.$$

Returning with the result of the real integral in the expression for the second member and rearranging, we can write

$$\text{II}_M = \frac{x}{\pi} \sum_{k=0}^{\infty} \frac{i^k}{\Gamma(\alpha k + 1)} \frac{\pi}{x}(ix)^k = \sum_{k=0}^{\infty} \frac{(-x)^k}{\Gamma(\alpha k + 1)}.$$

Since the remaining summation is exactly the power series representation for the classical Mittag-Leffler function, we obtain

$$\frac{2x}{\pi} \int_0^\infty \frac{E_{2\alpha}(-t^2)}{x^2 + t^2}\, dt = E_\alpha(-x)$$

which is the desired result   ◇

40. We start by writing this particular Mittag-Leffler function with four parameters

$$E_{k,\beta}^{\gamma,q}(z) = \sum_{n=0}^{\infty} \frac{(\gamma)_{qn}}{\Gamma(kn + \beta)} \frac{z^n}{n!}$$

or even using the definition of the generalized Pochhammer symbol

$$E_{k,\beta}^{\gamma,q}(z) = \frac{1}{\Gamma(\beta)} \sum_{n=0}^{\infty} \frac{(\gamma)_{qn}}{(\beta)_{kn}} \frac{z^n}{n!}.$$

Explaining the Pochhammer symbols in terms of a product,

$$(\gamma)_{qn} = \prod_{i=1}^{q} \left(\frac{\gamma + i - 1}{q}\right)_n q^{qn} \quad \text{and} \quad (\beta)_{kn} = \prod_{i=1}^{k} \left(\frac{\beta + j - 1}{k}\right)_k k^{kn}$$

we can write for the Mittag-Leffler function with four parameters

$$E_{k,\beta}^{\gamma,q}(z) = \frac{1}{\Gamma(\beta)} \sum_{n=0}^{\infty} \frac{1}{n!} \frac{\displaystyle\prod_{i=1}^{q} \left(\frac{\gamma + i - 1}{q}\right)_n}{\displaystyle\prod_{i=1}^{k} \left(\frac{\beta + j - 1}{k}\right)_k} \left(\frac{q^q}{k^k} z\right)^n.$$

Using the notation for the generalized hypergeometric function, according to Eq. (2.5), we can write

$$E_{k,\beta}^{\gamma,q}(z) = \frac{1}{\Gamma(\beta)} \, {}_qF_k \left(\begin{array}{c} \dfrac{\gamma}{q}, \dfrac{\gamma+1}{q}, \ldots, \dfrac{\gamma+q-1}{q} \\[2ex] \dfrac{\beta}{k}, \dfrac{\beta+1}{k}, \ldots, \dfrac{\beta+k-1}{k} \end{array} \, \middle| \, \dfrac{q^q}{k^k} z \right).$$

Entering the notation with q-uplas and k-uplas

$$\Delta(q;\gamma) = \frac{\gamma}{q}, \frac{\gamma+1}{q}, \ldots, \frac{\gamma+q-1}{q} \quad \text{and} \quad \Delta(k;\beta) = \frac{\beta}{k}, \frac{\beta+1}{k}, \ldots, \frac{\beta+k-1}{k}$$

we obtain the following expression [17]

$$E_{k,\beta}^{\gamma,q}(z) = \frac{1}{\Gamma(\beta)} \, {}_qF_k \left(\begin{array}{c} \Delta(q;\gamma) \\[2ex] \Delta(k;\beta) \end{array} \, \middle| \, \frac{q^q}{k^k} z \right)$$

which is the desired result   ◇

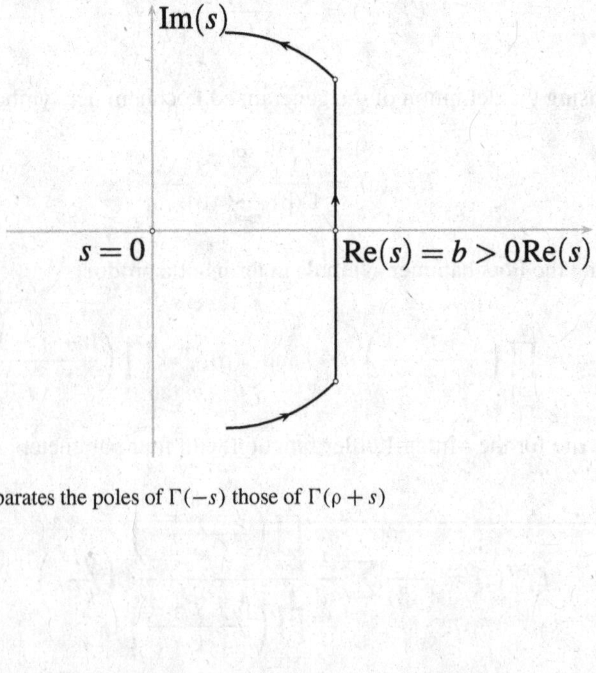

Fig. 3.2 $\mathcal{L}$ separates the poles of $\Gamma(-s)$ those of $\Gamma(\rho+s)$

3.3.4 Proposed Exercises

1. Let $\alpha > 0$. Show that $E_{1,\alpha}(x) = 1 + x E_{1,\alpha+1}(x)$.
2. Show that

$$E_2(-x^2) = \sum_{k=0}^{\infty} \frac{(-x^2)^k}{\Gamma(2k+1)} = \cos x \cdot$$

3. Using Proposed exercise (2), considering $x \rightarrow ix$ to obtain

$$E_2(x^2) = \sum_{k=0}^{\infty} \frac{(x^2)^k}{\Gamma(2k+1)} = \cosh x \cdot$$

4. Show that $E_{\frac{1}{2}}(x) = e^{x^2} \operatorname{erfc}(-x) \cdot$
5. Show that $x E_{2,2}(x^2) = \sinh x \cdot$
6. Show the relation

$$E_3(z^3) = \frac{1}{2}\left[e^z + 2 e^{-z/2} \cos\left(\frac{\sqrt{3}}{2} z\right) \right],$$

where $E_3(\cdot)$ is the Mittag-Leffler function of order three.
7. The generalized Laguerre function [18] with parameters α and n, denoted by $L_n^{(\alpha)}(x)$, is related with the confluent hypergeometric function by means of

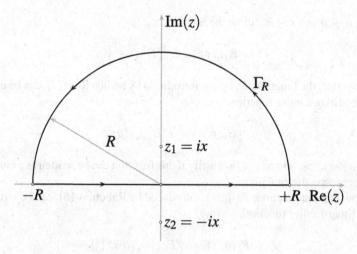

Fig. 3.3 Γ_R is a semicircle centered at the origin and radius $R > 0$ and a straight line segment of $-R$ to R, closed and counterclockwise oriented

$$L_n^{(\alpha)}(x) = \binom{n+\alpha}{n} {}_1F_1(-n; \alpha; x)$$

with $n = 0, 1, 2, \ldots$ and $\alpha \in \mathbb{R}$, being $\binom{n+\alpha}{n} = \dfrac{\Gamma(n+\alpha+1)}{n!\,\Gamma(\alpha+1)}$. Show that

$$L_k^{(\alpha)}(x) = (k+1)_\alpha \, E_{1,\alpha}^{-k}(x).$$

8. Let $E_{\alpha,\beta}(\cdot)$ be a Mittag-Leffler function with two parameters. Show the relation

$$\frac{d}{dx} E_{\alpha,\beta}(x) = x \frac{d}{dx} E_{\alpha,\alpha+\beta}(x) + E_{\alpha,\alpha+\beta}(x).$$

9. Let $\mathrm{Re}(\beta) > 1$. Show the contiguous relation in parameter β

$$\alpha x \frac{d}{dx} E_{\alpha,\beta}(x) = E_{\alpha,\beta-1}(x) + (1-\beta) E_{\alpha,\beta}(x).$$

10. Let $\alpha, \beta, \rho > 0$. Show that $E_{\alpha,\beta}(x^{\alpha/\gamma}) = E_{\gamma,\beta}(x)$.
11. Let $\alpha, \beta \in \mathbb{R}$ and $k = 0, 1, 2, \ldots$. Show that the k-th derivative of the Mittag-Leffler function with two parameters is equal to the product of the factorial of $k!$ by a Mittag-Leffler function with three parameters, that is,

$$\left(\frac{d}{dx}\right)^k E_{\alpha,\beta}(x) = k! \, E_{\alpha,\beta+\alpha k}^{k+1}(x).$$

12. Let $\alpha > 0$ and $x \in \mathbb{R}$. Show the relation

$$E_\alpha(x) = x \, E_{\alpha,\alpha+1}(x) + 1.$$

13. Show that, the function, $E_x(\nu, a)$, introduced by Miller-Ross [7], can be express as a Mittag-Leffler function,

$$E_x(\nu, a) = x^\nu E_{1,\nu+1}(ax)$$

with $\mathrm{Re}(\nu) > -1$ and $a \in \mathbb{R}$. Justify if this function can be written as a confluent hypergeometric function.
14. Show that the function, $R_\alpha(\beta, x)$, introduced by Rabotnov [8], can be written as a Mittag-Leffler function,

$$R_\alpha(\beta, x) = x^\alpha E_{\alpha+1,\alpha+1}(\beta x^{\alpha+1})$$

with $\mathrm{Re}(\alpha) > -1$. Justify if this function can be written as a confluent hypergeometric function.
15. Let $E_{1,3}(x)$ be a Mittag-Leffler function with two parameters, show that

$$E_{1,3}(x) = \frac{e^x - x - 1}{x^2}$$

and the particular result $\lim\limits_{x \to 0} E_{1,3}(x) = 1/2$.

16. In the Exercise (22), Chap. 6, the solution was written as a Mittag-Leffler function with two parameters. Show that

$$E_{2,2-\alpha}(-\omega^2 t^2) = i\omega^{\alpha-1} \sin\left(\omega t - \frac{\pi}{2}\alpha\right) \exp\left[i\frac{\pi}{2}(2\alpha - 1)\right]$$

with $0 < \alpha < 1$ and ω a positive constant.

17. Using the Exercise (21) discuss the case $\alpha = 2$.

18. Let $\mu > 0$. Show that the classical Mittag-Leffler function can be written as

$$E_\mu(x) = {}_1\Psi_1\left[\begin{matrix}(1, 1) \\ (1, \mu)\end{matrix}\bigg| x\right]$$

where ${}_1\Psi_1(\cdot)$ is a Wright function.

19. Let $\mu > 0$. Show that the classical Mittag-Leffler function can be written as

$$E_\mu(x) = H_{1,2}^{1,1}\left[-x \bigg| \begin{matrix}(1, 1) \\ (1, 1), (0, \mu)\end{matrix}\right]$$

where $H_{1,2}^{1,1}(\cdot)$ is a Fox's H-function.

20. Let $\alpha > 0$. Show the called duplication formula

$$E_{2\alpha}(x^2) = \frac{1}{2}[E_\alpha(x) + E_\alpha(-x)]$$

for the classical Mittag-Leffler function.

21. Using the result of the Exercise (22) and the duplication formula for the classical Mittag-Leffler function, show that

$$E_\alpha(x) = E_{2\alpha}(x^2) + x E_{2\alpha,\alpha+1}(x^2).$$

22. Let $\alpha > 0$. Show that the following relation is worth

$$\int_0^x E_\alpha(\xi^\alpha) \, d\xi = x E_{\alpha,2}(x^\alpha).$$

23. As in Exercise (17) and using partial fraction, show that

$$E_{\alpha,\beta}(x) = \frac{1}{2\pi i} \int_{C_\gamma} \frac{e^\xi \xi^{\alpha-\beta}}{\xi^\alpha - x} \, d\xi$$

with $\alpha > \beta > 0$.

24. Let $\alpha > 0$ and $x \neq 0$. Show that

$$E_{-\alpha}(x) + E_\alpha\left(\frac{1}{x}\right) = 1.$$

25. Show that

$$E_{-2,2}\left(-\frac{1}{x^2}\right) = 1 - \frac{\sin x}{x}.$$

26. Show that

$$\frac{d^k}{dx^k}\left[E_k(\lambda x^k)\right] = \lambda E_k(\lambda x^k)$$

with λ a positive constant.

27. Consider the parameters such that $\alpha, \beta, \delta, \rho \in \mathbb{C}, p, q > 0, \mathrm{Re}(\alpha) > 0, \mathrm{Re}(\beta) > 0, \mathrm{Re}(\rho) > 0, \mathrm{Re}(\delta) > 0, \mathrm{Re}(\alpha) + p \geq q$ and $\delta \neq 1$, show that [2]

$$E_{\alpha,\beta,p}^{\rho,\delta,q}(x) = \beta E_{\alpha,\beta+1,p}^{\rho,\delta-1,q}(x) + \alpha x \frac{d}{dx} E_{\alpha,\beta+1,p}^{\rho,\delta,q}(x).$$

28. Assume the conditions as in Proposed exercise (27), the Mittag-Leffler function with five parameters, was introduced by means of the following power series [19]

$$E_{\alpha,\beta,\delta}^{\rho,q}(x) = \sum_{k=0}^{\infty} \frac{(\rho)_{kq}}{\Gamma(\alpha k + \beta)(\delta)_k} \frac{x^k}{}.$$

Using the result given in Proposed exercise (27) and a particular value of the parameter to obtain the Mittag-Leffler function with five parameters, that is, a particular case of a Mittag-Leffler function with six parameters.

29. Assume the conditions as in Proposed exercise (27), show the relation involving Mittag-Leffler functions with five parameters

$$E_{\alpha,\beta,\delta}^{\rho,q}(x) - E_{\alpha,\beta,\delta-1}^{\rho,q}(x) = \frac{x}{1-\delta} \frac{d}{dx} E_{\alpha,\beta,\delta}^{\rho,q}(x).$$

30. Consider the parameters such that $\alpha, \beta, \rho \in \mathbb{C}, \mathrm{Re}(\alpha) > 0, \mathrm{Re}(\beta) > 0, \mathrm{Re}(\rho) > 0$, and $q \in (0, 1) \cup \mathbb{N}$. The Mittag-Leffler function with four parameters was introduced by means of the following power series [17]

$$E_{\alpha,\beta}^{\rho,q}(x) = \sum_{k=0}^{\infty} \frac{(\rho)_{kq}}{\Gamma(\alpha k + \beta)} \frac{x^k}{k!}.$$

Show the relation [2]

$$E_{\alpha,\beta-\alpha}^{\rho,q}(x) - E_{\alpha,\beta-\alpha}^{\rho-1,q}(x) = qx \sum_{k=0}^{\infty} \frac{(\rho)_{kq+q-1}}{\Gamma(\alpha k + \beta)} \frac{x^k}{k!}.$$

31. Consider the conditions as in the precedent one and taking $q = 1$ to show that

$$E_{\alpha,\beta-\alpha}^{\rho}(x) - E_{\alpha,\beta-\alpha}^{\rho-1}(x) = x E_{\alpha,\beta}^{\rho}(x),$$

with $E_{a,b}^{c}(\cdot)$ a Mittag-Leffler function with three parameters, a particular case of the Mittag-Leffler function with four parameters.

32. Use Exercise (39) to discuss particular cases of α. Consider three particular values $\alpha = 0$, $\alpha = 1/2$ and $\alpha = 1$.

33. Let $E_{\alpha,\beta}^{\gamma,q}(z)$ be a Mittag-Leffler function with four parameters with $q \in (0,1)$. Use the definition of generalized Wright function, as in Definition 2.4.2 to show the relation

$$E_{\alpha,\beta}^{\gamma,q}(z) = \frac{1}{\Gamma(\gamma)} \,{}_1\Psi_1 \left[\begin{array}{c} (\gamma, q) \\ \\ (\beta, \alpha) \end{array} \; z \right]$$

with ${}_1\Psi_1(\cdot)$ a Wright function.

34. Let $E_{\alpha,\beta}^{\gamma,q}(z)$ be a Mittag-Leffler function with four parameters with $q \in (0,1)$. Use the definition of Fox's H-function, as in Definition 2.4.4, to show that

$$E_{\alpha,\beta}^{\gamma,q}(z) = \frac{1}{\Gamma(\gamma)} H_{1,2}^{1,1} \left[-z \left| \begin{array}{c} (1-\gamma, q) \\ (0,1), (1-\beta, \alpha) \end{array} \right. \right]$$

with $H_{1,2}^{1,1}(\cdot)$ a Fox's H-function.

35. With the data of precedent one, show the following Mellin-Barnes representation for the Mittag-Leffler function with four parameters,

$$\Gamma(\gamma) E_{\alpha,\beta}^{\gamma,q}(z) = \frac{1}{2\pi i} \int_L \frac{\Gamma(s)\Gamma(\gamma - qs)}{\Gamma(\beta - \alpha s)} (-z)^{-s} \, ds$$

with L a contour that separates the poles from the gamma functions into the numerator of the integrand, according to Appendix A.

36. Using Exercise (39) evaluate the integral

$$\Omega \equiv \frac{2x}{\pi} \int_0^{\infty} \frac{\cos(i\omega)}{x^2 + \omega^2} \, d\omega.$$

37. In a recent paper was introduced the function [20]

$$\mathbb{E}_{\lambda,\mu}(z) = \Gamma(\mu) z E_{\lambda,\mu}(z)$$

with $z, \lambda, \mu \in \mathbb{C}$, $\operatorname{Re}(\lambda) > 0$ and $\mu \neq 0, -1, -2, \ldots$, being $E_{\lambda,\mu}(\cdot)$ a Mittag-Leffler function with two parameters. Show that

$$\mathbb{E}_{1,4}(z) = \frac{6(e^z - z - 1) - 3z^3}{z^3}.$$

38. Let $\alpha, \beta \in \mathbb{R}_+$. Show that

$$(1 + \delta_x)\, E_{\alpha,\beta}(x) = E_{\alpha,\beta}^2(x)$$

with $\delta_x \equiv x\dfrac{d}{dx}$.

39. Let $\alpha \in \mathbb{R}_+$. Show that

$$E_\alpha(x) = (1 + \alpha\,\delta_x)\, E_{\alpha,2}(x)$$

with $\delta_x \equiv x\dfrac{d}{dx}$. Discuss the particular case $\alpha = 1$.

40. Let $\mathrm{Re}(\beta) > 0$. Show that

$$(\beta - \alpha\rho) E_{\alpha,\beta+1}^\rho(x) = E_{\alpha,\beta}^\rho(x) - \alpha\rho E_{\alpha,\beta+1}^{\rho+1}(x)$$

so that the classical Mittag-Leffler functions are well defined.

References

1. Mittag-Leffler, G.M.: Sur la nouvelle fonction $E_\alpha(x)$. C. R. Acad. Sci. Paris **137**, 554–558 (1903)
2. Sales Teodoro, G.: Fractional Calculus and the Mittag-Leffler Functions. Master Thesis, Imecc-Unicamp, Campinas (2014). (in Portuguese)
3. Sales Teodoro, G., Capelas de Oliveira, E.: Laplace transform and the Mittag-Leffler function. Int. J. Math. Educ. Sci. Technol. **45**, 595–604 (2014)
4. Agarwal, R.P.: A propos d'une note de M. Pierre Humbert. C. R. Acad. Sci. Paris **236**, 2031–2032 (1953)
5. Prabhakar, T.R.: A singular integral equation with generalized Mittag-Leffler function in the kernel. Yokohama Math. J. **19**, 7–15 (1971)
6. Figueiredo Camargo, R., Capelas de Oliveira, E.: Fractional Calculus. Editora Livraria da Física, São Paulo (2015). (in Portuguese)
7. Miller, K.S., Ross, B.: An Introduction to the Fractional Calculus and Fractional Differential Equations. Wiley, New York (1993)
8. Rabotnov, Y.N.: Creep Problems in Structural Members. North-Holland, Amsterdam (1969)
9. Erdélyi, A. (ed.): Higher Tanscendental Functions, vol. 3. McGraw-Hill, New York (1955)
10. Wright, E.M.: On the coefficients of power series having exponential singularities. J. Lond. Math. Soc. **8**, 71–79 (1933)
11. Podlubny, I.: Fractional Differential Equations. Mathematical in Sciences and Engineering, vol. 1, p. 98. Academic Press, San Diego (1999)
12. Mainardi, F.: Fractional Calculus and Waves in Linear Viscoelasticity, An Introduction to Mathematical Models. Imperial College, London (2010)
13. Mainardi, F., Pagnini, G.: Space-time fractional diffusion: exact solutions and probability interpretation. In: Proceedings WASCOM-2001. World Scientific, Singapore (2002)

14. Mainardi, F., Pagnini, G.: The fundamental solutions of the time-fractional diffusion equation. In: Fabrizio, M., Lazzari, B., Morro, A. (eds.) Mathematical Models and Methods for Smart Material, vol. 62, pp. 207–224 (2002)
15. Abramowitz, M., Stegun, I.A.: Handbook of Mathematical Functions. Dover Publications Inc., New York (1972)
16. Salim, T.O., Faraj, A.W.: A generalization of Mittag-Leffler functions and integral operator associated with fractional calculus. J. Fract. Cal. Appl. **3**, 1–13 (2012)
17. Shukla, A.K., Prajapati, J.C.: On a generalization of Mittag-Leffler function and its properties. J. Math. Anal. Appl. **336**, 797–811 (2007)
18. Capelas de Oliveira, E.: Special Functions and Applications, 2nd edn. Livraria Editora da Física, São Paulo (2012). (in Portuguese)
19. Khan, M.A., Ahmed, S.: On some properties of fractional calculus operators associated with generalized Mittag-Leffler functions. Thai J. Math. **11**, 645–654 (2013)
20. Bansal, D., Orhan, H.: Partial sums of Mittag-Leffler functions. J. Math. Ineq. **13**, 423–431 (2018)

Chapter 4
Integral Transforms

After presenting the main functions that emerge naturally in integer order calculus and in fractional calculus —respectively the exponential and the classical Mittag-Leffler functions—, we present here the methodology of integral (Laplace, Fourier, and Mellin) transforms as a tool to be used for the resolution of differential equations, integral equations and integrodifferential equations, all of them in the fractional sense.

In general, the problems that are presented are accompanied by an equation (system of equations) and, possibly, with conditions arising either from the geometry of the problem and/or from the physics of the problem, among others, the so-called boundary conditions and, in many cases, the initial condition (initial velocity), when the temporal variable is present.

In relation to the equations, there are many possibilities related to a specific classification, since they can be, among others, differential or integral or integrodifferential. Moreover, its may be linear or non-linear, homogeneous or non-homogeneous, and ordinary or partial, so that we do not extend ourselves further. Once the equation is known, it is the turn of the conditions. Here too, we have several possibilities, but we will restrict ourselves to the initial conditions and boundary conditions. Then, given the problem, that is, equation and conditions, we should look for the best methodology to solve it. Here the range of options grows almost uncontrollably, because there are methods appropriate for each type of equation.

Regarding the methodology we will make a distinction, from the beginning, because we will only concern ourselves with analytical methods, not concerned with numerical and computational methods. Moreover, only with the called integral transformation methodology and, in this case, focus only on the Laplace, Fourier and Mellin transforms, since these will be used to solve a particular exercise.

© Springer Nature Switzerland AG 2019
E. Capelas de Oliveira, *Solved Exercises in Fractional Calculus*, Studies in Systems, Decision and Control 240,
https://doi.org/10.1007/978-3-030-20524-9_4

However, once the methodology is set, let us go back to the equations, since there is already a restriction to be imposed, namely: the methodology of restricts to linear equations, since the operators associated with them are linear operators. Let us, for now, explain the equations we are going to approach. With this restriction, we will be concerned with linear equations, as well as only with the differentials, integrals and integrodifferentials. Once this restriction is imposed, let us turn to the methodology of the integral transform.

The methodology (or method) of the integral transforms, in general lines, lies in the fact of converting the original problem into another, transformed, seemingly simpler to solve. The transformed problem is solved and, by operating with the respective inverse transform, the solution of the original problem is recovered.

An integral transform is characterized by the definition interval, for example, for the Fourier transform, the entire real axis; by the kernel, for example, for the Mellin transform, x^{s-1}, the variable s being associated with the inverse transform; and by the class of functions that will be transformed, for example, for the Laplace transform, functions of an exponential order. Once the transform has been characterized, we can discuss the properties associated with it. Finally, it is important to emphasize that we are focused on working with integral transforms in the sense of having an operative method so that we can have a tool to solve problems arising from the fractional calculus.

This chapter is arranged as follows: we begin by formalizing, in general lines, what we describe as the methodology of the integral transform; we specify the kernel, the definition interval and the class of functions for the Fourier transform, then the Laplace transform, and finally for the Mellin transform. In general, this is not the order that is usually presented in this topic, but we chose to do it in that order, because it seems to be more natural. For each of these transforms we present some properties, as well as the respective inverse transform. In the course of the text, we present some examples, but only at the end is a list of solved exercises, as well as some others proposed for the student to solve.

4.1   Methodology

In this section we present the general case of an integral transform so that in the following sections we discuss the particular integral transform that, although there are others, we will only work with the Fourier, Laplace and Mellin [1, 2].

Definition 4.1.1 *(Integral transform)* Let $f(x)$ be a real function of real variable and I be an interval. The integral transform of $f(x)$, denoted by $\mathscr{T}[f(x)]$ or by $F(\cdot)$, where $(\cdot)$ is the transformed variable, is defined by

$$\mathscr{T}[f(x)] = F(\cdot) = \int_I \mathscr{N}(x, \cdot) f(x) \, dx$$

with $\mathcal{N}(x, \cdot)$, a given function of two variables, the kernel of the transform. The $f(\cdot)$ function is called inverse transform, denoted by $f(x) = \mathcal{T}^{-1} F(\cdot)(x)$.

It should be mentioned that both the integral transform and its respective inverse transform are linear operators and, even more, operating first with the direct operator and then with the respective inverse, or vice versa, we obtain the same result, that is, results in the identity, that is, they are worth the following equalities

$$\mathcal{T} \mathcal{T}^{-1} = \mathbb{I} = \mathcal{T}^{-1} \mathcal{T}$$

where $\mathbb{I}$ is the identity operator.

Further, we can prove the uniqueness of the transform, that is, if $\mathcal{T}[f(x)] = \mathcal{T}[g(x)]$ then, for proper conditions, we have $f(x) = g(x)$ [1].

4.2 Fourier Transform

Before we define the Fourier transform, we will justify its use in engineering. In general, in this integral transform, the function $f(t)$ of the original problem can represent a signal (electric, acoustic, etc...) as a function of time, t. Then, the Fourier transform, denoted by $F(\omega)$, of $f(t)$, ω being a frequency, represents the frequency spectrum of this signal and is as useful as the signal itself. We also point out that the Fourier transform is associated with functions that are absolutely integrable in the interval $(-\infty, +\infty)$, as well as satisfy the so-called Dirichlet conditions [1], for the continuity of the function, in the interval $(-\infty, +\infty)$, so that it is worth the integral Fourier theorem,

$$f(x) = \frac{1}{\sqrt{2\pi}} \int_{-\infty}^{\infty} \left\{ \frac{1}{\sqrt{2\pi}} \int_{-\infty}^{\infty} f(\xi) \, e^{\pm i k \xi} \, d\xi \right\} e^{\mp i k x} \, dk.$$

Since we are going to introduce the direct and inverse Fourier transforms (hereafter referred to as the pair of transforms) from this theorem, two observations must be clear. First, in the denominator we write, separately, $\sqrt{2\pi}$, whose product provides 2π. So this is the symmetrical way of writing, but we could consider 2π outside the keys or inside the keys, the important thing is that the product is 2π. The other observation is relative to the plus or minus sign. Note that inside the keys we take $\pm$ while outside $\mp$, this means the following: when considering a signal inside, the other must be opposite. In both observations there is no obligation, however, it must be obeyed that the product is equal to 2π, in the denominator, and signals exchanged in the exponents of the kernel.

Finally, we mention that, in analogy to the Fourier series, parity of the function plays an important role, since it simplifies several calculations. In this sense, we usually define the called Fourier transform in sine (odd functions) and the Fourier transform in cosine (even functions).

Definition 4.2.1 (*Fourier transform*) The Fourier transform of the function $f(t)$, denoted by $\mathscr{F}[f(t)] = F(\omega)$, with $f : \mathbb{R} \to \mathbb{R}$ and $F(\omega) : \mathbb{R} \to \mathbb{C}$ being $\omega \in \mathbb{R}$, the transform variable, is defined by the integral

$$\mathscr{F}[f(t)] = F(\omega) = \frac{1}{\sqrt{2\pi}} \int_{-\infty}^{\infty} e^{-i\omega t} f(t)\, dt.$$

We note that from this definition, the kernel is equal to $e^{-i\omega t}$ (a function of two variables), the interval is $(-\infty, \infty)$ and the functions must be absolutely integrable. Note that we are considering the minus sign in the expression that lies between the keys in the expression of the integral Fourier theorem.

Definition 4.2.2 (*Inverse Fourier transform*) The inverse Fourier transform of the function $F(\omega)$, denoted by $\mathscr{F}^{-1}[F(\omega)] = f(t)$, is defined by the integral

$$\mathscr{F}^{-1}[F(\omega)] = f(t) = \frac{1}{\sqrt{2\pi}} \int_{-\infty}^{\infty} e^{i\omega t} F(\omega)\, d\omega.$$

As a consequence of the definition of the Fourier transform with the minus sign in the kernel, here the signal has been switched, i.e., the forward with the minus sign and the inverse with the plus sign. Further, the inverse Fourier transform retrieves the function $f(t)$.

4.2.1   Properties

Let us present some properties of the Fourier transform, in particular, by playing an important role in the definitions of integral and fractional derivatives, the Fourier convolution product. Several others are presented as solved or proposed exercises.

Property 4.2.1 (Linearity) *Let $a, b \in \mathbb{R}$. The Fourier transform is linear,*

$$\mathscr{F}(af_1 \pm bf_2) = a\mathscr{F}(f_1) \pm b\mathscr{F}(f_2)$$

provided that f_1 and f_2 support the respective Fourier transforms.

Property 4.2.2 (Fourier convolution product) *We define the convolution product, also called the Fourier convolution product, denoted by $\star$, between two well-behaved functions, in order to satisfy the Dirichlet conditions $f(t)$ and $g(t)$ with $t \in \mathbb{R}$, by means of the integral*

$$f(t) \star g(t) := \int_{-\infty}^{\infty} f(\tau)g(t - \tau)\, d\tau = \int_{-\infty}^{\infty} g(\tau)f(t - \tau)\, d\tau$$

since the commutative property is valid, that is, $f(t) \star g(t) = g(t) \star f(t)$.

Property 4.2.3 (Convolution Fourier transform) *The Fourier transform of the convolution product of two functions is equal to the product of the respective Fourier transforms*

$$\mathscr{F}[f_1 \star f_2](\omega) = \mathscr{F}[f_1](\omega)\mathscr{F}[f_2](\omega).$$

Property 4.2.4 (Fourier transform of multiplication) *The Fourier transform of the multiplication of two functions is given by*

$$\mathscr{F}[f_1 \cdot f_2](\omega) = \frac{1}{2\pi}\mathscr{F}[f_1](\omega) \star \mathscr{F}[f_2](\omega).$$

Property 4.2.5 (Fourier transform of the derivative) *Let $n \in \mathbb{N}$. The Fourier transform and the inverse Fourier transform of the derivative of order n, are given by*

$$\mathscr{F}[f^{(n)}](\omega) = (i\omega)^n \mathscr{F}[f](\omega) \quad and \quad \mathscr{F}^{-1}[f^{(n)}](\omega) = (it)^n f(t),$$

respectively.

4.3 Laplace Transform

In analogy to the Fourier transform, let's talk a little about the importance of the Laplace transform. The Laplace transform is a very useful tool, for example, to obtain a particular solution of a fractional differential equation. Further, the Fourier transform associated with a signal always exists, which does not occur with the Laplace transform, since exponential growth can be induced, as we shall see, immediately after the definition. Faced with this, we must be concerned about the called convergence region.

Definition 4.3.1 (*Bilateral Laplace transform*) Let $t \in \mathbb{R}$ and $f(t)$ be a signal. The bilateral Laplace transform of $f(t)$, denoted by $\mathscr{L}[f](s) = \mathsf{F}(s)$, is defined by the integral

$$\mathscr{L}[f](s) = \mathsf{F}(s) := \int_{-\infty}^{\infty} e^{-st} f(t)\,dt$$

being s the parameter such that $s = \sigma + i\tau$ with $\sigma, \tau \in \mathbb{R}$.

Before turning to the presentation of the properties involving the Laplace transform, it is worth highlighting the study of the convergence region, due to its relevance. We will discuss and justify this relevance through a specific example.

Example 4.1 Convergence region. Let $t \in \mathbb{R}$. Consider the function $f_1(t) = e^{-at}$ $u(t)$ with $a > 0$, being $u(t)$ the unitary step function, defined by

$$u(t) = \begin{cases} 1, & if\ t \geq 0, \\ 0, & if\ t < 0. \end{cases}$$

We have, then, for the bilateral Laplace transform of $f_1(t)$

$$\mathscr{L}[f_1](s) = \int_{-\infty}^{\infty} e^{-st} e^{-at} u(t) \, dt = \int_{0}^{\infty} e^{-(a+s)t} \, dt = \frac{1}{s+a}$$

with $\text{Re}(s) > -a$. So, the convergence region is the complex semiplane $\text{Re}(s) = \sigma > -a$.

Let us now consider the following function $f_2(t) = -e^{-at} u(-t)$ with $a > 0$, being $u(t)$ the unitary step function. In this case, the bilateral Laplace transform of $f_2(t)$ is given by

$$\mathscr{L}[f_2](s) = \int_{-\infty}^{\infty} e^{-st} e^{-at} [-u(t)] \, dt = -\int_{-\infty}^{0} e^{-(a+s)t} \, dt = \frac{1}{s+a}$$

with $\text{Re}(s) < -a$. So, the convergence region is the complex semiplane $\text{Re}(s) = \sigma < -a$.

In short, what is clear in this example is the fact that we have $\mathscr{L}[f_1](s) \equiv \mathscr{L}[f_2](s)$, the two bilateral Laplace transforms are equal, although we have $f_1(t) \neq f_2(t)$, the two functions are distinct, or else, what differentiates the two cases is the convergence region.

After emphasizing the importance of the convergence region in the study of the Laplace transform, we will restrict ourselves to the fact that, from now on, we will consider a continuous time signal, still denoted by $f(t)$, now t is the time variable. Before we define the unilateral Laplace transform, or Laplace transform only, we will present the concepts of exponential order function, limited function and admissible function. Note that in the Fourier transform analogy where functions should be square-integrable, here we have other conditions to be imposed for the function so that its Laplace transform is defined.

Definition 4.3.2 (*Exponential order function*) A function $f(t)$, defined for every $t \geq 0$, is said to be an exponential order function if there are real constants $M > 0$, $t_0 > 0$ and $a \in \mathbb{R}$, such that

$$|f(t)| \leq M \, e^{at}, \qquad \text{for all} \quad t > t_0$$

or, equivalently,

$$e^{-at} |f(t)| \leq M \qquad \Longleftrightarrow \qquad \lim_{t \to \infty} |f(t) \, e^{-at}| = 0 \cdot$$

Definition 4.3.3 (*Limited function*) A function $f(t)$ is said to be a limited function if there exists $M > 0$ such that $|f(t)| \leq M$, for every t in the domain of f.

Thus, we conclude: let $a > 0$; if f is a limited function, then $e^{-at} |f(t)| \leq M$, for every $t \geq 0$, since $|e^{-at}| \leq 1$. Therefore, any limited function is an exponential order function.

Definition 4.3.4 (*Admissible function*) An exponential order function at infinity and continuous by parts over the closed interval $[0, A]$, with $A > 0$, is said to be admissible.

Definition 4.3.5 (*Laplace transform*) Let $t \in \mathbb{R}_+$ and $f(t)$ be an admissible function. We define the Laplace transform, denoted by $\mathscr{L}[f](s) = \mathsf{F}(s)$, of the function $f(t)$, by means of the integral

$$\mathscr{L}[f](s) = \mathsf{F}(s) := \int_0^\infty e^{-st} f(t) \, dt$$

where $s = \sigma + i\tau$, with $\sigma, \tau \in \mathbb{R}$ a parameter (transformed variable).

Theorem 4.3.1 (Existence of the Laplace transform) *If a function $f(t)$, defined for every $t \geq 0$, is continuous by parts in every closed interval $0 \leq t \leq c$, with $c > 0$, and exponential order, then there exists $a \in \mathbb{R}$ so that $\mathscr{L}[f](s)$ exists for every* $\mathrm{Re}(s) = \sigma > a$.

4.3.1 Properties

Let us now present some properties of the Laplace transform, in particular, the one associated with the integer ordering due to its importance in the fractional differential equations.

Property 4.3.1 (Linearity) *Let $a, b \in \mathbb{R}$. The Laplace transform is linear,*

$$\mathscr{L}(af_1 \pm bf_2) = a\mathscr{L}(f_1) \pm b\mathscr{L}(f_2)$$

provided that f_1 and f_2 support the respective Laplace transforms.

Property 4.3.2 (Displacement in the independent variable) *Let $a \in \mathbb{R}$. If $\mathscr{L}[f](s) = \mathsf{F}(s)$, then*

$$\mathscr{L}[e^{-at} f(t)] = \mathsf{F}(s + a) \cdot$$

Property 4.3.3 (Displacement in function) *Let $a > 0$ and $H(t - a)$ be the unitary Heaviside function. If $\mathscr{L}[f](s)$, then*

$$\mathscr{L}[f(t - a)H(t - a)] = e^{-as}\mathsf{F}(s)$$

or, equivalently,

$$\mathscr{L}[f(t)H(t - a)] = e^{-as}\mathscr{L}[f(t)] \cdot$$

Property 4.3.4 (Laplace transform of a derivative) *Consider a function $f : \mathbb{R} \to \mathbb{R}$ whose derivatives up to order $(n - 1)$ are continuous in the closed interval $[0, c] \subset \mathbb{R}$*

and the derivative of order n is continuous by parts in the interval $[0, c] \subset \mathbb{R}$, *so that there exist constants* $M > 0$ *and* $t_0 > 0$ *such that*

$$|f(t)| \le M e^{bt}, \quad \left|\frac{\mathrm{d}}{\mathrm{d}t} f(t)\right| \le M e^{bt}, \quad \dots, \quad \left|\frac{\mathrm{d}^{n-1}}{\mathrm{d}t^{n-1}} f(t)\right| \le M e^{bt}$$

for every $t > t_0$. *Then, for* $\mathrm{Re}(s) > b$, *we have*

$$\mathscr{L}\left[\frac{\mathrm{d}^n}{\mathrm{d}t^n} f\right](s) = s^n \mathscr{L}[f](s) - \sum_{k=0}^{n-1} s^k \left[\left(\frac{\mathrm{d}^{n-1-k}}{\mathrm{d}t^{n-1-k}} f \bigg|_{t=0}\right)\right].$$

Property 4.3.5 (Laplace transform of the convolution) *Let the Laplace transforms of* $f(t)$ *and* $g(t)$, *denoted by* $\mathscr{L}[f](s) = \mathsf{F}(s)$ *and* $\mathscr{L}[g](s) = \mathsf{G}(s)$, *respectively, admitted. The Laplace transform of the convolution product, denoted by* $\mathscr{L}[f(t) \star g(t)](s)$, *is equal to the product of the Laplace transforms of the functions* $f(t)$ *and* $g(t)$

$$\mathscr{L}[f(t) \star g(t)](s) = \mathscr{L}[f](s)\mathscr{L}[g](s) = \mathsf{F}(s)\mathsf{G}(s),$$

where the Laplace convolution product is given by

$$f(t) \star g(t) = \int_0^t f(t - \tau)g(\tau) \, \mathrm{d}\tau = \int_0^t g(t - \tau)f(\tau) \, \mathrm{d}\tau.$$

Property 4.3.6 (Derivative of the Laplace transform) *Let* $n = 0, 1, 2, \dots$ *and* $\mathscr{L}[f](s) = \mathsf{F}(s)$ *be the Laplace transform of the function* $f(t)$, *admitted. Then, we have*

$$\mathscr{L}[t^n f(t)](s) = (-1)^n \frac{\mathrm{d}^n}{\mathrm{d}s^n} \mathsf{F}(s).$$

Property 4.3.7 (Integral of the Laplace transform) *If* $\mathscr{L}[f](s) = \mathsf{F}(s)$ *is the Laplace transform of the function* $f(t)$, *then,*

$$\mathscr{L}\left[\frac{f(t)}{t}\right](s) = \int_s^\infty \mathsf{F}(\xi) \, \mathrm{d}\xi.$$

Property 4.3.8 (Laplace transform of an integral) *If* $\mathscr{L}[f](s) = \mathsf{F}(s)$ *is the Laplace transform of the function* $f(t)$, *then,*

$$\mathscr{L}\left\{\int_0^t f(\tau) \, \mathrm{d}\tau\right\}(s) = \frac{\mathsf{F}(s)}{s}.$$

Before we introduce the inverse Laplace transform, we will discuss the called Bromwich and Hankel contours in the complex plane.

Fig. 4.1 Bromwich contour

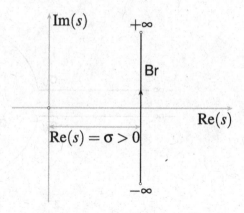

4.3.2 Bromwich and Hankel Contours

As we have already mentioned, the integral transform methodology transfers the initial problem to another one, supposedly simpler to be solved, in order to finally recover the solution of the initial problem through the inverse transform which is done with the help of the complex plane, $s = \sigma + i\tau$ with $\sigma, \tau \in \mathbb{R}$. In the particular case of the Laplace transform, the contour is the called Bromwich contour, also denoted by Br, as in Fig. 4.1.

This contour is composed by a straight line parallel to the vertical axis, $\text{Re}(s) = \sigma > 0$ and the immaginary part, denoted by $\text{Im}(s) = \tau$, such that $-\infty < \tau < +\infty$, so that the singular points are positioned to the left of the vertical straight line.

In practice, since we must have a closed contour, we conduct the Bromwich contour, through a continuous deformation, in the called Hankel contour, denoted by $\text{Ha}(+)$, begining in $-\infty$, along the lower side of the negative real semi-axis, by passing, in the positive direction, the circular disc $|s| = \sigma_0$, with $\sigma_0 < \sigma$ and ending in $-\infty$, along the upper side of the negative real half axis, as in Fig. 4.2.

In order to exemplify, we consider the function $f(s) = s^\mu/g(s)$ with $0 < \mu < 1$ and $g(s)$ being a polynomial with independent term other than zero, that is, $s = 0$ is not polynomial root of $g(s) = 0$. Consider a Hankel contour, denoted by $\text{Ha}(\varepsilon)$ being $\varepsilon > 0$, that is, a *loop* consisting of a circle $|s| = \varepsilon$ and two edges (lower and upper) of the cut line in the negative real semi-axis.

We close the contour with two arcs of circumference (in the lower and upper half planes). Using the residue theorem, the contributions are derived only from the singularities of the $g(s)$ function, since the branching point lies outside the Hankel contour.

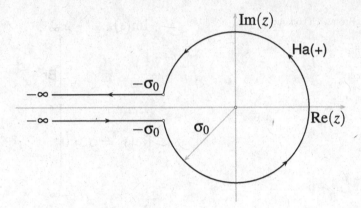

Fig. 4.2 Hankel contour

4.3.3 Inverse Laplace Transform

As we have already mentioned, the convergence region plays an important role in the definition of the bilateral Laplace transform, which will also occur in the respective inverse Laplace transform, denoted by $\mathscr{L}^{-1}[\mathsf{F}(s)] = f(t)$. In order to obtain an expression for the inverse Laplace transform, we introduce $s = \sigma + i\tau$, with $\sigma, \tau \in \mathbb{R}$ in the definition of the bilateral Laplace transform, we can write

$$\mathsf{F}(\sigma + i\tau) = \int_{-\infty}^{\infty} e^{-(\sigma + i\tau)t} f(t)\,dt$$

or in the following form

$$\mathsf{F}(s) = \int_{-\infty}^{\infty} e^{-\sigma t} f(t)\, e^{-i\tau t}\,dt$$

which can be interpreted as the Fourier transform of the function $e^{-\sigma t} f(t)$.

We can then obtain the function $e^{-\sigma t} f(t)$ through the respective inverse Fourier transform,

$$e^{-\sigma t} f(t) = \frac{1}{2\pi} \int_{-\infty}^{\infty} \mathsf{F}(\sigma + i\tau)\, e^{i\tau t}\,d\tau$$

or in the following suitable form

$$f(t) = \frac{1}{2\pi} \int_{-\infty}^{\infty} \mathsf{F}(\sigma + i\tau)\, e^{(\sigma + i\tau)t}\,d\tau.$$

Returning with the variable $s = \sigma + i\tau$, we obtain an integral representation for the inverse Laplace transform, in the complex plane, given by the expression

$$f(t) = \mathscr{L}^{-1}[F](t) = \frac{1}{2\pi i} \lim_{\tau \to \infty} \int_{\sigma-i\tau}^{\sigma+i\tau} e^{st} F(s) \, ds \, ,$$

where $\mathrm{Re}(s) = \sigma > 0$, so that all singularities of the function $F(s)$ are to the left of the straight line $\mathrm{Re}(s) = \sigma$, in the complex plane.

4.3.4 Theorems of Initial Value and Final Value

As we did with convergence, discussing it through an example, we will present two theorems that have their utility in engineering, in particular, in the discussion of asymptotic behavior in both zero and infinite, ends of the integral that defines the Laplace transform. Specifically, in the case of the lower limit, it must be taken into account, for example, if at this point we have a pulse type function, as well as the limits, sometimes from the left of the zero to the right of zero. The initial value theorem, as the name itself says, relates the limit for $t \to 0^+$ of the function $f(t)$ with the limit for $s \to \infty$ of the function sF, where F is the Laplace transform of the function $f(t)$. On the other hand, the final value theorem, as the name itself already states, relates the limit for $t \to \infty$ of function $f(t)$ with the limit for $s \to 0$ of function $sF(s)$, where, also, $F(s)$ is the Laplace transform of the function $f(t)$. It is also an observation in relation to the final value theorem, since it can not always be used.[1]

Finally, as mentioned, in both cases, we find the function $sF(s)$ which emerges directly from the Laplace transform of the first order derivative of the function $f(t)$. In what follows we will present and prove two theorems involving this result.

Theorem 4.3.2 (Theorem of the initial value) *Let $f : \mathbb{R} \to \mathbb{R}$ be a function whose derivative is continuous by parts in the closed interval of $\mathbb{R}$ and both, $f(t)$ and $f'(t)$, have the respective Laplace transform. If the limit $\lim_{s \to \infty} sF(s)$ does exists, then*

$$\lim_{t \to 0^+} f(t) = f(0^+) = \lim_{s \to \infty} sF(s)$$

being $F(s)$ the Laplace transform of $f(t)$ and s the parameter of the transform.

Proof To prove the theorem, we begin with the expression which gives the Laplace transform of the first derivative,

$$\mathscr{L}\left[\frac{d}{dt} f\right](s) = sF(s) - f(0^+)$$

where we use the notation $\mathscr{L}[f](s) = F(s)$.

---

[1] This is a problem which requires complex variables, in particular the study of the singularities and regions of the complex plane, and which is beyond the scope of this book [3].

Since the integration interval is $0^+ \le t \le \infty$, as $s \to \infty$, the kernel tends to zero, from where we can write, from the previous expression

$$\lim_{s \to \infty} \left\{ \int_{0^+}^{\infty} f'(t) \, e^{-st} \, dt \right\} = \lim_{s \to \infty} [sF(s) - f(0^+)] = 0$$

or, using the last equality, in the following form

$$f(0^+) = \lim_{t \to 0^+} f(t) = \lim_{s \to \infty} sF(s)$$

which is the desired result $\square$

This theorem guarantees that we can know the asymptotic behavior of the function $f(t)$ for $t \to 0^+$ from the $s \to \infty$ limit of the function $sF(s)$, as long as it exists.

Theorem 4.3.3 (Theorem of the final value) *Let $f : \mathbb{R} \to \mathbb{R}$ be a function whose derivative is continuous by parts in a closed interval of $\mathbb{R}$ and both, $f(t)$ and $f'(t)$, have the respective Laplace transform. If the limit $\lim_{t \to \infty} f(t)$ does exist, then*

$$\lim_{t \to \infty} f(t) = \lim_{s \to 0} sF(s)$$

being $F(s)$ the Laplace transform of $f(t)$ and s the parameter of the transform.

Proof To prove the theorem, we begin with the expression which gives the Laplace transform of the first derivative,

$$\mathscr{L}\left[\frac{d}{dt} f \right](s) = sF(s) - f(0)$$

where we use the notation $\mathscr{L}[f](s) = F(s)$, from where it follows, already writing $s \to 0$,

$$\lim_{s \to 0} \left\{ \int_{0}^{\infty} f'(t) \, e^{-st} \, dt \right\} = \lim_{s \to 0} [sF(s) - f(0)].$$

Taking the limit $s \to 0$, the kernel tends to one, from where we can write, from the previous expression and already integrating in the variable t,

$$\left\{ \int_{0}^{\infty} f'(t) \, dt \right\} = f(t) \Big|_{0}^{\infty} = f(\infty) - f(0).$$

Comparing the two results, we can write

$$\lim_{s \to 0} [sF(s) - f(0)] = f(\infty) - f(0)$$

or, simplifying, in the following form

$$f(\infty) = \lim_{t \to \infty} f(t) = \lim_{s \to 0} s F(s)$$

which is the desired result   □

Note that the condition $\lim_{t \to \infty} f(t)$, implies that the function tends to a constant value at the limit. Moreover, the behavior of $f(t)$, for $t \to \infty$, is the same as the behavior of $sF(s)$ for $s \to 0$. Thus, in short, we can obtain the value of $f(\infty)$ from the Laplace transform of $f(t)$, that is, $F(s)$.

4.4 Mellin Transform

In analogy to the Laplace transform, introduced through the Fourier transform, we will present the Mellin transform, here, too, from a convenient variable change in the Fourier transform. It should be mentioned that we could introduce the Mellin transform using the Laplace transform with another convenient variable change. Further, as we have introduced the respective inverse Fourier and Laplace transforms, let us consider a parallel between the Mellin transform, denoted by $\mathscr{M}[f(x)] = F(p)$ and the inverse Mellin transform, denoted by $\mathscr{M}^{-1}[F(p)] = f(x)$, which recovers the function $f(x)$. Finally, we assume that the function $f(x)$ satisfies conditions that result in the convergence of the integral.

Consider a function $g(\omega)$ that admits the Fourier transform, that is,

$$\mathscr{F}[g(\omega)] = G(k) = \frac{1}{\sqrt{2\pi}} \int_{-\infty}^{+\infty} e^{-i\omega k} g(\omega) \, d\omega$$

whose inverse Fourier transform is given by

$$\mathscr{F}^{-1}[G(k)] = f(x) = \frac{1}{\sqrt{2\pi}} \int_{-\infty}^{+\infty} e^{i\omega k} G(k) \, dk \cdot$$

From these two expressions, we introduce the following change of variable $e^{\omega} = x$ which implies $\omega = \ln x$, as well as defined $ik = c - p$, where c is a real constant and p, with $\mathrm{Re}(p) > 0$ the parameter associated with the Mellin transform.

Before we continue, it is worth an observation regarding the pair of Laplace transforms, direct and inverse. If we had considered the bilateral Laplace transform,

$$\mathscr{L}[g(t)] = \int_{-\infty}^{\infty} g(t) e^{-st} \, dt$$

we should introduce the following variable change $t = -\ln x$ and defined $g(-\ln x) = f(x)$, converting the kernel of the Laplace transform, e^{-st} in the kernel of the Mellin transform, that is, x^s. Further, it is immediate to note that the Mellin transform of $f(x)$ is the Laplace transform of $g(t)$ with $t = -\ln x$, so that it converges absolutely and is analytical in the band region $a < \text{Re}(s) < b$ with $b > a$. Let's go back with the Fourier transform. Then, with the new variables we can write to the Fourier transform pair

$$\mathscr{F}[g(\ln x)] = G(ip - c) = \frac{1}{\sqrt{2\pi}} \int_0^\infty x^{p-c-1} g(\ln x)\, dx$$

and

$$\mathscr{F}^{-1}[G(ip - c)] = g(\ln x) = \frac{1}{\sqrt{2\pi}} \int_{c-i\infty}^{c+i\infty} x^{c-p} G(ip - ic)\, dp.$$

First, we redefine the functions so that we have

$$\frac{1}{\sqrt{2\pi}} g(\ln x) = x^c f(x) \quad \text{and} \quad G(ip - ic) = G(p)$$

which, substituted in the two previous expressions, allow to write, respectively

$$\mathscr{F}[g(\ln x)] = F(p) = \int_0^\infty x^{p-1} f(x)\, dx$$

and

$$\mathscr{F}^{-1}[F(p)] = f(x) = \frac{1}{2\pi i} \int_{c-i\infty}^{c+i\infty} x^{-p} F(p)\, dp.$$

Definition 4.4.1 (*Mellin transform*) Let $f(x)$ be a real function with real variable, defined in a open interval $(0, \infty)$ and $p \in \mathbb{C}$. We define the Mellin transform of $f(x)$ by means of the integral

$$\mathscr{M}[f(x)] = F(p) = \int_0^\infty x^{p-1} f(x)\, dx$$

being p the parameter of the transform.

Definition 4.4.2 (*Inverse Mellin transform*) Let $p \in \mathbb{C}$ and c be a real constant in the convergence region, ensuring that $f(x)$ is continuous. We define the inverse Mellin transform, through the contour integral in the complex plane

$$\mathscr{M}^{-1}[F(p)] = f(x) = \frac{1}{2\pi i} \int_{c-i\infty}^{c+i\infty} x^{-p} F(p)\, dp,$$

which recovers the function $f(x)$.

In analogy to the Fourier and Laplace transforms, we will state properties of the Mellin transform, in particular, to justify the importance of the Mellin transform form of a derivative, which is very useful in solving differential equations involving the Laplace operator in cylindrical coordinates, for example.

Property 4.4.1 (Scale) *Let $a > 0$ and $\mathcal{M}[f(x)] = F(p)$ denoting the Mellin transform of the function $f(x)$. The property is valid*

$$\mathcal{M}[f(ax)] = \frac{F(p)}{a^p}.$$

Property 4.4.2 (Displacement) *Let $a \in \mathbb{R}$ and $\mathcal{M}[f(x)] = F(p)$ denoting the Mellin transform of the function $f(x)$. The property is valid*

$$\mathcal{M}[x^a f(x)] = F(p + a).$$

Property 4.4.3 (Power of x) *Let $a \neq 0$ and $\mathcal{M}[f(x)] = F(p)$ denotaing the Mellin transform of the function $f(x)$. The property is valid*

$$\mathcal{M}[f(x^a)] = \frac{1}{a} F\left(\frac{p}{a}\right).$$

Property 4.4.4 (Mellin transform of the derivative) *Let $\mathcal{M}[f(x)] = F(p)$ be the Mellin transform of the function $f(x)$. Imposing the validity of the limits*

$$\lim_{x \to 0}[x^{p-1} f(x)] \to 0 \quad and \quad \lim_{x \to \infty}[x^{p-1} f(x)] \to 0$$

we have the property

$$\mathcal{M}[f'(x)] = -(p - 1)F(p - 1).$$

With the extended imposition for the derivatives, terms that are outside the integral, in the integration by parts, we can obtain the expression for the k-order derivative

$$\mathcal{M}[f^{(k)}(x)] = (-1)^k \frac{\Gamma(p)}{\Gamma(p - k)} F(p - k)$$

that in the particular case where $k = 2$, useful in problems involving the derivative of order two, can be written in the form

$$\mathcal{M}[f''(x)] = (p - 1)(p - 2)F(p - 2).$$

Property 4.4.5 (Mellin transform of the function $x^k f^{(k)}(x)$) *Let $\mathcal{M}[f(x)] = F(p)$ be the Mellin transform of the function $f(x)$ and $k = 0, 1, 2, \ldots$. Imposing the validity of the limits*

$$\lim_{x \to 0}[x^p f(x)] \to 0 \quad and \quad \lim_{x \to \infty}[x^p f(x)] \to 0$$

extended to the functions, terms that are outside the integral, in the integration by parts, we have the expression for the product of x^k by the derivative of $f(x)$ of order k

$$\mathscr{M}[x^k f^{(k)}(x)] = (-1)^k \frac{\Gamma(p+k)}{\Gamma(p)} F(p).$$

Property 4.4.6 (Mellin transform of the function $\left(x\dfrac{d}{dx}\right)^k f(x)$) *If $\mathscr{M}[f(x)] = F(p)$ is the Mellin transform of the function $f(x)$, then*

$$\mathscr{M}\left[\left(x\frac{d}{dx}\right)^k f(x)\right] = (-1)^k p^k F(p)$$

with $k = 0, 1, 2, \ldots$

Property 4.4.7 (Convolution) *Let $\mathscr{M}[f(x)] = F(p)$ and $\mathscr{M}[g(x)] = G(p)$ the Mellin transforms of the functions $f(x)$ and $g(x)$, respectively, both with parameter p. Denoting by $\star$ the Mellin convolution product, we have*

$$\mathscr{M}[f(x) \star g(x)] = \mathscr{M}\left[\int_0^\infty f(\xi) g\left(\frac{x}{\xi}\right) \frac{d\xi}{\xi}\right] = F(p)G(p).$$

4.5 Exercises

We separate this section into four subsections. The first one contains only the statements of the exercises (List of exercises), the second account with a suggestion (Suggestions) for the respective solution, while the third contains the resolutions (Solutions) itself. The fourth subsection presents exercises (Proposed exercises) to be solved by the students, all of them similar to those discussed in the text.

4.5.1 Exercise List

1. Let $z \in \mathbb{C}$ and $z = x + iy$ with $x, y \in \mathbb{R}$. Consider the following integral in the complex plane

$$\int_{\gamma^+} e^{-z^2} dz$$

where the contour γ^+ is the closed and simple path, oriented in the positive direction, joining the points $A(-\rho, 0)$, $B(\rho, 0)$, $C(\rho, a)$ and $D(-\rho, a)$, vertices of a rectangle, with $a \in \mathbb{R}$ and $\rho \in \mathbb{R}$. Take the limit $\rho \to \infty$ to show that

$$\int_{-\infty}^\infty e^{-(x \pm ia)^2} dx = \int_{-\infty}^\infty e^{-x^2} dx.$$

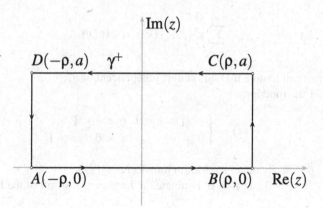

Fig. 4.3 Contour for Exercise (1)

2. Evaluate the Fourier transform of $f(x) = e^{-|t|}$.
3. Evaluate the Fourier transform of $f(t) = e^{-t^2}$.
4. (Displacement.) Let $a \in \mathbb{R}_+$. Show that $\mathscr{F}[f(t-a)](\omega) = e^{i\omega a} F(\omega)$, where $F(\omega)$ is the Fourier transform of $f(t)$.
5. (Scale.) Let $a \neq 0$. Show that $\mathscr{F}[f(at)](\omega) = \frac{1}{|a|} F\left(\frac{\omega}{a}\right)$, where $F(\omega)$ is the Fourier transform of $f(t)$.
6. (Translation.) Let $a \in \mathbb{R}$. Show that $\mathscr{F}[e^{iat} f(t)](\omega) = F(\omega - a)$, where $F(\omega)$ is the Fourier transform of $f(t)$.
7. (Duality.) Consider $F(\omega)$ being the Fourier transform of the function $f(t)$. Show that $f(-\omega) = \mathscr{F}[F(t)](\omega)$.
8. (Composition.) Consider $F(\omega) = \mathscr{F}[f(t)](\omega)$ and $G(\omega) = \mathscr{F}[g(t)](\omega)$ the Fourier transform of the functions $f(t)$ and $g(t)$, respectively. Show that

$$\int_{-\infty}^{\infty} F(\omega)g(\omega)\, e^{i\omega t}\, d\omega = \int_{-\infty}^{\infty} f(\xi)G(\xi - t)\, d\xi.$$

Also, consider the particular case, $t = 0$.
9. (Convolution.) Consider $F(\omega) = \mathscr{F}[f(t)](\omega)$ and $G(\omega) = \mathscr{F}[g(t)](\omega)$ the Fourier transform of the functions $f(t)$ and $g(t)$, respectively. Show that

$$\int_{-\infty}^{\infty} f(-\xi)g(\xi)\, d\xi = \int_{-\infty}^{\infty} F(\omega)G(\omega)\, d\omega.$$

10. Let $t \in \mathbb{R}$. Evaluate the Laplace transform of $t^{b-1} {}_1F_1(a, c; t)$. Show that the constraints for parameters a, b and c are the same as the series of the hypergeometric function.
11. Let $n \in \mathbb{N}$ and $E_{n,k}(\cdot)$ be a Mittag-Leffler function with two parameters. Show that

$$\sum_{k=1}^{n} x^{k-1} E_{n,k}(x^n) = \exp(x).$$

It is important to note that this sum is independent of n.

12. Consider the function

$$f(x) = \begin{cases} x^\alpha (1-x)^{\beta-1} & 0 \le x \le 1 \\ 0 & x < 0, \quad x > 1 \end{cases}$$

with $\alpha, \beta \in \mathbb{R}$. Evaluate the Mellin transform of $f(x)$.

13. Let $a \in \mathbb{R}$, $\alpha > 0$ and $\beta > 0$. Evaluate the Laplace transform of the function

$$\Lambda(\alpha, \beta) = t^{\beta-1} E_{\alpha,\beta}(at^\alpha)$$

where $E_{\alpha,\beta}(\cdot)$ is a Mittag-Leffler function with two parameters.

14. From the result of the previous one, enter $a \to -a$ and $\beta \to \gamma$ to show that

$$\mathscr{L}[t^{\gamma-1} E_{\alpha,\gamma}(at^\alpha)] = \frac{s^{\alpha-\gamma}}{s^\alpha - a}$$

with $a \in \mathbb{R}$, $\alpha > 0$ and $\gamma > 0$.

15. Let $a \in \mathbb{R}$, $\alpha > 0$, $\beta > 0$ and $\gamma > 0$. Show that

$$\mathscr{L}^{-1}\left[\frac{s^{2\alpha-\beta-\gamma}}{s^{2\alpha} - a^2}\right] = t^{\beta+\gamma-1} E_{2\alpha,\beta+\gamma}(a^2 t^{2\alpha}).$$

16. As a special case of the previous exercise, discuss the case $\gamma = \beta = \alpha$.

17. Evaluate the Laplace transform of $f(t) = e^{2t} \cos t$.

18. Let $a \in \mathbb{R}_+$. Evaluate the Laplace transform of the function $f(t) = (1 - at) e^{-at}$.

19. Evaluate the Laplace transform of the function $f(t)) = \dfrac{\sin t}{t}$.

20. Let $a, b \in \mathbb{R}$ with $a \ne b$. Evaluate the inverse Laplace transform

$$\mathscr{L}^{-1}\left[\frac{s}{(s+a)(s+b)}\right].$$

21. Evaluate the inverse Laplace transform $\mathscr{L}^{-1}\left[\dfrac{1}{(s-2)^2}\right]$.

22. Use the Laplace transform to solve the problem composed by a linear and homogeneous ordinary differential equation,

$$\frac{d^2}{dt^2} x(t) + x(t) = t$$

whose solution satisfies two initial conditions $x(0) = 0$ and $x'(0) = 0$, where the 'line' denotes derivative with respect to the independent variable, t.

23. With the Laplace transform methodology, solve the following linear and nonho-mogeneous ordinary differential equation

$$\frac{d^2}{dt^2}x(t) - 5\frac{d}{dt}x(t) + 6x(t) = e^{2t}$$

satisfying the inicial conditions $x(0) = 0 = x'(0)$. Note that the function in the second member is a solution of the respective homogeneous differential equation.

24. Use the Laplace transform methodology to solve the first-order system composed by linear and non-homogeneous differential equations

$$\begin{cases} \dfrac{d}{dt}u(t) + v(t) = 1 & \text{(a)} \\[3mm] \dfrac{d}{dt}v(t) - u(t) = -1 & \text{(b)} \end{cases}$$

and satisfying the initial conditions $u(0) = 2 = v(0)$.

25. Let $t \in \mathbb{R}_+$ and $x(t) \in \mathbb{R}_+$. Use the Laplace transform methodology to solve the integral equation

$$\frac{d}{dt}x(t) = 1 - \int_0^t x(\xi)\,d\xi$$

satisfying the initial condition $x(0) = 0$.

26. Let $x \in \mathbb{R}$ and $\varepsilon > 0$. Consider the function

$$f_\varepsilon(x) = \begin{cases} \dfrac{1}{2\varepsilon} & |x| \le \varepsilon \\[3mm] 0 & |x| > \varepsilon \end{cases}$$

Evaluate the Fourier transform of $f(x)$.

27. Evaluate the Mellin transform of the function $f(x) = (1 + x^2)^{-1}$.

28. Let $x > 1$ and $0 < \mu < 1$. Evaluate the Mellin transform of the function $f(x) = (x - 1)^{-\mu}$.

29. Dirichlet problem in the upper half-plane, $y > 0$. Consider the Laplace equation in upper half-plane $y > 0$, that is, solve the partial differential equation

$$\frac{\partial^2}{\partial x^2}u(x, y) + \frac{\partial^2}{\partial y^2}u(x, y) = 0, \qquad -\infty < x < \infty$$

satisfying the condition

$$u(x, 0) = f(x) \qquad -\infty < x < \infty$$

admitting $u(x, y) \to 0$ and $u_x(x, y) \to 0$ as $x \to \pm\infty$ and the function $f(x)$ is absolutely integrable. Show that

$$u(x, y) = \frac{y}{\pi} \int_{-\infty}^{\infty} \frac{f(\xi)}{y^2 + (x - \xi)^2} \, d\xi.$$

30. Let $k \in \mathbb{N}$ and $\alpha > 0$. Evaluate the Laplace transform of the function $t^k E_{\alpha,k+1}(-t^\alpha)$ where $E_{\alpha,k+1}(\cdot)$ is the Mittag-Leffler function with two parameters.

31. Let $\alpha > 0$. Show that the Laplace transform of the first derivative of the classical Mittag-Leffler function is given by

$$\mathscr{L}\left[\frac{d}{dt} E_\alpha(-t^\alpha)\right](s) = -\frac{1}{s^\alpha + 1}$$

with s the parameter of Laplace transform.

32. Let $1 \le \alpha \le 2$. Evaluate the inverse Laplace transform

$$\mathscr{L}^{-1}\left[\frac{s^{\alpha-1}}{s^\alpha + 1}\right].$$

Discuss the particular cases: $\alpha = 1$ and $\alpha = 2$.

33. Let $\alpha > \beta > 0$ be the double inequality and $\sigma \in \mathbb{R}_+$. Show that

$$\mathscr{L}^{-1}\left[\frac{s^{\rho-1}}{s^\alpha + 2\sigma s^\beta + 1}\right] = \tau^{\alpha-\rho} \sum_{r=0}^{\infty} (-2\sigma)^r \tau^{(\alpha-\beta)r} E_{\alpha,\alpha+1-\rho+(\alpha-\beta)r}^{r+1}(-\tau^\alpha)$$

with $E_{\alpha,\beta}^\rho(\cdot)$ being the Mittag-Leffler function with three parameters.

34. Let $\tau > 0$, ω and σ be two positive real constants satisfying the relation $\omega^2 + \sigma^2 = 1$ and $E_{\alpha,\beta}^\rho(\cdot)$ a Mittag-Leffler function with three parameters. Show that

$$\sum_{\ell=0}^{\infty} (-2\sigma)^\ell \tau^{\ell+1} E_{2,\ell+2}^{\ell+1}(-\tau^2) = e^{-\sigma\tau} \frac{\sin \omega\tau}{\omega}.$$

35. Evaluate the Mellin transform of the function $f(x) = (e^x - 1)^{-1}$ in order to relate it to the called Riemann zeta function [4], that is,

$$\zeta(p) = \sum_{k=1}^{\infty} \frac{1}{k^p}$$

with $\text{Re}(p) > 1$.

36. Prove the Mellin convolution theorem, as in Property 4.4.7.

37. Let $J_0(t)$ be the zero order Bessel function. Show that

$$\mathcal{L}[J_0(t)] = \frac{1}{\sqrt{s^2+1}}.$$

38. Let $t > 0$. Use the definition of the Laplace transform to show that

$$\mathcal{L}\left[\int_0^t \frac{\sin\xi}{\xi}\,d\xi\right] = \mathcal{L}\left[\int_0^1 \frac{\sin t\xi}{\xi}\,d\xi\right] = \frac{1}{s}\arctan\left(\frac{1}{s}\right)$$

where s is the parameter of Laplace transform.

39. Use the methodology of Mellin transform to show that the integral equation

$$f(x) = e^{-x} + \int_0^\infty \exp\left(-\frac{x}{\xi}\right) f(\xi)\frac{d\xi}{\xi}$$

has solution given by

$$f(x) = \frac{1}{2\pi i}\int_{c-i\infty}^{c+i\infty} x^{-p}\left[\frac{\Gamma(p)}{1-\Gamma(p)}\right]dp$$

with $c > 0$.

40. Show that

$$\mathcal{M}\left[\frac{1}{e^x+1}\right] = (1-2^{1-p})\Gamma(p)\zeta(p),$$

where s is the parameter of Mellin transform.

4.5.2 Sugestions

1. Use the integrand to be an integer function and the residue theorem.
2. Use the definition and separate into two different intervals.
3. Use the definition, write an adequate perfect square and use the result

$$\int_0^\infty e^{-\xi^2}\,d\xi = \frac{\sqrt{\pi}}{2}.$$

4. Use the definition and a suitable variable change.
5. Use the definition and separate in two cases, $a > 0$ and $a < 0$.
6. Use the definition and a suitable variable change.
7. From the inverse Fourier transform use the $t \rightleftarrows \omega$ and then make the exchange $\omega \to -\omega$.
8. Use the Fourier transform definition and rearrange it adequately.
9. Use the convolution product definition and, at the end, take $t = 0$.

10. Express the confluent hypergeometric function through your series, use the definition of the gamma function and the Pochhammer symbol. Identify the resulting series with the hypergeometric function.

11. Take the Laplace transform of the first member, introduce the series representation of the Mittag-Leffler function, and use the sum of the geometric progression. Evaluate the respective inverse Laplace transform.

12. Direct from the definition, using the integral representation for the beta function.

13. Use the series representation for the Mittag-Leffler function and change the order with the integral. Use a variable change to reduce the remaining integral in an integral identified as a gamma function and use the result of the geometric series.

14. Direct substitution.

15. Use the result of the two previous exercises and identify with the result.

16. Direct substitution.

17. Use the identity $\cos x = \text{Re}(e^{ix})$, integrate the exponential and take the real part.

18. Use direct integration by parts in the definition.

19. Use the identity $\sin x = \text{Im}(e^{ix})$, derives in relation to the parameter of the transform, take the imaginary part and integrate it. It is necessary to impose a condition to determine the constant that emerges from the integration.

20. Use partial fractions direct from the definition. From the residue theorem obtains, in the complex plane, the integral of an exponential.

21. Use the residue theorem in the case of a pole of order two.

22. Take the Laplace transform of both members, use the initial conditions and solve the algebraic equation in the transformed variable. Use partial fractions and the residue theorem to evaluate the inverse Laplace transform.

23. Proceed exactly as in the previous exercise.

24. Take the Laplace transform in both equations, use the initial conditions and solve the resulting system involving algebraic equations. Solve this algebraic system, use partial fractions and the residue theorem to recover the solution of the starting system.

25. Take the Laplace transform in both members, use the convolution theorem, and solve the algebraic equation. With the inverse Laplace transform, recover the solution.

26. Use the definition and expression for the sine in terms of exponentials.

27. Use the definition of the Mellin transform and the residue theorem to evaluate the integral and rearranging.

28. Note that the lower bound on the integral is unity. Take the variable change $x = 1/t$ and identify with the integral representation for the beta function.

29. From the Fourier transform in the x variable, obtain and solve the resulting ordinary differential equation, in the variable y. Use the conditions and the Fourier convolution theorem to obtain the solution as an integral.

30. Use the Laplace transform definition and the series representation of the Mittag-Leffler function with two parameters, perform a variable change, and use the geometric series.

31. Enter the series representation for the classical Mittag-Leffler function and change the order with integration. From a suitable change of variable and the geometric series, obtain the result.

32. Use the inversion formula, the geometric series, and the definition of the classical Mittag-Leffler function in terms of the power series.

33. Manipulate the quotient, imposing conditions, to use the geometric series and the series expansion of the Mittag-Leffler function with three parameters.

34. Enter the series representation for the Mittag-Leffler function with three parameters, take the Laplace transform of the first member and integrate. From the definition of the gamma function, rearrange and evaluate the sum. Finally, use the sum of the geometric series, take the inverse Laplace transform and compare with the second member.

35. Use the definition of Mellin transform, the geometric series, and the definition of the gamma function.

36. Direct from the definition with a suitable change of variable.

37. Use the series expansion for the zero order Bessel function

$$J_0(t) = \sum_{k=0}^{\infty} \frac{(-1)^k}{k!k!} \left(\frac{t}{2}\right)^{2k}$$

as well as the relation of the Laplace transform of a power function.

38. Use the definition of the Laplace transform and the expression that relates the sine function to complex exponentials. Integrate in order to obtain another integral that is calculated with a change of variable involving the tangent function.

39. Use the definition of the Mellin transform, its linearity, and the convolution theorem. Solve the resulting algebraic equation and invert the Mellin transform.

40. Use the results of Exercise (35), Proposed exercise (22) and the identity

$$\frac{1}{e^x + 1} = \frac{1}{e^x - 1} - \frac{2}{e^{2x} - 1}.$$

4.5.3 Solutions

1. We will show the result considering only the case where $a > 0$, since the other case is similar. Then, through the integration contour, as in Fig. 4.3 we can write, in a simplistic notation, which we will explain in the sequence

$$\int_{\gamma^+} e^{-z^2} \, dz = \int_A^B + \int_B^C + \int_C^D + \int_D^A = 0,$$

where the second equality comes from the residue theorem, since the integral on the left contains in the integrand an integer function.

In the first integral we use the parametrization $x = t$ and $y = 0$, in the second $x = \rho$ and $y = it$, in the third $x = t$ and $y = ai$ and in the fourth $x = -\rho$ and $y = it$, respectively. Then, we have

$$\int_{-\rho}^{\rho} e^{-t^2}\, dt + \int_0^a e^{-(\rho+it)^2}\, dt + \int_\rho^{-\rho} e^{-(t+ia)^2}\, dt + \int_a^0 e^{-(-\rho+it)^2}\, dt = 0.$$

Taking the limit $\rho \to \infty$ the second and the fourth integrals go to zero, so

$$\int_{-\infty}^{\infty} e^{-(x+ia)^2}\, dx = \int_{-\infty}^{\infty} e^{-x^2}\, dx,$$

which is the desired result   ◇

2. From the definition we can write, separating into two integrals

$$\mathscr{F}[e^{-|t|}](\omega) = \frac{1}{\sqrt{2\pi}} \int_{-\infty}^0 e^{t-i\omega t}\, dt + \frac{1}{\sqrt{2\pi}} \int_0^{\infty} e^{-t-i\omega t}\, dt.$$

Integrating, we obtain

$$\mathscr{F}[e^{-|t|}](\omega) = \frac{1}{\sqrt{2\pi}} \frac{e^{t(1-i\omega)}}{1-i\omega}\bigg|_{-\infty}^0 + \frac{1}{\sqrt{2\pi}} \frac{e^{-t(1+i\omega)}}{-(1+i\omega)}\bigg|_0^{\infty}.$$

Replacing the extremes and simplifying, we have

$$\mathscr{F}[e^{-|t|}](\omega) = \sqrt{\frac{2}{\pi}} \frac{1}{1+\omega^2},$$

which is the desired result   ◇

3. Using the definition we have

$$\mathscr{F}[e^{-t^2}](\omega) = \frac{1}{\sqrt{2\pi}} \int_{-\infty}^{\infty} e^{-t^2} e^{-it\omega}\, dt.$$

Separately, write as a perfect square, that is,

$$-t^2 - i\omega t = -\left(t^2 + i\omega t - \frac{\omega^2}{4} + \frac{\omega^2}{4}\right) = -\frac{\omega^2}{4} - \left(t + \frac{i\omega}{2}\right)^2.$$

Returning to the expression for the Fourier transform, we obtain

$$\mathscr{F}[e^{-t^2}](\omega) = \frac{1}{\sqrt{2\pi}} e^{-\frac{\omega^2}{4}} \int_{-\infty}^{\infty} e^{-(t+\frac{i\omega}{2})^2}\, dt$$

which, with the change of variable $t + i\omega/2 = \xi$, provides

$$\mathscr{F}[e^{-t^2}](\omega) = \frac{1}{\sqrt{2\pi}} e^{-\frac{\omega^2}{4}} \int_{-\infty}^{\infty} e^{-\xi^2}\, d\xi.$$

Using the parity of the function (the integrand is a even function and the interval is symmetric), we can write

$$\mathscr{F}[e^{-t^2}](\omega) = \sqrt{\frac{2}{\pi}} e^{-\frac{\omega^2}{4}} \int_{0}^{\infty} e^{-\xi^2}\, d\xi,$$

from where it follows, using the result of the integral and simplifying,

$$\mathscr{F}[e^{-t^2}](\omega) = \frac{1}{\sqrt{2}} e^{-\frac{\omega^2}{4}},$$

which is the desired result $\diamond$

4. From the definition of the Fourier transform, we have

$$\mathscr{F}[f(t-a)](\omega) = \frac{1}{\sqrt{2\pi}} \int_{-\infty}^{\infty} f(t-a)\, e^{-i\omega t}\, dt.$$

Introducing a change of variable $t - a = \xi$, we can write, rearranging

$$\mathscr{F}[f(t-a)](\omega) = \frac{e^{-i\omega a}}{\sqrt{2\pi}} \int_{-\infty}^{\infty} f(\xi)\, e^{-i\omega\xi}\, d\xi.$$

The resulting integral is the Fourier transform of the function $f(t)$, then

$$\mathscr{F}[f(t-a)](\omega) = e^{-i\omega a} F(\omega),$$

which is the desired result $\diamond$

5. First, we started imposing $a > 0$. From the definition we have

$$\mathscr{F}[f(at)](\omega) = \frac{1}{\sqrt{2\pi}} \int_{-\infty}^{\infty} f(at)\, e^{-i\omega t}\, dt.$$

Entering the change of variable $at = \xi$ we can write

$$\mathscr{F}[f(at)](\omega) = \frac{1}{a} \frac{1}{\sqrt{2\pi}} \int_{-\infty}^{\infty} f(\xi)\, e^{-i\omega\xi/a}\, d\xi$$

which, from the Fourier transform definition, provides

$$\mathscr{F}[f(at)](\omega) = \frac{1}{a} F\left(\frac{\omega}{a}\right).$$

Let us now consider $a < 0$. Being $b = -a > 0$, we can write the same expression for the Fourier transform, i.e., it suffices that we consider the module, so

$$\mathscr{F}[f(at)](\omega) = \frac{1}{|a|} F\left(\frac{\omega}{|a|}\right)$$

which is the desired result   ◇

6. From the definition of the Fourier transform, we have

$$\mathscr{F}[e^{iat} f(t)](\omega) = \frac{1}{\sqrt{2\pi}} \int_{-\infty}^{\infty} e^{iat} f(t) e^{-i\omega t} \, dt.$$

Rearranging, we can write

$$\mathscr{F}[e^{iat} f(t)](\omega) = \frac{1}{\sqrt{2\pi}} \int_{-\infty}^{\infty} f(t) e^{-i(\omega-a)t} \, dt.$$

which, from the definition of the Fourier transform, provides

$$\mathscr{F}[e^{iat} f(t)](\omega) = F(\omega - a),$$

which is the desired result   ◇

7. From the definition of the inverse Fourier transform, we have

$$f(t) = \frac{1}{\sqrt{2\pi}} \int_{-\infty}^{\infty} e^{i\omega t} F(\omega) \, d\omega.$$

Considering the change $t \rightleftarrows \omega$ we get

$$f(\omega) = \frac{1}{\sqrt{2\pi}} \int_{-\infty}^{\infty} e^{i\omega t} F(t) \, dt$$

and now taking the $\omega \rightarrow -\omega$, we can write

$$f(-\omega) = \frac{1}{\sqrt{2\pi}} \int_{-\infty}^{\infty} e^{-i\omega t} F(t) \, dt = \mathscr{F}[F(t)](\omega),$$

which is the desired result   ◇

8. Denotating by I_M the first member of equality and using the definition of the Fourier transform, we can write

$$I_M = \int_{-\infty}^{\infty} F(\omega) g(\omega) e^{i\omega t} \, d\omega$$

$$= \int_{-\infty}^{\infty} g(\omega) e^{i\omega t} \, d\omega \frac{1}{\sqrt{2\pi}} \int_{-\infty}^{\infty} e^{-i\omega\xi} f(\xi) \, d\xi.$$

Exchanging the order of integration and rearranging, we obtain

$$I_M = \int_{-\infty}^{\infty} f(\xi)\,d\xi \; \frac{1}{\sqrt{2\pi}} \int_{-\infty}^{\infty} g(\omega)\,e^{-i\omega(\xi-t)}\,d\omega.$$

The remaining integral is the Fourier transform of $g(\omega)$,

$$I_M = \int_{-\infty}^{\infty} f(\xi)G(\xi - t)\,d\xi$$

which is exactly the second member of equality, which proves the result. Now, taking $t = 0$ we get

$$\int_{-\infty}^{\infty} F(\omega)g(\omega)\,d\omega = \int_{-\infty}^{\infty} f(\xi)G(\xi)\,d\xi,$$

which is the desired result   ◇

9. The Fourier transform of the convolution product is given by

$$\mathscr{F}[f(t) \star g(t)](\omega) = \frac{1}{2\pi} \int_{-\infty}^{\infty} e^{-i\omega t}\,dt \int_{-\infty}^{\infty} f(t-\xi)g(\xi)\,d\xi,$$

which, rearranging adequately, can be written in the form

$$\mathscr{F}[f(t) \star g(t)](\omega) = \frac{1}{2\pi} \int_{-\infty}^{\infty} g(\xi)\,e^{-i\omega\xi}\,d\xi \int_{-\infty}^{\infty} f(t-\xi)e^{-i\omega(t-\xi)}\,dt.$$

Introducing the change of variable $t\xi = \eta$ in the second integral and rearranging, we have

$$\mathscr{F}[f(t) \star g(t)](\omega) = \frac{1}{\sqrt{2\pi}} \int_{-\infty}^{\infty} g(\xi)\,e^{-i\omega\xi}\,d\xi \; \cdot \; \frac{1}{\sqrt{2\pi}} \int_{-\infty}^{\infty} f(\eta)e^{-i\omega\eta}\,d\eta.$$

Therefore, using the definition of the Fourier transform, we obtain

$$\mathscr{F}[f(t) \star g(t)](\omega) = F(\omega)G(\omega).$$

From the previous expression, we can write, through the inverse Fourier transform

$$f(t) \star g(t) = \mathscr{F}^{-1}[F(\omega)G(\omega)](t)$$

which can be interpreted as: the inverse Fourier transform of a product of Fourier transforms is the convolution of the original functions. In order to show the desired result, we write the previous one in the form

$$\int_{-\infty}^{\infty} f(t-\xi)g(\xi)\,\mathrm{d}\xi = \int_{-\infty}^{\infty} e^{i\omega t} F(\omega)G(\omega)\,\mathrm{d}\omega,$$

which, for $t = 0$ in the previous expression, provides

$$\int_{-\infty}^{\infty} f(-\xi)g(\xi)\,\mathrm{d}\xi = \int_{-\infty}^{\infty} F(\omega)G(\omega)\,\mathrm{d}\omega,$$

which is the desired result   ◇

10. Introducing the series for confluent hypergeometric function and rearranging ,
    we have

$$\mathscr{L}[t^{b-1}\,{}_1F_1(a,c;t)] = \int_0^{\infty} t^{b-1}\,e^{-st}\,{}_1F_1(a,c;t)\,\mathrm{d}t$$

$$= \sum_{k=0}^{\infty} \frac{1}{k!}\frac{(a)_k}{(c)_k} \int_0^{\infty} e^{-st} t^{b+k-1}\,\mathrm{d}t.$$

Entering the variable change $st = u$ we have

$$\mathscr{L}[t^{b-1}\,{}_1F_1(a,c;t)] = \sum_{k=0}^{\infty} \frac{1}{k!}\frac{(a)_k}{(c)_k}\frac{1}{s^{b+k}} \int_0^{\infty} e^{-u} u^{b+k-1}\,\mathrm{d}u.$$

The remaining integral on the second member is a gamma function,

$$\mathscr{L}[t^{b-1}\,{}_1F_1(a,c;t)] = \sum_{k=0}^{\infty} \frac{1}{k!}\frac{(a)_k}{(c)_k}\frac{1}{s^{b+k}}\Gamma(b+k).$$

Multiplying numerator and denominator by $\Gamma(b)$, using the definition of the
Pochhammer symbol and rearranging, we can write

$$\mathscr{L}[t^{b-1}\,{}_1F_1(a,c;t)] = \frac{\Gamma(b)}{s^b} \sum_{k=0}^{\infty} \frac{(a)_k(b)_k}{(c)_k k!}\frac{1}{s^k}.$$

The summation in the second member is nothing more than the series representation of the hypergeometric function,

$$\mathscr{L}[t^{b-1}\,{}_1F_1(a,c;t)] = \frac{\Gamma(b)}{s^b}\,{}_2F_1\left(a,b;c;\frac{1}{s}\right)$$

which is interpreted as: the Laplace transform of a confluent hypergeometric
function equals a multiple of the hypergeometric function. Then, the conditions
to be imposed for the parameters are exactly the same as the definition of the
hypergeometric series, which is the desired result   ◇

11. Introducing the notation $I_M = \sum_{k=1}^{n} x^{k-1} E_{n,k}(x^n)$. Taking the Laplace transform of both members we have

$$\mathcal{L}[I_M] = \mathcal{L}\left\{\sum_{k=1}^{n} x^{k-1} E_{n,k}(x^n)\right\}.$$

Using the series representation for the classical Mittag-Leffler function and rearranging, we can write

$$\mathcal{L}[I_M] = \sum_{k=1}^{n} \sum_{\ell=0}^{\infty} \frac{1}{\Gamma(n\ell+k)} \int_{0}^{\infty} x^{n\ell+k-1} e^{-sx}\, dx.$$

By changing the variable $sx = t$, considering the definition of the gamma function and simplifying, we obtain

$$\mathcal{L}[I_M] = \sum_{k=1}^{n} \frac{1}{s^k} \sum_{\ell=0}^{\infty} \left(\frac{1}{s^n}\right)^{\ell}.$$

The second sum, in ℓ, is the sum of the infinite terms of a geometric progression of unitary first term and ratio s^{-n}, while the sum in k is the sum of the first n terms of a geometric progression of first term and ratio equal to s^{-1}, so

$$\mathcal{L}[I_M] = \frac{1}{s-1}.$$

Taking the inverse Laplace transform of both members, we can write

$$\sum_{k=1}^{n} x^{k-1} E_{n,k}(x^n) = e^x$$

which is the desired result   ◊

12. Take the Mellin transform on both sides, changing the upper bound of integration, instead of infinity, we can write

$$\mathcal{M}[f(x)] = \int_{0}^{1} x^{\alpha}(1-x)^{\beta-1} x^{s-1}\, dx$$

where s is the parameter of Mellin transform.

Using the integral representation that defines the beta function we get

$$\mathcal{M}[f(x)] = B(\alpha+s, \beta).$$

From the relation between the gamma and beta functions, we can write

$$\mathscr{M}[f(x)] = \frac{\Gamma(\alpha + s)\Gamma(\beta)}{\Gamma(\alpha + \beta + s)}$$

which is the desired result   ◇

13. Taking the Laplace transform of both members and introducing the series representation for the Mittag-Leffler function with two parameters, we have

$$\mathscr{L}[\Lambda(\alpha, \beta)] = \int_0^\infty t^{\beta-1} e^{-st} \sum_{k=0}^\infty \frac{(at^\alpha)^k}{\Gamma(\alpha k + \beta)} dt.$$

By changing the order of integration and rearranging, we can write

$$\mathscr{L}[\Lambda(\alpha, \beta)] = \sum_{k=0}^\infty \frac{a^k}{\Gamma(\alpha k + \beta)} \int_0^\infty t^{\alpha k + \beta - 1} e^{-st} dt$$

that, after changing the variable $st = u$ provides

$$\mathscr{L}[\Lambda(\alpha, \beta)] = \sum_{k=0}^\infty \frac{a^k}{\Gamma(\alpha k + \beta)} \frac{1}{s^{\alpha k + \beta}} \int_0^\infty u^{\alpha k + \beta - 1} e^{-u} du.$$

The remaining integral is nothing more than a representation for the gamma function, thus, already simplifying, we can write

$$\mathscr{L}[\Lambda(\alpha, \beta)] = \sum_{k=0}^\infty \frac{a^k}{s^{\alpha k + \beta}} = \frac{1}{s^\beta} \sum_{k=0}^\infty \left(\frac{a}{s^\alpha}\right)^k.$$

Using the sum of the geometric series with infinite terms, we obtain, simplifying

$$\mathscr{L}[t^{\beta-1} E_{\alpha,\beta}(at^\alpha)] = \frac{s^{\alpha - \beta}}{s^\alpha - a}$$

which is the desired result   ◇

14. By directly replacing $a \to -a$ and $\beta \to \gamma$ we obtain

$$\mathscr{L}[t^{\gamma-1} E_{\alpha,\gamma}(-at^\alpha)] = \frac{s^{\alpha - \gamma}}{s^\alpha + a}$$

which is the desired result   ◇

15. Using the result of the two previous exercises, we can write

$$\frac{s^{\alpha - \beta}}{s^\alpha - a} \cdot \frac{s^{\alpha - \gamma}}{s^\alpha + a} = \frac{s^{2\alpha - \beta - \gamma}}{s^{2\alpha} - a^2}.$$

From the result of the previous exercise and the inverse Laplace transform, we obtain directly

$$\mathscr{L}^{-1}\left[\frac{s^{2\alpha-\beta-\gamma}}{s^{2\alpha}-a^2}\right] = t^{\beta+\gamma-1}E_{2\alpha,\beta+\gamma}(a^2t^{2\alpha})$$

which is the desired result   ◇

16. By directly substituting $\gamma = \beta = \alpha$ we obtain the expression

$$\mathscr{L}^{-1}\left[\frac{1}{s^{2\alpha}-a^2}\right] = t^{2\alpha-1}E_{2\alpha,2\alpha}(a^2t^{2\alpha})$$

which is the desired result   ◇

17. We must calculate the integral $\mathscr{L}[e^{2t}\cos t] = \int_0^\infty e^{-st}e^{2t}\cos t\, dt = \int_0^\infty e^{-t(s-2)}\cos t\, dt$.

Before the resolution, an observation should be made. We could calculate the integral through the piecewise integration methodology, as well as explain the cosine as a semisome of two exponentials, but here we will choose to use the identity $\text{Re}(e^{it}) = \cos t$. With this, we exchange the operations, integrate the exponentials and, in the end, consider the real part, only.

Using the identity $\text{Re}(e^{it}) = \cos t$ for the Laplace transform, we have

$$\mathscr{L}[e^{2t}\cos t] = \text{Re}\left\{\int_0^\infty e^{-(s-2-i)t}\, dt\right\}.$$

By performing the (exponential) integration we obtain

$$\mathscr{L}[e^{2t}\cos t] = \text{Re}\left\{\frac{e^{-(s-2-i)t}}{-(s-2-i)}\bigg|_0^\infty\right\} = \text{Re}\left\{\frac{1}{s-2-i}\right\}.$$

Multiplying numerator and denominator by denominator conjugate we can write

$$\mathscr{L}[e^{2t}\cos t] = \text{Re}\left\{\frac{s-2+i}{(s-2)^2+1}\right\}$$

which, considering the real part, allows us to write

$$\mathscr{L}[e^{2t}\cos t] = \frac{s-2}{(s-2)^2+1}$$

which is the desired result   ◇

18. We must evaluate the integral $\mathscr{L}[(1-at)e^{-at}] = \int_0^\infty (1-at)e^{-at}e^{-st}\, dt$.

Separating the integral into two others and rearranging, we have

$$\mathscr{L}[(1-at)\,e^{-at}] = \int_0^\infty e^{-(s+a)t}\,dt - a\int_0^\infty t\,e^{-(s+a)t}\,dt\cdot$$

The first integral is immediate (exponential) while the second, we use integration by parts, from where it follows

$$\mathscr{L}[(1-at)\,e^{-at}] = \left.\frac{e^{-(s+a)t}}{-(s+a)}\right|_0^\infty - a\left\{-\frac{t}{s+a}e^{-(a+st)}\Big|_0^\infty + \int_0^\infty \frac{e^{-(s+a)t}}{s+a}\,dt\right\}\cdot$$

Simplifying and rearranging, we obtain

$$\mathscr{L}[(1-at)\,e^{-at}] = \frac{1}{s+a} - \frac{a}{s+a}\int_0^\infty e^{-(s+a)t}\,dt\cdot$$

In short, integrating and simplifying, we have

$$\mathscr{L}[(1-at)\,e^{-at}] = \frac{s}{(s+a)^2}$$

which is the desired result   ◇

19. We begin by introducing the following notation $J(s) = \int_0^\infty \frac{\sin t}{t}e^{-st}\,dt$. Using the identity $\text{Im}(e^{it}) = \sin t$, we have

$$J(s) = \text{Im}\left\{\int_0^\infty \frac{e^{it}}{t}e^{-st}\,dt\right\}\cdot$$

We derive in relation to the parameter s, the previous expression, then

$$\frac{d}{ds}J(s) = -\text{Im}\left\{\int_0^\infty e^{-(s-i)t}\,dt\right\},$$

with $\text{Re}(s) > 0$, so that the integral is convergent.
Integrating we obtain

$$\frac{d}{ds}J(s) = -\text{Im}\left\{\left.\frac{e^{-(s-i)t}}{-(s-i)}\right|_0^\infty\right\} = -\text{Im}\left\{\frac{1}{s-i}\right\}\cdot$$

Multiplying numerator and denominator by conjugate of denominator and considering the imaginary part, we can write

$$\frac{d}{ds}J(s) = -\frac{1}{1+s^2},$$

whose integration provides

$$J(s) = -\arctan s + C$$

where C is an integration constant. In order to determine this constant, we impose the following condition $\lim_{s\to\infty} J(s) = 0$, we can write $C = \pi/2$. With this constant value, we obtain

$$\mathscr{L}\left[\frac{\sin t}{t}\right] = \frac{\pi}{2} - \arctan s = \arctan(1/s)$$

which is the desired result $\diamond$

20. We begin by determining A and B (partial fractions) so that

$$\frac{s}{(s+a)(s+b)} = \frac{A}{s+a} + \frac{B}{s+b}.$$

Reducing to the same denominator and using polynomial identity, we obtain

$$-A = \frac{a}{b-a} \qquad \text{and} \qquad B = \frac{b}{b-a},$$

that, returning in the expression for the inverse Laplace transform and using the fact that it is a linear transform, allows us to write

$$\mathscr{L}^{-1}\left[\frac{s}{(s+a)(s+b)}\right] = -\frac{a}{b-a}\mathscr{L}\left[\frac{1}{s+a}\right] + \frac{b}{b-a}\mathscr{L}\left[\frac{1}{s+b}\right].$$

In order to solve the remaining integrals, we use the residue theorem, that is,

$$\mathscr{L}^{-1}\left[\frac{1}{s+a}\right] = \frac{1}{2\pi i}\int_{C_\gamma}\frac{e^{st}}{s+a}\,ds = \operatorname*{Res}_{s=-a}\left(\frac{e^{st}}{s+a}\right),$$

with

$$\operatorname*{Res}_{s=-a}\left(\frac{e^{st}}{s+a}\right) = \lim_{s\to -a}\left\{(s+a)\frac{e^{st}}{s+a}\right\} = e^{-at}.$$

Since the two integrals are similar, we obtain, already rearranging

$$\mathscr{L}^{-1}\left[\frac{s}{(s+a)(s+b)}\right] = \frac{b\,e^{-bt} - a\,e^{-at}}{b-a}$$

which is the desired result $\diamond$

21. From the definition, we must calculate the following integral in the complex plane

$$\mathscr{L}^{-1}\left[\frac{1}{(s-2)^2}\right] = \frac{1}{2\pi i}\int_{C_\gamma}\frac{e^{st}}{(s-2)^2}\,ds.$$

Since the integrand has a pole of order two, the residue theorem allows us to write

$$\mathscr{L}^{-1}\left[\frac{1}{(s-2)^2}\right] = \operatorname*{Res}_{s=2}\left[\frac{1}{(s-2)^2}\right] = \lim_{s\to 2}\frac{d}{ds}\left[(s-2)^2\frac{e^{st}}{(s-2)^2}\right].$$

Evaluating the derivative and calculating the limit, we obtain

$$\mathscr{L}^{-1}\left[\frac{1}{(s-2)^2}\right] = t\,e^{2t},$$

which is the desired result   ◇

22. Taking the Laplace transform of both members and using the linearity of the transform, we have

$$\int_0^\infty \frac{d^2}{dt^2}x(t)\,e^{-st}\,dt + \int_0^\infty x(t)\,e^{-st}\,dt = \int_0^\infty t\,e^{-st}\,dt.$$

Introducing the notation $F(s) = \int_0^\infty x(t)\,e^{-st}\,dt$ and integrate by parts the first integral in the first member and the integral in second member, we can write

$$s^2 F(s) - sx(0) - x'(0) + F(s) = \frac{1}{s}$$

which, using the initial conditions and solving for $F(s)$, provides

$$F(s) = \frac{1}{s(s^2+1)}.$$

Using partial fractions and the linearity of the inverse Laplace transform, we have

$$x(t) = \mathscr{L}[F(s)] = \mathscr{L}\left[\frac{1}{s}\right] - \mathscr{L}\left[\frac{s}{s^2+1}\right]$$

from where, using the residue theorem, we obtain

$$x(t) = \lim_{s\to 0}\left\{s\cdot\frac{e^{st}}{s}\right\} - \lim_{s\to i}\left\{(s-i)\cdot\frac{s}{s^2+1}e^{st}\right\} - \lim_{s\to -i}\left\{(s+i)\cdot\frac{s}{s^2+1}e^{st}\right\}$$

which, by calculating the limits and rearranging, allows us to write

$$x(t) = 1 - \cos t$$

which is the desired result   ◇

23. Taking the Laplace transform of both members and using the linearity of the transform, we have

$$\int_0^\infty \frac{d^2}{dt^2}x(t)\,e^{-st}\,dt - 5\int_0^\infty \frac{d}{dt}x(t)\,e^{-st}\,dt + 6\int_0^\infty x(t)\,e^{-st}\,dt = \int_0^\infty e^{2t}\,e^{-st}\,dt.$$

Introducing the notation $F(s) = \int_0^\infty x(t)\,e^{-st}\,dt$ and integrating by parts first and second integrals in the first member and directly the integral in the second member, we get

$$s^2 F(s) - sx(0) - x'(0) - 5sF(s) - 5x(0) + 6F(s) = \frac{1}{s-2}$$

which, using the initial conditions and solving for $F(s)$, provides

$$F(s) = \frac{1}{(s-2)^2(s-3)}.$$

Using partial fractions, we have

$$F(s) = \frac{1}{s-3} - \frac{1}{2}\frac{s}{(s-2)^2} - \frac{1}{2}\frac{1}{s-2}.$$

It should be noted that this is one of the forms (suitable for knowing other results) of writing the partial fractions. Using the result of Exercise (20) adequately and rearranging, we can write

$$x(t) = e^{3t} - (1+t)\,e^{2t}$$

which is the desired result   ◇

24. Introducing the notation $F(s) = \int_0^\infty u(t)\,e^{-st}\,dt$ and $G(s) = \int_0^\infty v(t)\,e^{-st}\,dt$ and taking the Laplace transform on both equations, we can write the following algebraic system

$$\begin{cases} sF(s) - u(0) + G(s) = 1/s \\ sG(s) - v(0) - F(s) = -1/s \end{cases}$$

or, in the following way, already using the initial conditions

$$\begin{cases} sF(s) + G(s) = 1/s + 2 \\ sG(s) - F(s) = -1/s + 2. \end{cases}$$

In order to solve this algebraic system, we multiply the second equation by s, add to the first one and solve to explicit $G(s)$, that is,

$$G(s) = \frac{2s^2 + s + 1}{s(s^2 + 1)}.$$

Using partial fractions, we can write

$$G(s) = \frac{1}{s} + \frac{s}{s^2 + 1} + \frac{1}{s^2 + 1}$$

and, taking the inverse transform, we obtain

$$v(t) = \mathscr{L}^{-1}\left[\frac{1}{s}\right] + \mathscr{L}^{-1}\left[\frac{s}{s^2 + 1}\right] + \mathscr{L}^{-1}\left[\frac{1}{s^2 + 1}\right].$$

These results are already known, from where it follows

$$v(t) = 1 + \sin t + \cos t.$$

To obtain $u(t)$ we can proceed in many ways, in particular, for simplicity, here we derive $v(t)$ and substitute directly in the second equation, from where we obtain

$$u(t) = 1 - \sin t + \cos t.$$

The last two equations are the desired solutions　　　　　　　　　　◇

25. Taking the Laplace transform of both members, we have

$$\int_0^\infty e^{-st} \frac{\mathrm{d}}{\mathrm{d}t} x(t)\, \mathrm{d}t = \int_0^\infty e^{-st}\, \mathrm{d}t - \mathscr{L}\left[\int_0^t x(\xi)\, \mathrm{d}\xi\right].$$

Integrating by parts the integral in the first member and using the product of Laplace convolution in the second integral in the second member, we can write

$$sF(s) - x(0) = \frac{1}{s} - \mathscr{L}[1] \cdot \mathscr{L}[x(t)]$$

or, in the following way

$$sF(s) - x(0) = \frac{1}{s} - \frac{F(s)}{s}.$$

Using the initial condition, solving for $F(s)$, rearranging and simplifying, we have

$$F(s) = \frac{1}{1 + s^2}.$$

Calculating the inverse Laplace transform, we have

$$x(t) = \mathscr{L}^{-1}[F(s)] = \mathscr{L}\left[\frac{1}{1+s^2}\right] = \sin t$$

which is the desired result   ⋄

26. From the definition of the Fourier transform, we have

$$\mathscr{F}[f_\varepsilon(x)] = \frac{1}{\sqrt{2\pi}} \int_{-\varepsilon}^{\varepsilon} \frac{1}{2\varepsilon} e^{ikx} \, dx.$$

Integrating and simplifying, we can write

$$\mathscr{F}[f_\varepsilon(x)] = \frac{1}{\sqrt{2\pi}k\varepsilon} \frac{e^{ik\varepsilon} - e^{-ik\varepsilon}}{2i} = \frac{1}{\sqrt{2\pi}} \frac{\sin k\varepsilon}{k\varepsilon}. \qquad (4.1)$$

Let's make one important observation. In the limit $\varepsilon \to 0$ we get

$$\mathscr{F}[f_0(x)] = \frac{1}{\sqrt{2\pi}}$$

which corresponds to the Fourier transform of the called Dirac delta function [5]. The result given by Eq.(4.1) is the desired result   ⋄

27. Using the definition of the Mellin transform, we must evaluate the integral

$$\mathscr{M}\left[\frac{1}{1+s^2}\right] = \int_0^\infty \frac{x^{s-1}}{1+x^2} \, dx.$$

In order to calculate the Mellin transform, let's consider the following integral in the complex plane, as in Fig. 4.4,

$$\oint_\gamma \frac{z^{s-1}}{1+z^2} \, dz$$

being $z = x + it$, with $x, t \in \mathbb{R}$. γ is a simple and closed contour, consisting of a semicircumference of radius $R > 0$, centered at the origin and the real axis $-R < x < R$. Within this contour, the only point that contributes to the residue theorem is $z = i$. Taking $R \to \infty$ the integral on the semicircumference goes to zero by the Jordan lemma [3].
Using the residue theorem we obtain

$$\oint_\gamma \frac{z^{s-1}}{1+z^2} \, dz = \int_{-\infty}^\infty \frac{x^{s-1}}{1+x^2} \, dx = 2\pi i \lim_{z \to i} (z-i) \frac{z^{s-1}}{(z+i)(z-i)}$$

which, by simplifying, allows to write for the Mellin transform

$$\mathcal{M}\left[\frac{1}{1+x^2}\right] = \left\{\cos\left[\frac{\pi}{2}(s-1)\right]\right\} \int_0^\infty \frac{x^{s-1}}{1+x^2}\,dx = \frac{\pi}{2}$$

or, in the following way

$$\mathcal{M}\left[\frac{1}{1+x^2}\right] = \int_0^\infty \frac{x^{s-1}}{1+x^2}\,dx = \frac{\pi}{2}\,\mathrm{cosec}\left(\frac{\pi s}{2}\right)$$

which is the desired result   ◇

28. From the definition of the Mellin transform, we must evaluate the integral

$$\mathcal{M}\left[\frac{1}{(x-1)^\mu}\right] = \int_1^\infty \frac{x^{s-1}}{(x-1)^\mu}\,dx.$$

Introducing the change of variable $x = 1/t$ and rearranging, we have

$$\mathcal{M}\left[\frac{1}{(x-1)^\mu}\right] = \int_0^1 t^{\mu-s-1}(1-t)^{-\mu}\,dt$$

which, by identifying with the integral representation of the beta function, allows

$$\mathcal{M}\left[\frac{1}{(x-1)^\mu}\right] = B(\mu - s, 1 - \mu).$$

Using the relation between beta and gamma functions, we obtain

$$\mathcal{M}\left[\frac{1}{(x-1)^\mu}\right] = \frac{\Gamma(\mu - s)\Gamma(1 - \mu)}{\Gamma(1 - s)}$$

which is the desired result   ◇

29. First, we introduce the following notation

$$\mathcal{F}[u(x, y)] = \overline{u}(\omega, y) = \frac{1}{\sqrt{2\pi}} \int_{-\infty}^\infty e^{-i\omega x} u(x, y)\,dx$$

and

$$\mathcal{F}[f(x)] = \overline{f}(\omega) = \frac{1}{\sqrt{2\pi}} \int_{-\infty}^\infty e^{-i\omega x} f(x)\,dx.$$

Taking the Fourier transform in the partial differential equation we have

$$\frac{\partial^2}{\partial y^2}\overline{u}(\omega, y) + \omega^2\overline{u}(\omega, y) = 0$$

as well as for the condition $\overline{u}(\omega) = \overline{f}(\omega)$. This differential equation is ordinary, whose general solution is given by

$$\overline{u}(\omega, y) = A\, e^{-|\omega|y} + B\, e^{|\omega|y}$$

being A and B constants. Since $u(x, y)$ is bounded, so is its transform, so we must consider the constant $B = 0$, then

$$\overline{u}(\omega, y) = A\, e^{-|\omega|y}.$$

From the transformed condition, we can write for the solution of the problem (equation and condition)

$$\overline{u}(\omega, y) = \overline{f}(\omega)\, e^{-|\omega|y} \equiv \overline{g}(\omega, y)\overline{f}(\omega)$$

in which we used the notation $\overline{g}(\omega, y) = e^{-|\omega|y}$. By the Fourier convolution theorem, taking the inverse Fourier transform, we can write

$$u(x, y) = \frac{1}{\sqrt{2\pi}} g(x, y) \star f(x) = \frac{1}{\sqrt{2\pi}} \int_{-\infty}^{\infty} g(x - \xi), y) f(\xi)\, \mathrm{d}\xi, \quad (4.2)$$

that is, we must, through the inverse Fourier transform, obtain $g(x, y)$ such that

$$g(x, y) = \frac{1}{\sqrt{2\pi}} \int_{-\infty}^{\infty} e^{i\omega x} e^{-|\omega|y}\, \mathrm{d}\omega.$$

In order to calculate the resulting integral, we separate into two others, that is,

$$g(x, y) = \frac{1}{\sqrt{2\pi}} \left(\int_{0}^{\infty} e^{i\omega x - \omega y}\, \mathrm{d}\omega + \int_{0}^{\infty} e^{-i\omega x - \omega y}\, \mathrm{d}\omega \right),$$

where it follows, after the integrations and simplification

$$g(x, y) = \frac{1}{\sqrt{2\pi}} \frac{2y}{x^2 + y^2}.$$

Returning in Eq.(4.2) we can write for the solution of the Dirichlet problem in the upper half-plane $y > 0$

$$u(x, y) = \frac{y}{\pi} \int_{-\infty}^{\infty} \frac{f(\xi)}{y^2 + (x - \xi)^2}\, \mathrm{d}\xi$$

which is the desired result   ◇

30. We must evaluate the integral

$$\mathscr{L}[t^k E_{\alpha,k+1}(-t^\alpha)](s) = \int_0^\infty e^{-st} t^k E_{\alpha,k+1}(-t^\alpha)\, dt.$$

Introducing the series representation for the Mittag-Leffler function with two parameters and changing the order of integration with the summation, we have

$$\mathscr{L}[t^k E_{\alpha,k+1}(-t^\alpha)](s) = \sum_{n=0}^\infty \frac{(-1)^n}{\Gamma(\alpha n + k + 1)} \int_0^\infty e^{-st} t^{\alpha n + k}\, dt.$$

From the change of variable $st = \xi$ and the integral representation for the gamma function, we can write

$$\mathscr{L}[t^k E_{\alpha,k+1}(-t^\alpha)](s) = \sum_{n=0}^\infty \frac{(-1)^n}{\Gamma(\alpha n + k + 1)} \frac{\Gamma(\alpha n + k + 1)}{s^{\alpha n + k + 1}}.$$

Simplifying the previous expression and rearranging we obtain

$$\mathscr{L}[t^k E_{\alpha,k+1}(-t^\alpha)](s) = \frac{1}{s^{k+1}} \sum_{n=0}^\infty \left(-\frac{1}{s^\alpha}\right)^n.$$

and, using the geometric series, in the following form

$$\mathscr{L}[t^k E_{\alpha,k+1}(-t^\alpha)](s) = \frac{s^{\alpha-k-1}}{s^\alpha + 1}$$

which is the desired result   ◇

31. By introducing the series representation for the classical Mittag-Leffler function and using the definition of the Laplace transform, we obtain the integral

$$\mathscr{L}\left[\frac{d}{dt} E_\alpha(-t^\alpha)\right](s) = \int_0^\infty e^{-st} \frac{d}{dt} \sum_{n=0}^\infty \frac{(-t^\alpha)^n}{\Gamma(\alpha n + 1)}\, dt.$$

Evaluating the derivative and changing the order of the sum, with the integral, we have

$$\mathscr{L}\left[\frac{d}{dt} E_\alpha(-t^\alpha)\right](s) = \sum_{n=1}^\infty \frac{(-1)^n}{\Gamma(\alpha n + 1)} (\alpha n) \int_0^\infty e^{-st} t^{\alpha n - 1}\, dt.$$

By doing the index change $n \to n+1$, using the relation $\Gamma(a+1) = a\,\Gamma(a)$ and introducing the change of variable $st = \xi$, we can write

$$\mathscr{L}\left[\frac{d}{dt} E_\alpha(-t^\alpha)\right](s) = -\sum_{n=0}^\infty \frac{(-1)^n}{\Gamma(\alpha n + 1)} \frac{1}{s^{\alpha n + \alpha}} \int_0^\infty e^{-\xi} \xi^{\alpha n + \alpha - 1}\, d\xi.$$

From the integral representation for the gamma function and simplifying, we have

$$\mathscr{L}\left[\frac{d}{dt}E_\alpha(-t^\alpha)\right](s) = -\frac{1}{s^\alpha}\sum_{n=0}^{\infty}\left(-\frac{1}{s^\alpha}\right)^n$$

which, using the geometric series, allows to write

$$\mathscr{L}\left[\frac{d}{dt}E_\alpha(-t^\alpha)\right](s) = -\frac{1}{s^{\alpha+1}}$$

which is the desired result   ⋄

32. Note that the interval is closed. At the extreme of the interval where $\alpha = 1$ we retrieve the case where the function is the exponential function while at the other end, $\alpha = 2$, we recover the case in which the function is the cosine function. It is convenient to emphasize that in these two cases the inversion formula provides directly, through the residue theorem, the respective functions, since integrands have only simple poles. On the other hand, for the parameter α different from these two limit cases, contrary to the simple poles, we have branching points which entails the use of the called modified Bromwich contour [3]. In order to circumvent this situation, in this particular example, we use the geometric series and the concept of gamma function.

Through the inversion formula we can write

$$\mathscr{L}^{-1}\left[\frac{s^{\alpha-1}}{s^\alpha+1}\right] = \frac{1}{2\pi i}\int_{\gamma-i\infty}^{\gamma+i\infty}e^{st}\frac{s^{\alpha-1}}{s^\alpha+1}\,ds$$

or in the following form, after simplification,

$$\mathscr{L}^{-1}\left[\frac{s^{\alpha-1}}{s^\alpha+1}\right] = \frac{1}{2\pi i}\int_{\gamma-i\infty}^{\gamma+i\infty}\frac{e^{st}}{s}\frac{ds}{1+1/s^\alpha}.$$

Using the geometric series and since the geometric series is uniformly convergent, we exchange the integral signal with the summation signal, so that we can write the previous expression in the form

$$\mathscr{L}^{-1}\left[\frac{s^{\alpha-1}}{s^\alpha+1}\right] = \sum_{k=0}^{\infty}(-1)^k\underbrace{\frac{1}{2\pi i}\int_{\gamma-i\infty}^{\gamma+i\infty}\frac{e^{st}}{s^{k\alpha+1}}ds}_{*}.$$

We identify $(*)$ as the inverse Laplace transform of the function $s^{-k\alpha-1}$, which allows us to write

$$\mathscr{L}^{-1}\left[\frac{s^{\alpha-1}}{s^\alpha+1}\right] = \sum_{k=0}^{\infty}(-1)^k\mathscr{L}^{-1}\left[\frac{1}{s^{k\alpha+1}}\right]$$

or by replacing the factorial by the gamma function, as follows

$$\mathscr{L}^{-1}\left[\frac{s^{\alpha-1}}{s^{\alpha}+1}\right] = \sum_{k=0}^{\infty}(-1)^k \frac{t^{\alpha k}}{\Gamma(\alpha k+1)} = \sum_{k=0}^{\infty}\frac{(-t^{\alpha})^k}{\Gamma(\alpha k+1)}.$$

The remaining summation is nothing more than the series representation of powers of the classic Mittag-Leffler function,

$$\mathscr{L}^{-1}\left[\frac{s^{\alpha-1}}{s^{\alpha}+1}\right] = E_{\alpha}(-t^{\alpha}).$$

As a direct consequence of this result we recover the identities

$$E_1(-t) = e^{-t} \quad \text{and} \quad E_2(-t^2) = \cos t$$

relative to the two extremes of the interval, $\alpha = 1$ and $\alpha = 2$, respectively. This result allows us to write that the classical Mittag-Leffler function interpolates the exponential function and the cosine trigonometric function ◇

33. We start by writing the first member in the form

$$\mathscr{L}^{-1}\left[\frac{s^{\rho-1}}{s^{\alpha}+2\sigma s^{\beta}+1}\right] = \mathscr{L}^{-1}\left[\frac{s^{\rho-1}}{s^{\alpha}+1}\frac{1}{1+\frac{2\sigma s^{\beta}}{s^{\alpha}+1}}\right]$$

where, by imposing the restriction $|2\sigma s^{\beta}/(s^{\alpha}+1)| < 1$ and using the geometric series we have

$$\mathscr{L}^{-1}\left[\frac{s^{\rho-1}}{s^{\alpha}+2\sigma s^{\beta}+1}\right] = \mathscr{L}^{-1}\left[\frac{s^{\rho-1}}{s^{\alpha}+1}\sum_{k=0}^{\infty}\left(-\frac{2\sigma s^{\beta}}{s^{\alpha+1}}\right)^k\right].$$

By changing the order of the summation with the inverse Laplace transform and rearranging, we can write

$$\mathscr{L}^{-1}\left[\frac{s^{\rho-1}}{s^{\alpha}+2\sigma s^{\beta}+1}\right] = \sum_{k=0}^{\infty}(-2\sigma)^k\mathscr{L}^{-1}\left[\frac{s^{\rho+\beta k-1}}{(s^{\alpha}+1)^{k+1}}\right].$$

Using the Proposed exercise (14), we obtain

$$\mathscr{L}^{-1}\left[\frac{s^{\rho-1}}{s^{\alpha}+2\sigma s^{\beta}+1}\right] = \tau^{\alpha-\rho}\sum_{r=0}^{\infty}(-2\sigma)^r\tau^{(\alpha-\beta)r}E_{\alpha,\alpha+1-\rho+(\alpha-\beta)r}^{r+1}(-\tau^{\alpha})$$

which is the desired result ◇

34. First, we introduce the notation $\Omega \equiv \sum_{\ell=0}^{\infty}(-2\sigma)^{\ell}\tau^{\ell+1}E_{2,\ell+2}^{\ell+1}(-\tau^2)$. By introducing the series expression for the Mittag-Leffler function with three parameters, taking the Laplace transform on both sides and rearranging, we have

$$\mathscr{L}[\Omega] = \sum_{\ell=0}^{\infty}(-2\sigma)^{\ell}\sum_{n=0}^{\infty}\frac{(\ell+1)_n}{\Gamma(2n+\ell+2)}\frac{(-1)^n}{n!}\int_0^{\infty}e^{-s\tau}\tau^{2n+\ell+1}\,d\tau$$

where s is the parameter of the Laplace transform and $(\cdot)_k$ is the Pochhammer symbol.

By changing the variable $s\tau = u$, we can write

$$\mathscr{L}[\Omega] = \sum_{\ell=0}^{\infty}(-2\sigma)^{\ell}\sum_{n=0}^{\infty}\frac{(\ell+1)_n}{\Gamma(2n+\ell+2)}\frac{(-1)^n}{n!}\frac{1}{s^{2n+\ell+2}}\int_0^{\infty}e^{-u}u^{2n+\ell+2-1}\,du\cdot$$

The resulting integral is the gamma function $\Gamma(2n+\ell+2)$ from where, by rearranging, we obtain

$$\mathscr{L}[\Omega] = \sum_{\ell=0}^{\infty}\frac{(-2\sigma)^{\ell}}{s^{\ell+2}}\sum_{n=0}^{\infty}\frac{(\ell+1)_n}{n!}\left(-\frac{1}{s^2}\right)^n.$$

Using the following result [6]

$$\sum_{k=0}^{\infty}\frac{(m+k)!}{k!}x^k = \frac{m!}{(1-x)^{m+1}}$$

which is valid for $m \in \mathbb{N}$ and $|x| < 1$ and simplifying, we get

$$\mathscr{L}[\Omega] = \frac{1}{s^2+1}\sum_{\ell=0}^{\infty}\left(-\frac{2\sigma s}{s^2+1}\right)^{\ell}.$$

In order to calculate the remaining sum, we use the geometric series, whence it follows, rearranging

$$\mathscr{L}[\Omega] = \frac{1}{s^2+2\sigma s+1}$$

which, by taking the inverse Laplace transform on both sides, provides

$$\Omega = \mathscr{L}^{-1}\left[\frac{1}{(s+\sigma)^2+\omega^2}\right]$$

where we have introduced the notation $1-\sigma^2 = \omega^2$.

Using the well-known result

$$\mathscr{L}[e^{-bt}\sin at] = \left[\frac{a}{(s+b)^2 + a^2}\right]$$

we obtain, through the respective inverse,

$$\sum_{\ell=0}^{\infty}(-2\sigma)^{\ell}\tau^{\ell+1}E_{2,\ell+2}^{\ell+1}(-\tau^2) = e^{-\sigma\tau}\frac{\sin\omega\tau}{\omega}$$

which is the desired result   ◇

35. Consider the function $f(x) = e^{-kx}$ with $k > 0$. Using the geometric series we
can write

$$\sum_{k=0}^{\infty}e^{-kx} = \sum_{k=0}^{\infty}(e^{-x})^k = \frac{1}{1-e^{-x}}$$

because $|e^{-x}| < 1$. Evidenciating the first term of the summation, we have

$$\sum_{k=0}^{\infty}e^{-kx} = 1 + \sum_{k=1}^{\infty}e^{-kx} = \frac{1}{1-e^{-x}}$$

from where it follows, rearranging, the result

$$\sum_{k=1}^{\infty}e^{-kx} = \frac{1}{e^x - 1}.$$

Taking the Mellin transform of both members of the preceding one and changing
the order of the summation with the integral in the second member, we obtain

$$\mathscr{M}\left[\frac{1}{e^x - 1}\right] = \sum_{k=1}^{\infty}\int_0^{\infty}e^{-kx}x^{p-1}\,\mathrm{d}x$$

where p is the parameter of the Mellin transform.
Introducing the variable change $kx = u$ and rearranging, we have

$$\mathscr{M}\left[\frac{1}{e^x - 1}\right] = \sum_{k=1}^{\infty}\frac{1}{k^p}\int_0^{\infty}e^{-u}u^{p-1}\,\mathrm{d}u$$

from where, using the definition of the gamma function, we can write

$$\mathscr{M}\left[\frac{1}{e^x - 1}\right] = \sum_{k=1}^{\infty}\frac{\Gamma(p)}{k^p} = \Gamma(p)\sum_{k=1}^{\infty}\frac{1}{k^p} = \Gamma(p)\zeta(p)$$

which is the desired result ◇

36. From the definition of the Mellin transform, we can write

$$\mathscr{M}[f(x) \star g(x)] = \int_0^\infty x^{p-1}dx \underbrace{\int_0^\infty f(\xi)g\Big(\frac{x}{\xi}\Big)\frac{d\xi}{\xi}}_{f(x)\star g(x)}.$$

Changing the integration order and introducing the change $x = \xi\, t$, we have

$$\mathscr{M}[f(x) \star g(x)] = \int_0^\infty f(\xi)\frac{d\xi}{\xi} \int_0^\infty (t\xi)^{p-1}g(t)\xi\, dt$$

rearranging provides

$$\mathscr{M}[f(x) \star g(x)] = \int_0^\infty \xi^{p-1} f(\xi)d\xi \int_0^\infty t^{p-1}g(t)\, dt$$

from where it follows, using the definition of the Mellin transform

$$\mathscr{M}[f(x) \star g(x)] = F(p)G(p)$$

which is the desired result ◇

37. Using the series expansion for the zero order Bessel function

$$J_0(t) = \sum_{k=0}^\infty \frac{(-1)^k}{k!k!}\Big(\frac{t}{2}\Big)^{2k}$$

we can write for the Laplace transform, already rearranging adequately

$$\mathscr{L}[J_0(t)] = \mathscr{L}\left(1 - \frac{t^2}{2^2} + \frac{t^4}{2^2\cdot 4^2} - \frac{t^6}{2^2\cdot 4^2\cdot 6^2} + \cdots\right).$$

Using the linearity of the Laplace transform and the expression that provides the Laplace transform of the power function

$$\mathscr{L}[t^k] = \frac{k!}{s^{k+1}}$$

we can write as follows

$$\mathscr{L}[J_0(t)] = \frac{1}{s}\left[1 - \frac{1}{2}\Big(\frac{1}{s^2}\Big) + \frac{1\cdot 3}{2\cdot 4}\Big(\frac{1}{s^4}\Big) - \frac{1\cdot 3\cdot 5}{2\cdot 4\cdot 6}\Big(\frac{1}{s^6}\Big) + \cdots\right].$$

Identifying the expression in brackets as the Maclaurin series expansion of the inverse of the square root, we have

$$\mathscr{L}[J_0(t)] = \frac{1}{s}\left[\frac{1}{\sqrt{1+\frac{1}{s^2}}}\right]$$

or simplifying, in the following way

$$\mathscr{L}[J_0(t)] = \frac{1}{\sqrt{1+s^2}}$$

which is the desired result   ◇

38. For the first equality, consider a change of variable $\xi \to t\xi$ so that we will work directly with the second expression. Using the definition of the Laplace transform we have

$$\mathscr{L}\left[\int_0^1 \frac{\sin t\,\xi}{\xi}\,d\xi\right] = \int_0^\infty e^{-st}\int_0^1 \frac{\sin t\,\xi}{\xi}\,d\xi\,dt\cdot$$

Exchanging the order of integration and expressing the sine function in terms of the complex exponentials, we obtain

$$\mathscr{L}\left[\int_0^1 \frac{\sin t\,\xi}{\xi}\,d\xi\right] = \int_0^1 \frac{d\xi}{2i\xi}\int_0^\infty \left[e^{-t(s-i\xi)} - e^{-t(s+i\xi)}\right]\,dt\cdot$$

By integrating into t, rearranging and simplifying, we have

$$\mathscr{L}\left[\int_0^1 \frac{\sin t\,\xi}{\xi}\,d\xi\right] = \int_0^1 \frac{d\xi}{s^2+\xi^2}\cdot$$

The remaining integral is calculated by changing the variable $\xi = s\tan\theta$ from where it follows, already returning in the variable s

$$\mathscr{L}\left[\int_0^1 \frac{\sin t\,\xi}{\xi}\,d\xi\right] = \frac{1}{s}\arctan\left(\frac{1}{s}\right)$$

which is the desired result   ◇

39. Taking the Mellin transform from both members of the equation we have

$$\mathscr{M}[f(x)] = \mathscr{M}[e^{-x}] + \mathscr{M}\left[\int_0^\infty \exp(-x/\xi)\,f(\xi)\frac{d\xi}{\xi}\right]\cdot$$

Using the definition of the convolution product we obtain

$$F(p) = \mathscr{M}[e^{-x}] + F(p)\mathscr{M}[e^{-x}] \tag{4.3}$$

with $F(p) = \mathscr{M}[f(x)]$. Calculating the remaining Mellin transform,

$$\mathcal{M}[e^{-x}] = \int_0^\infty e^{-x} x^{p-1}\,\mathrm{d}x = \Gamma(p)$$

we can write, already solving the algebraic equation in $F(p)$,

$$F(p) = \frac{\Gamma(p)}{1 - \Gamma(p)}$$

whose inverse Mellin transform, provides

$$f(x) = \frac{1}{2\pi i} \int_{c-i\infty}^{c+i\infty} x^{-p} \left[\frac{\Gamma(p)}{1 - \Gamma(p)} \right] \mathrm{d}p$$

which is the desired result   ◇

40. Consider the Mellin transform on both sides of identity

$$\frac{1}{e^x + 1} = \frac{1}{e^x - 1} - \frac{2}{e^{2x} - 1}$$

from where we can write, due to the linearity of the Mellin transform,

$$\mathcal{M}\left[\frac{1}{e^x + 1} \right] = \mathcal{M}\left[\frac{1}{e^x - 1} \right] - \mathcal{M}\left[\frac{2}{e^{2x} - 1} \right].$$

Using the results of Exercise (35) and Proposed exercise (22), we obtain

$$\mathcal{M}\left[\frac{1}{e^x + 1} \right] = \Gamma(p)\zeta(p) - 2^{1-p}\Gamma(p)\zeta(p)$$

which, by rearranging and simplifying, allows us to write

$$\mathcal{M}\left[\frac{1}{e^x + 1} \right] = (1 - 2^{1-p})\Gamma(p)\zeta(p)$$

which is the desired result   ◇

4.5.4 *Proposed Exercises*

1. Let A and σ be two real constants. Determine the constants A and σ, if any, so that the Fourier transform of a Gaussian, $f(x) = A\,e^{-\sigma x^2}$ is equal to its inverse.
2. Demonstrate the Theorem 4.3.1 [7].
3. Let $f(t) = t^n$, for $t \geq 0$ and $n \in \mathbb{N}$ and $a > 0$. Use the l'Hôpital rule n times to show that $\lim_{t \to \infty} e^{-at} t^n = 0$, from which we conclude that f is of exponential

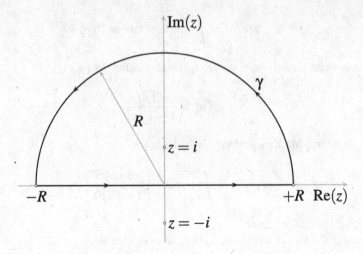

Fig. 4.4 Contour for Exercise (27)

order. If f is continuous, there exists the Laplace transform and is given by

$$F(s) = \frac{n!}{s^{n+1}}$$

with $\text{Re}(s) > 0$.

4. Obtain the result of the Proposed exercise (3) by simulating a derivative in the parameter of the Laplace transform.
5. Demonstrate Properties 4.3.1 and 4.3.2.
6. Demonstrate Properties 4.3.3 and 4.3.4.
7. Demonstrate Properties 4.3.5 and 4.3.6.
8. Demonstrate Properties 4.3.7 and 4.3.8.
9. Let $0 < \text{Re}(b) < \text{Re}(a)$ be the restrictions. Show that

$$\int_0^\infty t^{b-1} {}_1F_1(a; c; -t)\, dt = \frac{\Gamma(b)\Gamma(c)\Gamma(a - b)}{\Gamma(a)\Gamma(c - b)}.$$

Verify that this result can be obtained as a result limit as obtained in Exercise (10).

10. Consider the result of Exercise (20) and discuss the case $b \to a$.
11. Evaluate the inverse Laplace transform

$$\mathcal{L}^{-1}\left[\frac{s}{(s + 1)^2}\right].$$

12. Solve the Dirichlet problem in the semiplane $y > 0$ in the particular case where we have a even function, for example, $f(x) = \cos x$.

13. Let $0 < \nu < 1$. Consider the Mainardi function, denoted by $M_\nu(z)$, and the Mittag-Leffler function, denoted by $E_\nu(z)$. Show that the Laplace transform of the Mainardi function is a Mittag-Leffler function, that is,

$$\mathscr{L}[M_\nu(z)] = E_\nu(-s)$$

where s is the parameter of the Laplace transform.

14. In analogy to Exercise (30) show that

$$\mathscr{L}[t^{\beta-1} E_{\alpha,\beta}^\rho(at^\alpha)] = \frac{s^{\alpha\rho-\beta}}{(s^\alpha - a)^\rho}$$

with $\mathrm{Re}(s) > 0, \mathrm{Re}(\beta) > 0, a \in \mathbb{C}$ and $|as^{-\alpha}| < 1$, whose corresponding inverse transform is given by

$$\mathscr{L}^{-1}\left[\frac{s^{\alpha\rho-\beta}}{(s^\alpha - a)^\rho}\right] = t^{\beta-1} E_{\alpha,\beta}^\rho(at^\alpha). \tag{4.4}$$

15. Use the relation $\Gamma(n + \ell + 1) = n!(n+1)_\ell$ to show that

$$\sum_{\ell=0}^\infty (\mu\tau)^\ell E_{2,\ell+2}^{\ell+1}(-\tau^2) = \sum_{n=0}^\infty (-\tau^2)^n E_{1,2n+2}^{n+1}(\mu\tau)$$

for μ and τ positive reals.

16. Let $\mathscr{M}[f(x)] = F(p)$. Show that

$$\mathscr{M}\left[\frac{1}{x}f\left(\frac{1}{x}\right)\right] = F(1 - p)$$

where p is the parameter of the Mellin transform.

17. Let $\mathscr{M}[f(x)] = F(p)$. Use the result $\dfrac{d}{dp}x^{p-1} = (\ln x)\, x^{p-1}$ to obtain

$$\mathscr{M}[(\ln x)^n f(x)] = \frac{d^n}{dp^n} F(p)$$

valid for $n = 0, 1, 2, \ldots$

18. Let $\mathscr{M}[f(x)] = F(p)$. Show that

$$\mathscr{M}\left[\int_0^x f(\xi)\, d\xi\right] = -\frac{1}{p}F(p + 1)$$

where p is the parameter of the Mellin transform.

19. A different way to introduce the Mellin transform of another convolution product is as follows. Let $\mathscr{M}[f(x)] = F(p)$ and $\mathscr{M}[g(x)] = G(p)$ the Mellin trans-

forms of $f(x)$ and $g(x)$, respectively, both with parameter p. Denoting by $\star$ the Mellin convolution product, show that

$$\mathscr{M}[f(x) \star g(x)] = \mathscr{M}\left[\int_0^\infty f(x\xi)g(\xi)\,d\xi\right] = F(p)G(1-p).$$

20. Let $\mathscr{M}[f(x)] = F(p)$ and $\mathscr{M}[g(x)] = G(p)$ the Mellin transforms of $f(x)$ and $g(x)$, respectively, both with parameter p. Show that

$$\mathscr{M}[f(x)g(x)] = \frac{1}{2\pi i}\int_{c-i\infty}^{c+i\infty} F(\xi)G(p-\xi)\,d\xi$$

with p the parameter of the Mellin transform. In the particular case where $p = 1$, obtain the called Parseval formula for the Mellin transform,

$$\int_0^\infty f(x)g(x)\,dx = \frac{1}{2\pi i}\int_{c-i\infty}^{c+i\infty} F(\xi)G(1-\xi)\,d\xi.$$

21. Solve the Exercise (38) by means of the Maclaurin series expansion of the sine function.
22. Using the Exercise (35) show that

$$\mathscr{M}\left[\frac{2}{e^{2x}-1}\right] = 2^{1-p}\Gamma(p)\zeta(p)$$

where $\zeta(p)$ is the Riemann zeta function.
23. Using the Exercise (40) show that

$$\sum_{k=1}^\infty \frac{(-1)^{k-1}}{k^p} = (1-2^{1-p})\Gamma(p)\zeta(p)$$

where $\zeta(p)$ is the Riemann zeta function.
24. Assume a periodic function $f(t)$ with period T, that is, $f(t+T) = f(t)$. Show that the Laplace transform of $f(t)$ is given by

$$\mathscr{L}[f(t)] = \frac{1}{1-e^{-sT}}\int_0^\infty e^{-st}f(t)\,dt$$

where s is the parameter of the Laplace transform.
25. Let $\mathscr{L}[f(t) =]F(s)$ and $a > 0$, show that $\mathscr{L}[e^{at}f(t)] = F(s-a).$
26. Let $a > 0$. Show that

$$\mathscr{L}[\sinh at] = \frac{a^2}{s^2-a^2}.$$

27. Let $a > 0$. Show that
$$\mathscr{L}[\cosh at] = \frac{s}{s^2 - a^2}.$$

28. Knowing that $x(t) = J_0(t)$, where $J_0(t)$ is the zero order Bessel function, satisfies the ordinary differential equation
$$t^2 \frac{d^2}{dt^2} x(t) + t \frac{d}{dt} x(t) + t^2 x(t) = 0$$

and $J_0(0) = 1$ and $J'_0(0) = 0$, use the Laplace transform to obtain the same result as in Exercise (37).

29. Using the result
$$\arctan x = x - \frac{x^3}{3} + \frac{x^5}{5} - \frac{x^7}{7} + \cdots$$

valid for $|x| < 1$ show that
$$\mathscr{L}\left[\int_0^t \frac{\sin \xi}{\xi} d\xi\right] = \frac{1}{s} \arctan\left(\frac{1}{s}\right).$$

Compare with the result of Exercise (38).

30. Let $\theta(t - a)$, with $a > 0$, be the unitary Heaviside function. Show that
$$\mathscr{L}[\theta(t - a)] = \frac{e^{-as}}{s}$$

where s is the parameter of the Laplace transform.

31. Show that
$$\mathscr{L}\left[\frac{e^{-t} - e^{-2t}}{t}\right] = \ln\left(\frac{s + 2}{s + 1}\right)$$

where s is the parameter of the Laplace transform.

32. Let $\Gamma'(1) = -\gamma = -0,5772152\ldots$ be the Euler-Mascheroni constant. Show that
$$\mathscr{L}[\ln t] = -\frac{1}{s}(\gamma + \ln s)$$

where s is the parameter of the Laplace transform.

33. Let $t > 0$. Show that
$$\mathscr{L}^{-1}\left[\frac{1}{s^2(s^2 + 1)}\right] = t - \sin t.$$

34. Use the Laplace transform methodology to solve the integral equation
$$f(t) = t + f(t) \star \sin t$$

where $\star$ is the Laplace convolution product.

35. Let $\alpha > 0$ and $\beta > 0$. Evaluate the Fourier transform of the function

$$f(x) = \frac{1}{-\alpha x^2 + \beta x}.$$

36. The Fourier transform can be generalized to more than one dimension. For example, in three-dimensional space, the following expression is valid

$$\mathscr{F}[f(\mathbf{x})] \equiv F(\mathbf{k}) = \int d^3x \, f(\mathbf{x}) \, e^{-i\mathbf{k}\cdot\mathbf{x}}$$

with $\mathbf{x} = (x_1, x_2, x_3)$, the dot denotes the scalar product and $f(\mathbf{x})$ the called wave function. a) Introducing the spherical coordinates (r, θ, ϕ) and considering $\mathbf{k}$ in the z direction, then $\mathbf{k} \cdot \mathbf{x} = kr \cos \theta$, show that

$$\mathscr{F}[f(\mathbf{x})] = \int_0^{2\pi} \int_0^{\pi} \int_0^{\infty} r^2 \sin \theta \, f(r, \theta, \phi) \, e^{-ikr\cos\theta} dr \, d\theta \, d\phi. \qquad (4.5)$$

(b) Consider the particular case of the $2p$ electron state of the hydrogen atom, whose wave function is given by

$$f(r, \theta, \phi) \equiv f(r, \theta) = \beta r \cos \theta \, e^{-\alpha r}$$

with α and β positive constants. Determine the Fourier transform associated with this wave function, that is, replace $f(r, \theta)$ in Eq.(4.5) and compute the integrals.

37. A second kind Volterra integral equation is given by

$$y(x) = F(x) + \int_0^x K(x - \xi)y(\xi)\,d\xi.$$

Solve the Volterra equation, using the Laplace transform methodology, in the case where $F(x) = \sin x$ and $K(x) = -\sinh x$.

38. Consider the Gel'fand-Shilov function

$$\phi_\lambda(t) = \begin{cases} \dfrac{t^{\lambda-1}}{\Gamma(\lambda)}, & t \geq 0, \\ 0, & t < 0 \end{cases}$$

with $\text{Re}(\lambda > 0$. Let $\text{Re}(\mu) > 0$. Show that

$$\phi_\lambda(t) \star \phi_\mu(t) = \phi_{\lambda+\mu}(t)$$

where $\star$ denotes the Laplace convolution product.

39. Consider two power-type functions. Prove the following expression involving the Laplace convolution product

$$t^{p-1} \star t^{q-1} = \int_0^t \tau^{p-1}(t-\tau)^{q-1} \, d\tau = t^{p+q-1} B(p,q)$$

with $\text{Re}(p) > 0$, $\text{Re}(q) > 0$ and $B(p,q)$ denotes the beta function.

40. Consider the integral representation for the classical Mittag-Leffler function

$$E_\alpha(-t^\alpha) = \frac{1}{2\pi i} \int_{\text{Br}} e^{st} \frac{s^{\alpha-1}}{s^\alpha + 1} \, ds$$

with $0 < \alpha < 1$ and Br is the Bromwich contour. a) Show that we can write the following expression, interpreted as the Laplace transform of $K_\alpha(r)$,

$$E_\alpha(-t^\alpha) = \int_0^\infty e^{-rt} K_\alpha(r) \, dr$$

where $K_\alpha(r)$ is an algebraic function given by

$$K_\alpha(r) = -\frac{1}{\pi} \text{Im} \left\{ \left. \frac{s^{\alpha-1}}{s^\alpha + 1} \right|_{s=r\,e^{i\pi}} \right\}.$$

(b) Show that $K_\alpha(r)$ is always positive for every $0 < r < \infty$.

References

1. Sneddon, I.N.: Fourier Transforms. Dover Publications, New York (1995)
2. Debnath, L., Bhatta, D.: Integral Transform and Their Applications, 2nd edn. Chapman & Hall/CRC, Boca Raton (2007)
3. Capelas de Oliveira, E., Rodrigues Jr., W.A.: Analytical Functions with Applications. Editora Livraria da Física, São Paulo (2005). (in Portuguese)
4. Edwards, H.M.: Riemann's Zeta Function. Dover Publications Inc., Mineola, New York (1974)
5. Capelas de Oliveira, E.: Special Functions and Applications, 2nd edn. Livraria Editora da Física, São Paulo (2012). (in Portuguese)
6. Prudnikov, A.P., Brychkov, YuA, Marichev, O.I.: Integral and Series (Elementary Functions). Gordon and Breach Science Publishers, London (1986)
7. Figueiredo Camargo, R., Capelas de Oliveira, E.: Fractional Calculus. Editora Livraria da Física, São Paulo (2015). (in Portuguese)

Chapter 5
Fractional Derivatives

Just as the integer order calculus has integer order derivatives (and integrals), in fractional calculus we have fractional derivatives (and integrals), which should at least satisfy the condition of recovering the form of integer order derivatives (and integrals), for a suitable value of the parameter associated with the order of the fractional derivative. As we have already mentioned, there are several ways of approaching the study of fractional derivatives, since there are many distinct formulations of this concept, some of which should not even be denominated fractional. The proliferation of such definitions is due to the demand for an adequate derivative to describe each particular phenomenon. In general, different approaches to differential operators have been created because of the different domains in which they should be used. After presenting the special functions, in particular those associated with fractional calculus, as well as the methodology of the integral transforms, we are now able to introduce and work specifically with fractional calculus.

From the end of the last decade of the twentieth century and the beginning of the first decade of the twenty-first century there was a great proliferation of derivatives, always with the purpose of generalizing the classics derivative from the calculus of integer order. Many, although containing the fractional name, are only a multiplicative factor ahead of the derivative of order one, as well as others, proposing as kernel a nonsingular function.

As is well known, the integer-order derivative is defined from an adequate limit and has a well-consolidated interpretation, both physical and geometrical, differently from non-integer order. Here, we propose a possible classification of the derivatives so that, in the end, we highlight only those that will be presented in this chapter.

In the fractional calculus, in most definitions, the derivative is introduced from a generalization of the Cauchy integral of integer order [1]. Exception is the

© Springer Nature Switzerland AG 2019
E. Capelas de Oliveira, *Solved Exercises in Fractional Calculus*, Studies in Systems, Decision and Control 240,
https://doi.org/10.1007/978-3-030-20524-9_5

derivative, as proposed by Grünwald-Letnikov, which is the most direct generalization of the concept of derivative, since it is given in terms of a limit of an infinite series. We highlight that fractional derivatives are an excellent tool for the description of processes involving both the called memory effect and heredity properties, since in many cases they refine the solution.

Since the purpose of this chapter is to highlight only the derivatives that we will need for the resolution of fractional differential equations, let us go directly to our classification proposal, which is summarized in classifying the derivatives in only three classes, namely: (i) classical, which can also be subdivided; (ii) local and (iii) non-singular kernel. The first class includes the derivatives, already consolidated by the respective name, Grünwald-Letnikov, Riemann, Liouville, Riemann-Liouvile, Marchaud, Weyl, Riesz, Hadamard, Caputo, Hilfer, among others. It is important, right now, to highlight another possibility of classification, in the sense of deriving (integer order) an integral (fractional order) Riemann-Liouville type calls, as well as integrate (fractional order) a derivative (integer order), Caputo type calls. It is clear that the derivatives of the Caputo type are more restrictive, since they require the function to accept the derivative. Since the derivative is of integer order, the interpretation, in particular, of initial conditions, seems to be more suitable.

Local derivatives, introduced in the last decade of the twentieth century, are those that can be driven to a multiple of the integer-order derivative and, in our view, should not carry the fractional name, that is, they should be called local derivatives (in general, local operator) instead of local fractional derivatives. We emphasize that the conformal name comes from the possibility of the respective differential operator to admit different forms, as the need arises. Part of this class are the derivatives as proposed by Kolwankar-Gangal, Khalil, Katugampola and the called fractional alternative, among others. The third class of derivatives began in the second decade of the twenty-first century, with called Caputo-Fabrizio derivative which have the kernel a non-singular function. This class of derivatives, contains Caputo-Fabrizio derivatives, those that were proposed by Atangana, Atangana-Baleanu and Tenreiro Machado and Yang, among others. Finally, just to mention, there are the derivatives of an arbitrary order in order to derive a function in relation to another function of which we mention the derivatives as proposed by Almeida [2] and Sousa and Capelas de Oliveira [3].

In this book, for a brief introduction, through solved exercises, of fractional calculus concepts, involving fractional differential equations, we will take into account only those derivatives from Riemann-Liouville, Caputo and Hadamard, all belonging to the first class, according to our classification. These three formulations are presented in terms of an integral, which already characterizes nonlocal behavior, the first two in terms of the Riemann-Liouville integral and the third in terms of the Hadamard integral. Further, let's take into account only real functions of real variables, although several results can be extended to the complex field.

The chapter is arranged as follows: first we present the calculation of the derivative of order 1/2 of the function $f(x) = x$, through the Grünwald-Letnikov formulation, exclusively to obtain the result that led to the paradox, as the prophetic words of Leibniz, therefore the Riemann-Liouville and Caputo formulations are easier to handle. We now turn to the formulations given in terms of integrals, and to this end we introduce the concept of iterated integral, the called Cauchy's integral and its generalization, which brings us to the Riemann-Liouville integral and from a convenient change of variable, we introduced the called Hadamard's integral. We mention the importance of these two formulations in the sense that the former is associated with the translations and the latter associated with the dilations. After introducing the convenient spaces, we define the fractional derivatives, as proposed by Riemann-Liouville, Caputo and Hadamard.

5.1 Grünwald-Letnikov Formulation

As we have already mentioned, the formulation of the fractional derivative, as proposed by Grünwald-Letnikov, plays an important role in numerical problems, since it is given in terms of a summation, which can be "truncated" with a number of terms sufficient to satisfy the conditions and precision imposed on the particular problem being studied. We point out that this formulation, given in terms of a summation, is quite tedious as far as the calculations are concerned, but we will mention it solely and exclusively because of the history, in particular about the questioning of l'Hôpital that led Leibniz to prophesy about a possible paradox that would lead in the future to new and important results. In 1869, Sonin published an important work on arbitrary differentiation [4] which, together with Grünwald [5] and Letnikov [6] leds to introduce the formulation of the non-integer order derivative which is now known as the Grünwald-Letnikov fractional derivative, a natural generalization of the n-th integer-order derivative, given in terms of a sum, as presented in Application (25) of Chap. 6.

Finally, let's just explain the calculations, via Grünwald-Letnikov formulation, of the derivative of order $1/2$ of the function $f(x) = x$, to verify how laborious this type of calculation is, as well as for historical reasons, after all can this fact led Leibniz to utter the words by mentioning a possible paradox in response to l'Hôpital.

The derivative of order k with $k \in \mathbb{N}$ is given by

$$D^k f(x) = \lim_{h \to 0} \frac{1}{h^k} \sum_{j=0}^{n} (-1)^j \binom{k}{j} f(x - jh)$$

with $n \geq k$ and $\binom{k}{j} = \dfrac{k!}{j!(k-j)!}$ the binomial coefficient. For values of negative k (integration) we define

$$\binom{-k}{j} = (-1)^j \binom{k}{j}$$

where can we write the expression

$$D^{-k} f(x) = \lim_{h \to 0} \frac{1}{h^k} \sum_{j=0}^{n} \binom{k}{j} f(x - jh).$$

Then, in our case, with a view to Grünwald-Letnikov formulation, let us first consider $x_0 \in \mathbb{R}$ an arbitrary point (expansion around that point) and $h = (x - x_0)/n$ (step size). Let us assume for convenience and simplicity $x_0 = 0$ and replace h in the expression for the derivative, as well as take the limit $n \to \infty$, in order to obtain the derivative of Grünwald-Letnikov, denoted in this book by ${}^{GL}_{0}D^{\mu}_{x}$

$$^{GL}_{0}D^{\mu}_{x} f(x) = \lim_{n \to \infty} \left(\frac{n}{x} \right)^{\mu} \sum_{j=0}^{n} (-1)^j \binom{\mu}{j} f\left(x - \frac{x}{n} j \right)$$

where, in this notation, the subscript 0, on the left, means that we are considering the point $x_0 = 0$; the sub-index x, to the right, means that our derivative is in relation to x; being $\mu \in \mathbb{R}$ the order of the derivative while the superscript ^{GL} denotes that the derivative is taken in the Grünwald-Letnikov sense, to differentiate from other formulations.

In our case, $f(x) = x$ and $\mu = 1/2$, in which, after a simplification, we get

$$^{GL}_{0}D^{\frac{1}{2}}_{x} x = \sqrt{x} \lim_{n \to \infty} \sqrt{n} \sum_{j=0}^{n} (-1)^j \binom{1/2}{j} \left(1 - \frac{j}{n} \right).$$

We start by separating the summation into two others and simplifying the notation, so

$$D^{\frac{1}{2}} x = \sqrt{x} \left\{ \lim_{n \to \infty} \sqrt{n} \sum_{j=0}^{n} (-1)^j \binom{1/2}{j} - \lim_{n \to \infty} \frac{\sqrt{n}}{n} \sum_{j=1}^{n} (-1)^j \binom{1/2}{j} j \right\}.$$

By performing an index change, $j \to j + 1$ in the second sum and rearranging, we have

$$D^{\frac{1}{2}} x = \sqrt{x} \left\{ \lim_{n \to \infty} \sqrt{n} \sum_{j=0}^{n} (-1)^j \frac{\Gamma(3/2)}{j!\Gamma(-j+3/2)} + \lim_{n \to \infty} \frac{\Gamma(3/2)}{\sqrt{n}} \sum_{j=0}^{n-1} \frac{(-1)^j}{j!\Gamma(-j+1/2)} \right\}. \quad (5.1)$$

We calculate the sums separately. Let Ω_1 such that

$$\Omega_1 = \lim_{n \to \infty} \sqrt{n} \sum_{j=0}^{n} (-1)^j \binom{1/2}{j}.$$

Using the relation [7]

$$\sum_{j=0}^{n} (-1)^j \binom{a}{j} = (-1)^n \binom{a-1}{n}$$

which, in our case, $a = 1/2$, simplifying, provides

$$\Omega_1 = \lim_{n \to \infty} \sqrt{n} \, \frac{(-1)^n \Gamma(1/2)}{\Gamma(n+1)\Gamma(-n+1/2)}.$$

Using the reflection formula $\Gamma(-n+1/2)\Gamma(n+1/2) = \pi(-1)^n$, coming back into Ω_1 and rearranging, we can write

$$\Omega_1 = \frac{1}{\sqrt{\pi}} \lim_{n \to \infty} \sqrt{n} \, \frac{\Gamma(n+1/2)}{\Gamma(n+1)}.$$

For $z \to \infty$, the following asymptotic expansion is valid for the quotient of gamma functions [8]

$$\frac{\Gamma(z+\alpha)}{\Gamma(z+\beta)} \approx z^{\alpha-\beta}$$

which, in our case, $\alpha = 1/2$ and $\beta = 1$, from where it follows

$$\Omega_1 = \frac{1}{\sqrt{\pi}} \lim_{n \to \infty} \sqrt{n} n^{\frac{1}{2}-1} = \frac{1}{\sqrt{\pi}}.$$

By an analogous procedure to Ω_1, now, Ω_2 is given by

$$\Omega_2 = \lim_{n \to \infty} \frac{\Gamma(3/2)}{\sqrt{n}} \sum_{j=0}^{n-1} \frac{(-1)^j}{j!\Gamma(-j+1/2)}.$$

Multiplying the numerator and denominator of the precedent by $\Gamma(1/2)$ and using the notation for the binomial, we can write

$$\Omega_2 = \frac{\Gamma(3/2)}{\Gamma(1/2)} \lim_{n \to \infty} \frac{1}{\sqrt{n}} \sum_{j=0}^{n-1} (-1)^j \binom{-1/2}{j}.$$

Using the same expression as in the previous one, now, with $a = -1/2$ and $n \to n-1$, we obtain

$$\Omega_2 = \frac{1}{2} \lim_{n \to \infty} \frac{1}{\sqrt{n}} (-1)^{n-1} \binom{-3/2}{n-1}.$$

Now, after explaining the binomial coefficient in terms of gamma functions, using the reflection formula for the gamma functions, simplifying and rearranging, we can write

$$\Omega_2 = \frac{1}{\sqrt{\pi}} \lim_{n \to \infty} \frac{1}{\sqrt{n}} \frac{\Gamma(n + 1/2)}{\Gamma(n)}.$$

Finally, in analogy to Ω_1, using the asymptotic formula for the quotient of gamma functions, with $\alpha = 1/2$ and $\beta = 0$, follows

$$\Omega_2 = \frac{1}{\sqrt{\pi}} \lim_{n \to \infty} \frac{1}{\sqrt{n}} n^{\frac{1}{2}-0} = \frac{1}{\sqrt{\pi}}.$$

Returning with the results of Ω_1 and Ω_2 in Eq. (5.1) we obtain

$$\mathsf{D}^{\frac{1}{2}} x = \sqrt{x} \left(\frac{1}{\sqrt{\pi}} + \frac{1}{\sqrt{\pi}} \right) = 2\sqrt{\frac{x}{\pi}} \qquad (5.2)$$

which is the desired result, the same as that obtained by the fractional derivative formulations in the Riemann-Liouville and Caputo sense, as we will see later. It is necessary to frighten that, not always the definition is the best way to approach the problem, in particular, if we use either the representation integral or even, in this case, the derivative of a power, the result emerges in a much simpler way, besides of course, because we know how to evaluate the derivative, when the order of the derivative is an integer, a power-type function.

5.2 Integer Order Integral

In order to present the concept of integral of integer order, we begin by showing that an integral of order n, with $n \in \mathbb{N}$, of a function $f(x)$ with $x \in \mathbb{R}$, can be seen as a Laplace convolution product between the $f(x)$ function and the Gel'fand-Shilov function of order n, $\phi_n(x)$. This integer-order integral is also called a multiple or iterated integral. From the generalization of the concept of factorial, through the gamma function, we introduce the concept of non-integer order integral or fractional order integral.

Definition 5.2.1 (*Integer Order Integral*) The integer-order integral, through the operator $\mathcal{J}$, acting on the function $f(t)$, is given by

$$\mathcal{J} f(t) = \int_0^t f(t_1) \, \mathrm{d}t_1.$$

From the integer order integral, iterating, we get

$$\mathcal{J}^2 f(t) = \mathcal{J}[\mathcal{J} f(t)] = \int_0^t \int_0^{t_1} f(t_2)\, dt_2\, dt_1,$$

$$\mathcal{J}^3 f(t) = \mathcal{J}[\mathcal{J}^2 f(t)] = \int_0^t \int_0^{t_1} \int_0^{t_2} f(t_3)\, dt_3\, dt_2\, dt_1.$$

We define the integral of order n by the expression

$$\mathcal{J}^n f(t) = \int_0^t \int_0^{t_1} \int_0^{t_2} \cdots \int_0^{t_{n-2}} \int_0^{t_{n-1}} f(t_n)\, dt_n\, dt_{n-1} \cdots dt_3\, dt_2\, dt_1.$$

Let us express the integral of order n, with $n \in \mathbb{N}$, through a theorem involving the Gel'fand-Shilov function and the Laplace convolution product.

Theorem 5.2.1 (Integral of order n) *Let $n \in \mathbb{N}$, $t \in \mathbb{R}_+$ and $f(t)$ be an integrable function. The integral of order n is given by*

$$\mathcal{J}^n f(t) = \phi_n(t) \star f(t) := \int_0^t \phi_n(t - \tau) f(\tau)\, d\tau = \int_0^t \frac{(t - \tau)^{n-1}}{(n - 1)!} f(\tau)\, d\tau,$$

with $\star$ denoting the Laplace convolution product.

Proof Let us prove the theorem by induction in the n parameter. For $n = 1$ we have

$$\mathcal{J} f(t) = \int_0^t f(\tau)\, d\tau = \int_0^t \frac{(t - \tau)^{1-1}}{(1 - 1)!} f(\tau)\, d\tau = \phi_1(t) \star f(t).$$

Just show that: if $\mathcal{J}^n f(t) = \phi_n(t) \star f(t)$ then $\mathcal{J}^{n+1} f(t) = \phi_{n+1}(t) \star f(t)$.
  By the inductive hypothesis we obtain

$$\mathcal{J}^{n+1} f(t) = \mathcal{J}[\mathcal{J}^n f(t)] = \mathcal{J}[\phi_n(t) \star f(t)] = \int_0^t \phi_n(u) \star f(u)\, du$$

$$= \int_0^t \int_0^u \frac{(u - \tau)^{n-1}}{(n - 1)!} f(\tau)\, d\tau\, du.$$

By Goursat's theorem [9] it is possible to exchange the integration order, i.e.,

$$\mathcal{J}^{n+1} f(t) = \int_0^t \left[\int_\tau^t \frac{(u - \tau)^{n-1}}{(n - 1)!}\, du \right] f(\tau)\, d\tau.$$

Calculating the integral between brackets we can write

$$\mathcal{J}^{n+1} f(t) = \int_0^t \frac{(t - \tau)^n}{n!} f(\tau) \, d\tau = \phi_{n+1}(t) \star f(t),$$

which is precisely the desired result   □

Given the result obtained in Theorem 5.2.1 and using the concept of gamma function, a generalization of the concept of factorial, as discussed in Chap. 2, we will generalize the order of the integral to a number $v \in \mathbb{R}$.

Definition 5.2.2 (*Integral of Order v*) Let $f(t)$ be an integrable function. We define the integral of the order $v \in \mathbb{R}$, of the function $f(t)$, denoted by $\mathcal{J}^v f(t)$, through the expression

$$\mathcal{J}^v f(t) = \phi_v(t) \star f(t) = \int_0^t \frac{(t - \tau)^{v-1}}{\Gamma(v)} f(\tau) \, d\tau. \tag{5.3}$$

In the case where the parameter associated with the order, v, is such that $v = n + 1$ with $n \in \mathbb{N}$, we retrieve the result for the integral of integer order.

5.3   Riemann-Liouville and Hadamard Integrals

Since both fractional integrals and their fractional derivatives are considered to the left and right of a point, in this book we only present the formulation to the left. The respective formulation to the right, with simple modifications, follows the same steps of the formulation to the right. With this condition, we begin by defining the Riemann-Liouville fractional integral on the left, as well as the Hadamard fractional integral on the left. From now on, we will omit the nomenclature on the left, that is, we present the definitions of the fractional integrals of Riemann-Liouville and Hadamard.

Definition 5.3.1 (*Riemann-Liouville Integral*) Let $t \in \mathbb{R}$ and $\mathrm{Re}(v) > 0$. The Riemann-Liouville fractional integral of order v, acting in function $f \in L^p[a, b]$, $1 \le p < +\infty$, $-\infty < a < b < +\infty$, for $t \in [a, b]$, is defined by

$$(\mathcal{J}^v f)(t) \equiv \mathcal{J}^v f(t) := \frac{1}{\Gamma(v)} \int_a^t \frac{f(\tau)}{(t - \tau)^{1-v}} \, d\tau, \qquad t > a. \tag{5.4}$$

We mention that the Riemann-Liouville fractional integrals are characterized by the particular class of functions where this operator acts, as well as by the respective integration interval to be considered. For example, in the case where $a = 0$ in Eq. (5.4), we obtain the definition of the fractional integral, as proposed by Riemann, without the called complementary function [10, 11]. Also, if we consider $a = -\infty$, we obtain the Liouville fractional integral [11]. Furthermore, it should be stressed that expression given by Eq. (5.4) will be used to introduce the concept of fractional derivative in the Riemann-Liouville sense (integer order derivative of a fractional order integral) and Caputo sense (fractional order integral of an integer order derivative).

Definition 5.3.2 (*Hadamard Integral*) Let $t \in \mathbb{R}^+$ and $\mathrm{Re}(\nu) > 0$. The Hadamard fractional integral of order ν, acting on the function $f \in L^p[a,b]$, $1 \le p < +\infty$, $-\infty < a < b < +\infty$, for $t \in [a,b]$, is defined by

$$(\mathsf{H}\mathcal{J}^\nu f)(t) \equiv {}_\mathsf{H}\mathcal{J}^\nu f(t) := \frac{1}{\Gamma(\nu)} \int_a^t \left(\ln \frac{t}{\tau} \right)^{\nu-1} f(\tau) \frac{d\tau}{\tau}, \qquad t > a. \qquad (5.5)$$

We label with a subscript the Hadamard fractional integral, unlike the Riemann-Liouville fractional integral. We chose this notation, leaving without subscript, only the Riemann-Liouville fractional integral and placing in the others. With this, we use the same letter $\mathcal{J}$ to denote the integral, as well as the derivative denoting it with the letter $\mathcal{D}$. Finally, we mention that, by means of a suitable variable change of type $\eta = \ln \xi$, respecting the conditions of existence of the logarithm, we can drive Eq. (5.4) in Eq. (5.5) and vice versa.

5.4 Riemann-Liouville, Caputo and Hadamard Derivatives

Before we define the Riemann-Liouville, Caputo, and Hadamard derivatives, we will introduce a convenient function space, from the space of absolutely continuous functions, as well as a suitable notation for the operator $t \dfrac{d}{dt}$. Note that, since $k \in \mathbb{R}$ is a constant, the change of variable $t \to kt$ does not change this operator, that is, an invariant by such a change.

Definition 5.4.1 (*Absolutely Continuous Weighted Function Spaces*) Let $[a,b]$ be a finite interval such that $-\infty < a < b < +\infty$ and $AC[a,b]$ the space of the f absolutely continuous functions in $[a,b]$. Let $AC^n[a,b]$ be the space of the complex value functions, f which have continuous derivatives up to order $n-1$ in $[a,b]$ such that $f^{(n-1)}$ in $AC[a,b]$. Let $\delta := t \dfrac{d}{dt}$ be the operator. We introduce a weighted modification of the space $AC^n[a,b]$, in which the usual derivative $\mathrm{D} = \dfrac{d}{dt}$ is replaced by δ-derived. Such modification, which we denote by $AC^n_{\delta,\mu}[a,b]$ with $n \in \mathbb{N}$ and $\mu \in \mathbb{R}$, involves the measurable Lebesgue functions of complex value g in (a,b)

such that $t^{\mu}g(t)$ admits δ-derivatives up to order $n-1$ in $[a, b]$ and $\delta^{n-1}[t^{\mu}g(t)]$ is absolutely continuous in $[a, b]$. In particular,

$$AC_{\delta}^{n}[a, b] := \left\{ f : t \in [a, b] \to \mathbb{R} \text{ such that } \left(\delta^{n-1}f\right) \in AC[a, b] \right\}$$

is the called spaces of absolutely continuous weighted functions $(n-1)$-differentiable in $[a, b]$. For $n = 1$, we have $AC_{\delta}^{1}[a, b] = AC[a, b]$, that is, we recover the space of absolutely continuous functions.

Definition 5.4.2 (*Riemann-Liouville fractional derivative*) Let $t \in \mathbb{R}$, $\mathrm{Re}(v) > 0$, $n = [\mathrm{Re}(v)] + 1$, with $[\mathrm{Re}(v)]$ the integer part of $\mathrm{Re}(v)$, $n \in \mathbb{N}$ and $v \notin \mathbb{N}$. The Riemann-Liouville fractional derivative of order v, acting on the function $f \in L^p[a, b]$, $1 \le p < +\infty$, $-\infty < a < b < +\infty$, for $t \in [a, b]$, is defined by

$$(\mathcal{D}^{v}f)(t) \equiv \mathcal{D}^{v}f(t) := \left(\frac{\mathrm{d}}{\mathrm{d}t}\right)^{n}\left(\mathcal{J}^{n-v}f\right)(t)$$

$$= \frac{1}{\Gamma(n-v)}\left(\frac{\mathrm{d}}{\mathrm{d}t}\right)^{n}\int_{a}^{t}\frac{f(\tau)}{(t-\tau)^{v-n+1}}\mathrm{d}\tau. \qquad (5.6)$$

As we have already mentioned, the note on the left and right integrals is also valid for the derivatives, that is, we will consider only the formulations on the left. Further, from Eq. (5.6) we can say: the derivative of arbitrary order, according to the Riemann-Liouville formulation, is equivalent to the integer-order derivative of an arbitrary order integral.

Definition 5.4.3 (*Caputo Fractional Derivative*) Let $t \in \mathbb{R}$, $\mathrm{Re}(v) > 0$, $n = [\mathrm{Re}(v)] + 1$, with $[\mathrm{Re}(v)]$ the integer part of $\mathrm{Re}(v)$, $n \in \mathbb{N}$ and $v \notin \mathbb{N}$. Let $f(t) \in AC^{n}[a, b]$. The Caputo fractional derivative of order v, acting on the function f, with $-\infty < a < b < +\infty$, for $t \in [a, b]$, is defined by

$$(\mathrm{c}\mathcal{D}^{v}f)(t) \equiv \mathrm{c}\mathcal{D}^{v}f(t) := \left(\mathcal{J}^{n-v}\mathrm{D}^{n}f\right)(t)$$

$$= \frac{1}{\Gamma(n-v)}\int_{a}^{t}\frac{\mathrm{D}^{n}f(\tau)}{(t-\tau)^{v-n+1}}\mathrm{d}\tau \qquad (5.7)$$

where we use the notation $\mathrm{D}^{n} \equiv \left(\frac{\mathrm{d}}{\mathrm{d}t}\right)^{n}$.

In analogy to the one mentioned for the Riemann-Liouville formulation, Eq. (5.7) allows us to state: the arbitrary order derivative, according to the Caputo formulation, is equivalent to the arbitrary order integral of an integer-order derivative [12–14].

The Hadamard fractional derivative will be introduced from the Hadamard integral, according to Eq. (5.5). Moreover, taking into account that the integral and the derivative can commute in analogy to those Riemann-Liouville and Caputo derivatives, we shall call it a fractional derivative in the Riemann-Liouville-Hadamard sense, also known simply by the name of Hadamard fractional derivative, this nomenclature that we will use, while in the sense of Caputo, will be called the Hadamard-Caputo fractional derivative, also known as a Hadamard fractional derivative regularized Caputo type. Unlike the case of integrals, we introduced the subscript in the two formulations of the fractional derivatives of Hadamard and Hadamard-Caputo.

Definition 5.4.4 (*Hadamard Fractional Derivative*) Let $t \in \mathbb{R}$, $\mathrm{Re}(v) > 0$, $n = [\mathrm{Re}(v)] + 1$, with $[\mathrm{Re}(v)]$ the integer part of $\mathrm{Re}(v)$, $n \in \mathbb{N}$ and $\delta := t\dfrac{\mathrm{d}}{\mathrm{d}t}$. The Hadamard fractional derivative of order v, acting on the function $f \in AC^n_\delta[a, b]$, $0 \le a < b < \infty$, is defined by

$$(_H\mathcal{D}^v f)(t) \equiv {_H}\mathcal{D}^v f(t) = \frac{1}{\Gamma(n - v)} \left(\frac{\mathrm{d}}{\mathrm{d}t}\right)^n \int_a^t \left(\ln \frac{t}{\tau}\right)^{n-v-1} f(\tau)\frac{\mathrm{d}\tau}{\tau} \qquad (5.8)$$

or, using the δ operator, in the following form

$$(_H\mathcal{D}^v f)(t) \equiv {_H}\mathcal{D}^v f(t) = \delta^n (\mathcal{J}^{n-v} f)(t). \qquad (5.9)$$

Definition 5.4.5 (*Hadamard-Caputo Fractional Derivative*) Let $t \in [a, b]$, $0 \le a < b < \infty$ and $n - 1 < v \le n$, with $u \in \mathbb{N}$. Consider $\mathrm{Re}(v) > 0$, $n = [\mathrm{Re}(v)] + 1$, with $[\mathrm{Re}(v)]$ the integer part of $\mathrm{Re}(v)$, $n \in \mathbb{N}$ and $\delta := t\dfrac{\mathrm{d}}{\mathrm{d}t}$. The Hadamard-Caputo fractional derivative of order v, acting on the function $f \in AC^n_\delta[a, b]$, $0 \le a < b < \infty$, is defined by

$$(_{HC}\mathcal{D}^v f)(t) \equiv {_{HC}}\mathcal{D}^v f(t) = \frac{1}{\Gamma(n - v)} \int_a^t \left(\ln \frac{t}{\tau}\right)^{n-v-1} \left(\frac{\mathrm{d}}{\mathrm{d}\tau}\right)^n f(\tau)\frac{\mathrm{d}\tau}{\tau} \qquad (5.10)$$

or, using the δ operator, in the following form

$$(_{HC}\mathcal{D}^v f)(t) \equiv {_{HC}}\mathcal{D}^v f(t) = \mathcal{J}^{n-v}(\delta^n f)(t). \qquad (5.11)$$

Note that in analogy to the Riemann-Liouville and Caputo derivatives, the Hadamard fractional derivative, Eq. (5.9), derives an integral of fractional order and the Hadamard-Caputo fractional derivative, Eq. (5.11), integrates the integer-order derivative.

5.5 Exercises

We separate this section into four subsections. The first one contains only the statements of the exercises (List of exercises), the second account with a suggestion (Suggestions) for the respective solution, while the third contains the resolutions (Solutions) itself. The fourth subsection presents exercises (Proposed exercises) to be solved by the students, all of them similar to those discussed in the text.

5.5.1 Exercise list

1. Let $n, m \in \mathbb{N}$ with $m \geq n$. The derivative of order n of the function $f(x) = x^m$ is

$$D^n x^m = \frac{m!}{(n-m)!} x^{m-n}.$$

   Let $\nu, \mu \in \mathbb{R}$ with $\mu > -1$. Obtain an expression for the derivative of order ν of the $f(x) = x^\mu$ function. Retrieve the derivative of order $1/2$ the function $f(x) = x$, this is the result of the l'Hôpital questioning for Leibniz.

2. Let $a, \beta \in \mathbb{R}$. Show that $\mathcal{D}_x^\beta (\cos ax) = a^\beta \cos \left(ax + \frac{\pi \beta}{2} \right).$

3. Let $a \in \mathbb{R}$, $\beta \in \mathbb{C}$ with $\mathrm{Re}(\beta) > 0$. Show that: The Riemann-Liouville integral of a power function $f(t) = (t-a)^{\beta-1}$ is a power function, also.

4. Evaluate the integral of order k, $\mathcal{J}_{0+}^k$, of the classical Mittag-Leffler function.

5. Let $\mathrm{Re}(\beta) > \mathrm{Re}(\alpha) > 0$ and $0 < a < b < \infty$. Show that: The Hadamard integral $(_H\mathcal{J}_{a+}^\alpha f)(t)$ of the function $f(t) = \left(\ln \frac{t}{a} \right)^{\beta-1}$ is given by

$$\left[{}_H\mathcal{J}_{a+}^\alpha \left(\ln \frac{t}{a} \right)^{\beta-1} \right](x) = \frac{\Gamma(\beta)}{\Gamma(\beta+\alpha)} \left(\ln \frac{x}{a} \right)^{\beta+\alpha-1}.$$

6. Let $\alpha, \beta \in \mathbb{C}$ such that $\mathrm{Re}(\alpha) > 0$, $\mathrm{Re}(\beta) > 0$ and $1 \leq p \leq \infty$. If $c < 0$, then, for $f \in X_c^p(\mathbb{R}^+)$, the semigroup property is valid

$$\left[{}_H\mathcal{J}_{0+}^\alpha {}_H\mathcal{J}_{0+}^\beta f(x) \right](t) = \left[{}_H\mathcal{J}_{0+}^{\alpha+\beta} f(x) \right](t)$$

   being $_H\mathcal{J}_{0+}^\alpha$ the Hadamard integral on the left.

7. Consider the space $L^p[a, b]$ with $p \in [1, \infty]$ and $0 \leq a \leq b < \infty$. Show that: the Hadamard fractional derivative of order α on the left, with $0 < \alpha \leq 1$ is the inverse operator of the fractional integral on the left, that is,

$$({}_H\mathcal{D}^\alpha {}_H\mathcal{J}^\alpha f(t) = f(t),$$

for $a \le t \le b$.

8. Show that

$$\mathscr{L}^{-1}\left[\frac{F(s)}{s^2}\right] = \int_0^t \int_0^{t_1} f(t_2)\, dt_2\, dt_1$$

where $F(s)$ is the Laplace transform of $f(t)$, admitted to exist.

9. Use Exercise (8) to show that

$$\int_0^t \int_0^{t_1} f(t_2)\, dt_2\, dt_1 = \int_0^t (t - \xi) f(\xi)\, d\xi.$$

10. Show the result given by Eq. (5.4).

11. Let $f(t)$ be an analytical function and $\mu \in \mathbb{C}$. Show that: The Riemann-Liouville fractional derivative of order μ can be written as follows

$$(\mathcal{D}^\mu f)(t) \equiv \mathcal{D}^\mu f(t) := \sum_{k=0}^\infty \binom{\mu}{k} \frac{t^{k-\mu} f^{(k)}(t)}{\Gamma(k - \mu + 1)},$$

with $f^{(k)}(t)$ denoting the derivative of $f(x)$ of order k.

12. Let $\mu \in \mathbb{C}$ and $n \in \mathbb{N}$ such that $n - 1 < \mathrm{Re}(\mu) \le n$. Show that the relation between the fractional derivatives of Riemann-Liouville and Caputo is valid,

$$(_C\mathcal{D}^\mu f)(t) = (\mathcal{D}^\mu f)(t) - \sum_{k=0}^{n-1} f^{(k)}(0) \frac{t^{k-\mu}}{\Gamma(k - \mu + 1)}.$$

13. Set $t > 0$. Consider a function $f(t)$ and admit that the fractional derivatives of Riemann-Liouville and Caputo, of the same order, exist. What must be the condition in order for these two derivatives to coincide?

14. Let f defined for every $t > 0$. Show that the Laplace transform of the Riemann-Liouville fractional integral of order μ of f is given by

$$\mathscr{L}[J^\mu f(t)] = \frac{\mathscr{L}[f(t)]}{s^\mu}.$$

15. Let $\mathrm{Re}(\mu) > 0$ and $n \in \mathbb{N}$ such that $n - 1 < \mathrm{Re}(\mu) \le n$. Show that the Laplace transform of the Caputo derivative is given by

$$\mathscr{L}[_C\mathcal{D}^\mu f(t)] = s^\mu \mathscr{L}[f(t)] - \sum_{k=0}^{n-1} s^{\mu-1-k} f^{(k)}(0) \qquad (5.12)$$

where s is the parameter of the Laplace transform and $f^{(k)}(0) = \lim_{t \to 0} {_C\mathcal{D}^k f(t)}$.

16. Let $\nu \in \mathbb{R}$ such that $n - 1 < \nu \leq n$ with $n = 1, 2, 3, \ldots$ Calculate the Caputo fractional derivative of the Mittag-Leffler function. Discuss the case where the order of the derivative is equal to the parameter of the Mittag-Leffler function.

17. Let $n - 1 < \nu \leq n$ with $n = 1, 2, 3, \ldots$ Calculate the Riemann-Liouville fractional derivative of the Mittag-Leffler function. Compare with the result in the case where we have the Caputo fractional derivative.

18. Let $\nu > 0$. Calculate the Riemann-Liouville fractional integral of order ν of a particular Mittag-Leffler function with three parameters $E_{\alpha,1}^{\gamma}(-t^{\alpha})$ and retrieve the particular case in which $\gamma = 1$, that is, the classical Mittag-Leffler function.

19. Let $t > a$ and $\alpha > 0$. We define the classical Mittag-Leffler function with one parameter

$$E_{\alpha}\left[\left(\ln\frac{t}{a}\right)^{\alpha}\right] = \sum_{k=0}^{\infty} \frac{(\ln\frac{t}{a})^{\alpha k}}{\Gamma(\alpha k + 1)}.$$

Evaluate the Hadamard fractional integral of the classical Mittag-Leffler function.

20. Let $\nu > 0$, $\lambda \in \mathbb{R}$ and $t \geq 0$. Show that $x(t) = E_{\nu}(\lambda t^{\nu})$ is solution of the fractional differential equation

$$_{C}\mathcal{D}^{\nu} x(t) = \lambda x(t)$$

where the derivative is taken in the Caputo sense.

21. Let $m \in \mathbb{R}$. Calculate the Caputo fractional derivative of order ν, with $n - 1 < \nu \leq n$ with $n \in \mathbb{N}$ for the following function

$$f(t) = (t + 1)^{m}.$$

22. Let $\nu > 0$ and $t > a$. Denoting by $_{a}\mathcal{D}_{t}^{\nu}$ and $_{a}\mathcal{J}_{t}^{\nu}$ the Riemann-Liouville fractional derivative and the Riemann-Liouville fractional integral, respectively, both of order ν and initialized at $t = a$, show that

$$_{a}\mathcal{D}_{t}^{\nu}\left[_{a}\mathcal{J}_{t}^{\nu} f(t)\right] = f(t)$$

interpreted as: the Riemann-Liouville fractional derivative is the inverse to the left of the Riemann-Liouville fractional integral.

23. Show that the Riemann-Liouville fractional derivative of order $\nu = n > 1$ coincides with the usual derivative of order n.

24. Let $\mu > \nu \geq 0$ and m and n integers such that $0 \leq m - 1 \leq \mu < m$ and $0 \leq n \leq \mu - \nu < n$. Show that the following relation is valid

$$_{a}\mathcal{D}_{t}^{\mu}\left[_{a}\mathcal{J}_{t}^{\nu} f(t)\right] = {}_{a}\mathcal{D}_{t}^{\mu-\nu} f(t).$$

25. Let C be a constant. Show that the derivative of $f(t) = C$ is zero, using the Caputo fractional derivative and nonzero, using the Riemann-Liouville derivative.

26. Let $0 < v < 1$ and $a = -\infty$. Calculate the Fourier transform of the Riemann-Liouville fractional integral.

27. Calculate the Fourier transform of the Riemann-Liouville fractional derivative.

28. Let $v > 0$ and the initialization at $a = 0$. Show that the Mellin transform of the Riemann-Liouville fractional integral is

$$\mathcal{M}\left[{}_0\mathcal{J}_t^v f(t)\right] = \frac{\Gamma(1 - s - v)}{\Gamma(1 - s)} F(s + v)$$

where s is the parameter of the Mellin transform and $F(s) = \mathcal{M}[f(t)]$.

29. Let $s < 0$ be the parameter of Mellin transform. Show that the Mellin transform of the Hadamard fractional integral is given by

$$\mathcal{M}[{}_H\mathcal{J}_0^v f(t)] = (-s)^{-v} F(s)$$

with $F(s) = \mathcal{M}[f(t)]$.

30. Let μ, with $\mathrm{Re}(\mu) > -1$ the order of the Bessel function of the first kind, denoted by $J_\mu(\cdot)$. Show that the Riemann-Liouville fractional integral of the function

$$f(t) = t^{\mu/2} J_\mu(2\sqrt{t})$$

is given by

$${}_0\mathcal{J}_t^v \left[t^{\mu/2} J_\mu(2\sqrt{t})\right] = t^{\frac{\mu+v}{2}} J_{\mu+v}(2\sqrt{t})$$

being $v > 0$ the order of the fractional integral.

31. Let $v > 0$, $\alpha > 0$, $\beta > 0$, $\gamma > 0$ and $a \in \mathbb{R}$. Show that the Riemann-Liouville fractional integral of the function $f(t) = t^{\beta-1} E_{\alpha,\beta}^\gamma(at^\alpha)$ is given by

$${}_0\mathcal{J}_t^v \left[t^{\beta-1} E_{\alpha,\beta}^\gamma(at^\alpha)\right] = t^{v+\beta-1} E_{\alpha,v+\beta}^\gamma(at^\alpha)$$

being $E_{\alpha,\beta}^\gamma(\cdot)$ the Mittag-Leffler function with three parameters.

32. Let $\mu, \lambda \in \mathbb{R}$ and $0 \le \alpha < 1$. Show that the integral equation [15]

$$x(t) = \mu \frac{t^{1-\alpha}}{\Gamma(2-\alpha)} - \frac{\lambda}{\Gamma(1-\alpha)} \int_0^t (t - \xi)^{-\alpha} x(\xi)\, d\xi$$

admits as solution

$$x(t) = \frac{\mu}{\lambda} \left[1 - E_{1-\alpha}(-\lambda t^{1-\alpha})\right]$$

with $E_v(\cdot)$ the classical Mittag-Leffler function.

33. Consider the Riemann-Liouville fractional integral of order α

$$_0\mathcal{J}_t^\alpha f(t) = \frac{1}{\Gamma(\alpha)} \int_0^t (t-\tau)^{\alpha-1} f(\tau)\,d\tau$$

and $D = d/dt$ the differential operator of order one. Show that

$$D[_0\mathcal{J}_t^\alpha] - {}_0\mathcal{J}_t^\alpha[Df(t)] = f(0)\frac{t^{\alpha-1}}{\Gamma(\alpha)}.$$

34. Obtain a Rodrigues formula for the Legendre functions of first kind [16].
35. Let $J_0(\sqrt{x})$ the zero order Bessel function, show that

$$_0\mathcal{D}_x^{1/2} J_0(\sqrt{x}) = \frac{\cos(\sqrt{x})}{\sqrt{\pi x}}.$$

36. Calculate the fractional derivative of order $1/2$, in the Riemann-Liouville sense, initialized at zero, of the function $f(t) = \ln t$ [17].
37. Let $\alpha > 0$ and $h(r)$ be a known function. Solve the Poisson integral equation [18]

$$\int_0^{\pi/2} \phi(r\cos\theta)\,\sin^{2\alpha+1}\theta\,d\theta = h(r).$$

38. Let $0 < \alpha < 1$ and $f(t)$ be a continuous function. Solve the integral equation

$$\int_0^t (t-\tau)^{-\alpha} f(\tau)\,d\tau = 1.$$

39. Let $t > 0$ and $\nu > 0$ the order of the integral. Evaluate the Hadamard fractional, initialized at $a = 1$ of the function $f(t) = t$.
40. Let $\mu > -1$ and $\nu > 0$. Use the definition of the Caputo fractional derivative to get

$$_C\mathcal{D}_t^\nu t^\mu = \frac{\Gamma(\mu+1)}{\Gamma(\mu-\nu+1)} t^{\mu-\nu}.$$

5.5.2   Suggestions

1. Use the gamma function definition and take $\nu = 1/2$ and $\mu = 1$.
2. Use the Euler relation and $\mathcal{D}^\beta e^{iax} = (ia)^\beta e^{iax}$.
3. From the definition, perform a variable change to bring the integral into the integral representation for the beta function.

4. Use the definition, change the order of the integral with the summation and lead the resulting integral into an integral representation for the beta function. Identify the result with a Mittag-Leffler function with two parameters.

5. From the Hadamard integral, use a suitable change of variable to reduce the integral into an integral representation for the beta function.

6. Use the Hadamard integral definition on the left, the Dirichlet expression to change the integration order and use the integral representation for the beta function to calculate the remaining integral.

7. From the definitions of the fractional derivative and the Hadamard fractional integral, use the Dirichlet expression to change the order of the integrations, as well as use the Leibniz rule to derive the integral and simplify.

8. Enter the notation $g(t) = \int_0^t \int_0^{t_1} f(t_2) \, dt_2 \, dt_1$ and use the expression that gives the Laplace transform of the second derivative.

9. Use convolution theorem and Exercise (8).

10. Integrate the function once, then once more and once more. Use the gamma function.

11. Since the function is analytic, use the two-part Taylor series expansion for $\text{Re}(\mu) < 0$ and $\text{Re}(\mu) > 0$ and use the reflection formula for the gamma function and the expression for the Riemann-Liouville fractional integral. In the case where $\text{Re}(\mu) > 0$, it is necessary to use the expression [19]

$$\sum_{n=0}^{\infty} \binom{\alpha}{m-n} \binom{\beta}{n} = \binom{\alpha+\beta}{m}.$$

12. Use the linearity of the operators and show by induction in n that

$$\mathscr{I}^n \mathscr{D}^n f(t) = f(t) - \sum_{k=0}^{n-1} f^{(k)}(0) \frac{t^k}{k!}.$$

13. Direct from the relation between the two fractional derivatives, according to Exercise (12).

14. Take the Laplace transform of the fractional integral expression in terms of the convolution product and, to integrate, use the definition of the gamma function.

15. Express the Caputo derivative in terms of fractional integral and Exercise (14).

16. Use the definition of the classical Mittag-Leffler function and switch the order of integration with the derivative. Use the result of the Exercise (1) and the relation between the gamma and beta functions.

17. Use the definition of the classical Mittag-Leffler function and switch the order of integration to the summation. Use the result of the Exercise (1) and the relation between gamma and beta functions.

18. Use the series representation for this particular Mittag-Leffler function with three parameters and change the order of the integration with the summation. Use the definition of the beta function and the relationship between beta and gamma functions.

19. Enter the power-series representation of the Mittag-Leffler function in the Hadamard integral and change the order of the integral with the summation. Use a suitable change of variable and the relationship between gamma and beta functions.

20. Separate in two cases, $\lambda = 0$ and $\lambda \neq 0$. Introduce the series representation of the Mittag-Leffler function in the definition for the Caputo fractional derivative. Use the gamma function definition and the relation between gamma and beta functions.

21. To evaluate the resulting integral, use an adequate variable change and use the result of Exercise (15) in Chap. 2.

22. Divide in two cases $\nu = n \geq 1$, a positive integer and then ν other than an integer such that $n - 1 \leq \nu < n$ with $n = 1, 2, 3, \dots$ and use the semigroup property, according to Exercise (14) in Chap. 6.

23. Use the definition and the result $_a\mathcal{D}_t^0 f(t) = f(t)$.

24. Use the definition and result in Exercise (14) in Chap. 6.

25. Direct from the definition. In the case of the Riemann-Liouville fractional derivative, integrate first and then evaluate the derivative.

26. Write the fractional integral using the definition of convolution product, according to Eq. (5.3) and the Fourier transform of the Gel'fand-Shilov function.

27. Start from the definition of the Riemann-Liouville fractional derivative written in terms of the convolution product and use the result of the Fourier transform of the usual derivative of order n.

28. Enter the variable change $\tau = tx$ into the fractional integral and an adequate function so that the convolution theorem can be used, taking into account the Property 4.4.2, as well as the relation between gamma and beta functions.

29. Use the convolution product to be able to explicitly compute an integral, with the constraint $s < 0$, in terms of a gamma function.

30. Write the Bessel function in terms of the power series

$$J_\mu(z) = \sum_{k=0}^{\infty} \frac{(-1)^k}{k!} \frac{(z/2)^{\mu+2k}}{\Gamma(\mu + k + 1)}$$

and use the relationship between beta and gamma functions.

31. Enter the series representation of the Mittag-Leffler function with three parameters, change the variable to convert the integration into a representation for the beta function and use the relation between beta and gamma functions.

32. Use the Laplace transform and the convolution product to obtain an algebraic equation. Solve this algebraic equation and calculate the inverse Laplace transform, expressing the solution in terms of a Mittag-Leffler function with two parameters. Use the relationship $E_\mu(z) = z E_{\mu,\mu+1}(z) + 1$.

33. Use Leibniz's rule to derive an integral in the first parcel and integration by parts in the second, subtract one from the other.

34. Express the Legendre functions in terms of a hypergeometric function, use the expression that gives the derivative of order α of a power and identify with the power series of the hypergeometric function.
35. Use the power series expansion for the zero-order Bessel function and the Legendre duplication formula.
36. Use a change of variables and drive the integral to use the result

$$\int_0^1 \frac{\ln(1-x)}{\sqrt{x}} \, dx = -4(1 - \ln 2).$$

37. Introduce variable changes to drive the Poisson integral equation into an Abel integral equation from which the solution is obtained with the fractional derivative.
38. Multiply and divide the first member by $\Gamma(1 - \alpha)$ and identify with the Riemann-Liouville fractional integral.
39. Introduce suitable changes of variables in order to bring the integral into integral representation for the incomplete gamma function.
40. Calculate the derivative of the integer order n of the function $f(t) = t^\mu$ and introduce a variable change in order to drive the integral into the integral representation of the beta function and use the relation between the beta and gamma functions.

5.5.3 Solutions

1. Consider the parameters $n = \nu$ and $m = \mu$ in the expression for the derivative of order n of the function $f(x) = x^m$. Using the generalization of the factorial, that is, from the result $\Gamma(\ell + 1) = \ell!$ we can write

$$D^\nu x^\mu = \frac{\Gamma((\mu + 1))}{\Gamma(\mu - \nu + 1)} x^{\mu - \nu}.$$

This expression is valid for all values of μ and ν such that μ is different from a negative integer, the poles of the gamma function and that is in the numerator. If it occurs in the denominator, the result becomes zero. On the other hand, in the particular case where the parameters are such that $\mu = 1$ and $\nu = 1/2$, this is the derivative of order $1/2$ of the function $f(x) = x$, directly replacing these values in the preceding expression

$$D^{\frac{1}{2}} x^1 = \frac{\Gamma((1+1))}{\Gamma(1 - 1/2 + 1)} x^{1-1/2} = \frac{\Gamma(2)}{\Gamma(3/2)}.$$

Using the result $\Gamma(3/2) = \Gamma(1/2)/2 = \sqrt{\pi}/2$ and rearranging, we have

$$D^{\frac{1}{2}} x = 2\sqrt{\frac{x}{\pi}}.$$

This expression is the same as the result obtained in Eq. (5.2), that is, the derivative of order $1/2$ of the function $f(x) = x$, the questioning from l'Hôpital to Leibniz   ◊

2. Using the Euler relation $e^{iax} = \cos ax + i \sin ax$ and the relation involving the derivative of an exponential $\mathcal{D}^{\alpha} e^{iax} = (ia)^{\alpha} e^{iax}$, we can write

$$\mathcal{D}^{\beta} e^{iax} = \mathcal{D}^{\beta}(\cos ax + i \sin ax) = a^{\beta} \left(\cos \frac{\beta\pi}{2} + i \sin \frac{\beta\pi}{2} \right)(\cos ax + i \sin ax).$$

From the distributive property and rearranging, we have

$$\mathcal{D}^{\beta}(\cos ax + i \sin ax) = a^{\beta} \left[\cos \left(ax + \frac{\beta\pi}{2} \right) + i \sin \left(ax + \frac{\beta\pi}{2} \right) \right]$$

Identifying, real part with real part, we get

$$\mathcal{D}^{\beta}(\cos ax) = a^{\beta} \left[\cos \left(ax + \frac{\beta\pi}{2} \right) \right]$$

which is the desired result   ◊

3. Let $\mathrm{Re}(v) > 0$ and $t > a$. We must evaluate the integral

$$[\mathcal{J}_{a+}^{v}(t-a)^{\beta-1}](t) = \frac{1}{\Gamma(v)} \int_{a}^{t} \frac{(\tau-a)^{\beta-1}}{(t-\tau)^{1-v}} d\tau.$$

Note that this is the integral on the left. An analogous procedure is performed for the integral on the right, according to Proposed exercise (1). Consider the change of variable $\tau - a = x$ from where it follows to the integral

$$[\mathcal{J}_{a+}^{v}(t-a)^{\beta-1}](t) = \frac{(t-a)^{v-1}}{\Gamma(v)} \int_{0}^{t-a} x^{\beta-1} \left(1 - \frac{x}{t-a} \right)^{v-1} dx.$$

We now proceed with another change of variable $x = (t-a)y$ from where we get

$$[\mathcal{J}_{a+}^{v}(t-a)^{\beta-1}](t) = \frac{(t-a)^{v+\beta-1}}{\Gamma(v)} \int_{0}^{1} y^{\beta-1}(1-y)^{v-1} dy.$$

Finally, using the integral representation for the beta function, the relation between beta and gamma functions and simplifying, we obtain

$$[\mathcal{J}_{a+}^v (t-a)^{\beta-1}](t) = \frac{(t-a)^{v+\beta-1}}{\Gamma(v)} B(\beta, v) = \frac{\Gamma(\beta)}{\Gamma(v+\beta)}(t-a)^{v+\beta-1}$$

which is the desired result $\diamond$

4. Let $k \in \mathbb{N}$ and $\alpha > 0$. Using the definition, we must calculate

$$[\mathcal{J}_{0+}^k E_\alpha(-t^\alpha)](t) = \frac{1}{\Gamma(k)} \int_0^t E_\alpha(-\tau^\alpha) \frac{d\tau}{(t-\tau)^{1-k}}.$$

Introducing the series representation for the classical Mittag-Leffler function and permuting the order of this summation with the integral, we can write

$$[\mathcal{J}_{0+}^k E_\alpha(-t^\alpha)](t) = \frac{1}{\Gamma(k)} \sum_{n=0}^{\infty} \frac{(-1)^n}{\Gamma(n\alpha+1)} \int_0^t \tau^{n\alpha}(t-\tau)^{k-1} \, d\tau.$$

Consider the change of variable $\tau = \xi t$ from where it follows, already rearranging

$$[\mathcal{J}_{0+}^k E_\alpha(-t^\alpha)](t) = \frac{1}{\Gamma(k)} \sum_{n=0}^{\infty} \frac{(-1)^n}{\Gamma(n\alpha+1)} t^{n\alpha+k} \int_0^1 \xi^{n\alpha+1-1}(1-\xi)^{k-1} \, d\xi.$$

The resulting integral is nothing more than the integral representation for the beta function. Using the relation between beta and gamma functions and simplifying, we have

$$[\mathcal{J}_{0+}^k E_\alpha(-t^\alpha)](t) = t^k \sum_{n=0}^{\infty} \frac{(-t^\alpha)}{\Gamma(\alpha n + k + 1)}.$$

The remaining sum is identified with the Mittag-Leffler function with two parameters,

$$[\mathcal{J}_{0+}^k E_\alpha(-t^\alpha)](t) = t^k E_{\alpha,k+1}(-t^\alpha)$$

which is the desired result $\diamond$

5. Using the definition of the Hadamard integral given by Eq. (5.4), we have

$$\left[{}_H\mathcal{J}_{a+}^\alpha \left(\ln \frac{t}{a} \right)^{\beta-1} \right](x) = \frac{1}{\Gamma(\alpha)} \int_a^x \left(\ln \frac{x}{t} \right)^{\alpha-1} \left(\ln \frac{t}{a} \right)^{\beta-1} \frac{dt}{t}.$$

Introducing the change of variable ξ, defined by the relation

$$\xi = \frac{\ln \frac{t}{a}}{\ln \frac{x}{a}}$$

in the preceding equation and rearranging, we can write

$$\left[{}_H\mathcal{J}_{a+}^{\alpha} \left(\ln \frac{t}{a} \right)^{\beta-1} \right](x) = \frac{1}{\Gamma(\alpha)} \left(\ln \frac{x}{a} \right)^{\alpha+\beta-1} \int_0^1 (1-\xi)^{\alpha-1} \xi^{\beta-1} \, d\xi.$$

The remaining integral is an integral representation for the beta function, using the relation between beta and gamma functions, we obtain

$$\left[{}_H\mathcal{J}_{a+}^{\alpha} \left(\ln \frac{t}{a} \right)^{\beta-1} \right](x) = \frac{\Gamma(\beta)}{\Gamma(\alpha+\beta)} \left(\ln \frac{x}{a} \right)^{\alpha+\beta-1}$$

which is the desired result   ◇

6. From the definition of the Hadamard integral on the left (similar treatment is given for the Hadamard integral on the right), we have

$$\left[{}_H\mathcal{J}_{0+}^{\alpha} {}_H\mathcal{J}_{0+}^{\beta} f(x) \right](t) = \frac{1}{\Gamma(\alpha)} \int_a^t \left(\ln \frac{t}{\tau} \right)^{\alpha-1} \left[\frac{1}{\Gamma(\beta)} \int_a^\tau \left(\ln \frac{\tau}{\xi} \right)^{\beta-1} f(\xi) \frac{d\xi}{\xi} \right] \frac{d\tau}{\tau}.$$

In order to change the order of integration, we use Dirichlet's expression from where, by rearranging, we can write

$$\left[{}_H\mathcal{J}_{0+}^{\alpha} {}_H\mathcal{J}_{0+}^{\beta} f(x) \right](t) = \frac{1}{\Gamma(\alpha)\Gamma(\beta)} \int_a^t f(\xi) \frac{d\xi}{\xi} \left[\int_\xi^t \left(\ln \frac{t}{u} \right)^{\alpha-1} \left(\ln \frac{u}{\xi} \right)^{\beta-1} \frac{du}{u} \right].$$

The remaining integral, between brackets, was calculated in Exercise (5), then

$$\left[{}_H\mathcal{J}_{0+}^{\alpha} {}_H\mathcal{J}_{0+}^{\beta} f(x) \right](t) = \frac{1}{\Gamma(\alpha)\Gamma(\beta)} \int_a^t f(\xi) \left(\ln \frac{t}{\xi} \right)^{\alpha+\beta-1} \frac{\Gamma(\alpha)\Gamma(\beta)}{\Gamma(\alpha+\beta)} \frac{d\xi}{\xi}.$$

Simplifying, we can write

$$\left[{}_H\mathcal{J}_{0+}^{\alpha} {}_H\mathcal{J}_{0+}^{\beta} f(x) \right](t) = \frac{1}{\Gamma(\alpha+\beta)} \int_a^t f(\xi) \left(\ln \frac{t}{\xi} \right)^{\alpha+\beta-1} \frac{d\xi}{\xi}$$

or, in the following form

$$\left[{}_H\mathcal{J}_{0+}^{\alpha} {}_H\mathcal{J}_{0+}^{\beta} f(x) \right](t) = \left[{}_H\mathcal{J}_{0+}^{\alpha+\beta} f(x) \right](t)$$

which is the desired result   ◇

7. Using the definition of the Hadamard fractional derivative, given by Eq. (5.8), of order α, on the left with $0 < \alpha < 1$, implying $n = 1$ and the Hadamard fractional integral, given by Eq. (5.4), we have

$$_H\mathcal{D}^{\alpha} {}_H\mathcal{J}^{\alpha} f(t) = \frac{1}{\Gamma(1-\alpha)} \left(t\frac{d}{dt} \right) \int_a^t \left(\ln \frac{t}{\tau} \right)^{-\alpha} \left[\frac{1}{\Gamma(\alpha)} \int_a^\tau \left(\ln \frac{\tau}{\xi} \right)^{\alpha-1} f(\xi) \frac{d\xi}{\xi} \right] \frac{d\tau}{\tau}.$$

Using the Dirichlet condition to change the order of integrations and rearranging, we can write

$$_H\mathcal{D}^\alpha{}_H\mathcal{J}^\alpha f(t) = \frac{1}{\Gamma(1-\alpha)\Gamma(\alpha)} \left(t\frac{d}{dt}\right) \int_a^t f(\xi)\frac{d\xi}{\xi} \left[\int_\xi^t \left(\ln\frac{t}{u}\right)^{-\alpha} \left(\ln\frac{u}{\xi}\right)^{\alpha-1} \frac{du}{u} \right].$$

The remaining integral, between brackets, has already been calculated, then

$$_H\mathcal{D}^\alpha{}_H\mathcal{J}^\alpha f(t) = \frac{1}{\Gamma(1-\alpha)\Gamma(\alpha)} \left(t\frac{d}{dt}\right) \int_a^t f(\xi) \frac{\Gamma(\alpha)\Gamma(1-\alpha)}{\Gamma(\alpha+1-\alpha)} \left(\ln\frac{t}{\xi}\right)^{1-\alpha+\alpha-1} \frac{d\xi}{\xi},$$

rearranging provides

$$_H\mathcal{D}^\alpha{}_H\mathcal{J}^\alpha f(t) = t\frac{d}{dt} \int_a^t f(\xi)\frac{d\xi}{\xi}.$$

Using Leibniz's rule to derive an integral and simplifying, we have

$$_H\mathcal{D}^\alpha{}_H\mathcal{J}^\alpha f(t) = f(t)$$

which is the desired result   ◇

8. Consider the function

$$g(t) = \int_0^t \int_0^{t_1} f(t_2)\,dt_2\,dt_1$$

whose first derivative provides

$$g'(t) = \int_0^t f(t_2)\,dt_2$$

and the second one $g''(t) = f(t)$. Note that, also, $g(0) = 0 = g'(0)$.
From the expression which gives the Laplace transform of the second derivative

$$\mathcal{L}[g''(t)] = s^2 G(s) - sg(0) - g'(0)$$

being $G(s) = \mathcal{L}[g(t)]$ and conditions at the extreme $t = 0$, we obtain

$$\mathcal{L}[g''(t)] = s^2 \mathcal{L}[g(t)] = F(s) = \mathcal{L}[f(t)]$$

from where it follows

$$\mathcal{L}[g(t)] = \frac{F(s)}{s^2}.$$

Using the inverse Laplace transform in the previous one and the definition of $g(t)$ we have

$$\mathscr{L}^{-1}\left[\frac{F(s)}{s^2}\right] = g(t) = \int_0^t \int_0^{t_1} f(t_2)\, dt_2\, dt_1,$$

which is the desired result   ◇

9. Enter the notation $\Lambda = \int_0^t (t-\xi)f(\xi)\, d\xi$ and $F(s) = \mathscr{L}[f(t)]$. Taking the Laplace transform of Λ and using the definition of the convolution product we can write

$$\mathscr{L}\left[\int_0^t (t-\xi)f(\xi)\, d\xi\right] = \mathscr{L}[t]\mathscr{L}[f(t)] = \frac{F(s)}{s^2}.$$

Taking the inverse Laplace transform in the preceding one, we obtain

$$\int_0^t (t-\xi)f(\xi)\, d\xi = \mathscr{L}^{-1}\left[\frac{F(s)}{s^2}\right]$$

which, using the result of Exercise (8), provides

$$\int_0^t (t-\xi)f(\xi)\, d\xi = \int_0^t \int_0^{t_1} f(t_2)\, dt_2\, dt_1,$$

which is the desired result   ◇

10. We start by denoting the first integral of the function $f(t)$,

$$\mathcal{J}^1 f(t) = \frac{1}{\Gamma(1)} \int_a^t \frac{f(\tau)}{(t-\tau)^{1-1}}\, d\tau = \int_a^t f(\tau)\, d\tau.$$

In analogy to the previous one, integrating, we have for the second integral

$$\mathcal{J}^2 f(t) = \int_a^t \int_a^\tau f(\xi)\, d\xi\, d\tau = \int_a^t f(\xi)\, d\xi \int_\xi^t du = \int_a^t (t-\tau)f(\tau)d\tau$$

where, in the second equality, we change the order of integration through change of variable, according to Dirichlet. In analogy to the two previous ones we can write for the integral of order three

$$\mathcal{J}^3 f(t) = \int_a^t \int_a^x \int_a^y f(x)\, dz\, dy\, dx = \frac{1}{2}\int_a^t (t-\xi)^2 f(\xi)\, d\xi$$

from where it follows, using the notation for the gamma function and the principle of finite induction, after substituting the factorial by gamma function and $n \to \nu - 1$,

$$\mathcal{J}^\nu f(t) = \frac{1}{\Gamma(\nu)} \int_a^t \frac{f(\tau)}{(t-\tau)^{1-\nu}}\, d\tau$$

which is the desired result   ◇

11. Let $\mathrm{Re}(\mu) < 0$. Using the notation $(\mathscr{D}^\mu f)(t) = (\mathscr{I}^{-\mu} f)(t)$ and the Riemann-Liouville fractional integral, we can write

$$(\mathscr{D}^\mu f)(t) = (\mathscr{I}^{-\mu} f)(t) = (\mathscr{I}^\beta f)(t) = \frac{1}{\Gamma(\beta)} \int_0^t f(\tau)(t - \tau)^{\beta - 1}\, d\tau,$$

where we introduce the notation $\beta = -\mu$. Since $f(t)$ is an analytic function, we can expand it into a Taylor series

$$f(\tau) = \sum_{k=0}^\infty (\tau - t)^k \frac{f^{(k)}(t)}{k!}$$

which, substituted in the previous and rearranging, provides for the derivative

$$(\mathscr{D}^\mu f)(t) = \frac{1}{\Gamma(\beta)} \sum_{k=0}^\infty \frac{f^{(k)}(t)(-1)^k}{k!} \int_0^t (t - \tau)^{\beta - 1 + k}\, d\tau.$$

By integrating and introducing the gamma function, we can write

$$(\mathscr{D}^\mu f)(t) = \frac{1}{\Gamma(\beta)} \sum_{k=0}^\infty \frac{f^{(k)}(t)(-1)^k}{k!} \frac{\Gamma(\beta + k)}{\Gamma(\beta + k + 1)} t^{\beta + k}.$$

Using the reflection formula for the gamma function, we obtain

$$\frac{\Gamma(-\beta + 1)}{k!\,\Gamma(-\beta - k + 1)} = (-1)^k \frac{\Gamma(\beta + k)}{k!\,\Gamma(\beta)}$$

which, substituting in the expression for the derivative and simplifying, provides

$$(\mathscr{D}^\mu f)(t) = \sum_{k=0}^\infty \binom{\mu}{k} \frac{t^{k - \mu} f^{(k)}(t)}{\Gamma(k - \mu + 1)}$$

where, we have already returned in the parameter $\mu = -\beta$.

Let us now turn to the case $\mathrm{Re}(\mu) > 0$. Since the Riemann-Liouville derivative of order μ is equal to the derivative of order k of the Riemann-Liouville integral of order $k - \mu$ and using the result obtained previously, we can write

$$(\mathscr{D}^\mu f)(t) = \frac{d^k}{dt^k} \mathscr{I}^{k-\mu} f(t) = \frac{d^k}{dt^k} \sum_{n=0}^\infty \binom{\mu - k}{n} \frac{t^{n+k-\mu} f^{(n)}(t)}{\Gamma(k + n - \mu + 1)}.$$

Considering that the series converges uniformly and using the Leibniz rule for the integer case, we obtain

$$(\mathcal{D}^\mu f)(t) = \sum_{n=0}^{\infty}\sum_{i=0}^{\infty} \binom{k}{i} \frac{f^{(n+i)}(t)\dfrac{d^{k-i}}{dt^{k-i}}t^{n+k-\mu}}{\Gamma(n+k-\mu+1)} \binom{\mu-k}{n}.$$

Deriving $k-i$ times the power of t and rearranging, we obtain

$$(\mathcal{D}^\mu f)(t) = \sum_{n=0}^{\infty}\sum_{i=0}^{\infty} \binom{k}{i} f^{(n+i)}(t)\frac{t^{n+i-\mu}}{\Gamma(n+i-\mu+1)} \binom{\mu-k}{n}$$

which, with the change of index $i \to j-n$, allows to write

$$(\mathcal{D}^\mu f)(t) = \sum_{j=0}^{\infty}\sum_{n=0}^{\infty} \binom{k}{j-n} \binom{\mu-k}{n} f^{(j)}(t)\frac{t^{j-\mu}}{\Gamma(j-\mu+1)}.$$

Using the relation [19]

$$\sum_{n=0}^{\infty} \binom{\alpha}{m-n}\binom{\beta}{n} = \binom{\alpha+\beta}{m}$$

we can write the previous expression in the form

$$(\mathcal{D}^\mu f)(t) = \sum_{j=0}^{\infty} \binom{\mu}{j} f^{(j)}(t)\frac{t^{j-\mu}}{\Gamma(j-\mu+1)}$$

which is the desired result   ◇

12. Since the Riemann-Liouville fractional derivative is a linear operator, we have

$$(\mathcal{D}^\mu f)(t) - \sum_{k=0}^{n-1} f^{(k)}(0)\frac{t^{k-\mu}}{\Gamma(k-\mu+1)} = (\mathcal{D}^\mu f)(t) - \sum_{k=0}^{n-1} f^{(k)}(0)\frac{t^{k-\mu}}{k!}\frac{\Gamma(k+1)}{\Gamma(k-\mu+1)}$$

or, in the following form

$$(\mathcal{D}^\mu f)(t) - \mathcal{D}^\mu \sum_{k=0}^{n-1} f^{(k)}(0)\frac{t^k}{k!} = \mathcal{D}^\mu \left[f(t) - \sum_{k=0}^{n-1} f^{(k)}(0)\frac{t^k}{k!} \right].$$

Given this expression, it is enough to show the relationship

$$(_C\mathcal{D}^\mu f)(t) = \mathcal{D}^\mu \left[f(t) - \sum_{k=0}^{n-1} f^{(k)}(0)\frac{t^k}{k!} \right]. \tag{5.13}$$

In order to obtain this result, we first use induction in n to show that

$$\mathscr{I}^n \mathscr{D}^n f(t) = f(t) - \sum_{k=0}^{n-1} f^{(k)}(0) \frac{t^k}{k!}. \tag{5.14}$$

In fact, considering $n = 1$, we have

$$\mathscr{I}^1 \mathscr{D}^1 f(t) = \mathscr{I}^1 f'(t) = \int_0^t f'(\tau) \, d\tau = f(t) - f(0).$$

Now, we assume that the expression is valid for $n = m$, that is,

$$\mathscr{I}^m \mathscr{D}^m f(t) = f(t) - \sum_{k=0}^{m-1} f^{(k)}(0) \frac{t^k}{k!}.$$

being $m \in \mathbb{N}$ and we will show that it is valid for $n = m + 1$. So we have

$$\mathscr{I}^{m+1} \mathscr{D}^{m+1} f(t) = \mathscr{I}^1 \left(f'(t) - \sum_{k=0}^{m-1} f^{(k+1)}(0) \frac{t^k}{k!} \right)$$

or by operating with the integral and using linearity

$$\mathscr{I}^{m+1} \mathscr{D}^{m+1} f(t) = \int_0^t f'(\tau) \, d\tau - \int_0^t \sum_{k=0}^{m-1} f^{(k+1)}(0) \frac{t^k}{k!} d\tau$$

which, by integrating, can be written in the form

$$\mathscr{I}^{m+1} \mathscr{D}^{m+1} f(t) = f(t) - f(0) - \sum_{k=0}^{m-1} f^{(k+1)}(0) \frac{t^{k+1}}{(k+1)!}.$$

Considering the change of index $k \to k - 1$ and simplifying, we have

$$\mathscr{I}^{m+1} \mathscr{D}^{m+1} f(t) = f(t) - \sum_{k=0}^{m} f^{(k)}(0) \frac{t^k}{k!}.$$

Now, using Eqs. (5.13) and (5.14), we obtain

$$\mathscr{D}^\mu \left[f(t) - \sum_{k=0}^{n-1} f^{(k)}(0) \frac{t^k}{k!} \right] = \mathscr{D}^\mu [\mathscr{I}^n \mathscr{D}^n f(t)]$$

which, in turn, can be written in the form

$$\mathcal{D}^{\mu}\left[f(t) - \sum_{k=0}^{n-1} f^{(k)}(0)\frac{t^k}{k!}\right] = \mathcal{D}^n \mathcal{I}^{n-\mu} \mathcal{I}^n \mathcal{D}^n f(t)$$

or, finally

$$\mathcal{D}^{\mu}\left[f(t) - \sum_{k=0}^{n-1} f^{(k)}(0)\frac{t^k}{k!}\right] = \mathcal{I}^{n-\mu} \mathcal{D}^n f(t) = ({}_c\mathcal{D}^{\mu} f)(t)$$

which is the desired result   ◇

13. In order to have the two fractional derivatives equal, directly from Exercise (12), it suffices that the function and all its derivatives of order k, with $k = 1, 2, \ldots, n-1$, are zero, that is, $D^{(k)} f(t)\big|_{t=0} = 0$, from where it follows

$$\mathcal{D}^{\mu} f(t) = {}_c\mathcal{D}^{\mu} f(t)$$

which is the desired result   ◇

14. Since the fractional integral is given by means of a convolution product, by taking the Laplace transform, we can write

$$\mathcal{L}[\mathcal{J}^{\mu} f(t)] = \mathcal{L}[\phi_{\mu} * f(t)] = \mathcal{L}[\phi_{\mu}(t)]\mathcal{L}[f(t)]$$

where $*$ denotes the convolution product and $\phi_{\mu}(t)$ is the Gel'fand-Shilov function.

Since the Laplace transform of the Gel'fand-Shilov function is known, we have

$$\mathcal{L}[\mathcal{J}^{\mu} f(t)] = \frac{\mathcal{L}[f(t)]}{\Gamma(\mu)} \int_0^{\infty} t^{\mu-1} e^{-st}\, dt$$

where s is the Laplace transform parameter.

In order to calculate the resulting integral, we introduce the change of variable $\xi = st$ and using the definition of the gamma function, we obtain

$$\int_0^{\infty} t^{\mu-1} e^{-st}\, dt = \frac{\Gamma(\mu)}{s^{\mu}}$$

where, replacing in the previous and simplifying, we have

$$\mathcal{L}[\mathcal{J}^{\mu} f(t)] = \frac{\mathcal{L}[f(t)]}{\Gamma(\mu)}\frac{\Gamma(\mu)}{s^{\mu}} = \frac{\mathcal{L}[f(t)]}{s^{\mu}}$$

which is the desired result   ◇

15. Let $n - 1 < \text{Re}(\mu) \le n$. Taking the Laplace transform from the expression that gives the Caputo derivative in terms of the fractional integral, we can write

$$\mathscr{L}[{}_C\mathcal{D}^\mu f(t)] = \mathscr{L}[\mathcal{J}^{n-\mu}\mathcal{D}^n f(t)].$$

Entering the notation $g(t) = \mathcal{D}^n f(t)$ and replacing in the previous one, we have

$$\mathscr{L}[{}_C\mathcal{D}^\mu f(t)] = \mathscr{L}[\mathcal{J}^{n-\mu} g(t)]$$

which, from the result of Exercise (14), can be written in the form

$$\mathscr{L}[{}_C\mathcal{D}^\mu f(t)] = \frac{\mathscr{L}[g(t)]}{s^{n-\mu}}.$$

In order to calculate the Laplace transform of $g(t)$, we use the expression that provides the Laplace transform of the derivative of order $n \in \mathbb{N}$, then

$$\mathscr{L}[g(t)] = \mathscr{L}[\mathcal{D}^n f(t)] = s^n \mathscr{L}[f(t)] - \sum_{k=0}^{n-1} s^{n-k-1} f^{(k)}(0)$$

where s is the parameter of the Laplace transform. Compounding the last two results, we can write for the Laplace transform of the Caputo derivative

$$\mathscr{L}[{}_C\mathcal{D}^\mu f(t)] = \frac{1}{s^{n-\mu}} \left(s^n \mathscr{L}[f(t)] - \sum_{k=0}^{n-1} s^{n-k-1} f^{(k)}(0) \right)$$

which, by simplifying, provides

$$\mathscr{L}[{}_C\mathcal{D}^\mu f(t)] = s^\mu \mathscr{L}[f(t)] - \sum_{k=0}^{n-1} s^{\mu-k-1} f^{(k)}(0)$$

which is the desired result   ◇

16. We must evaluate the following derivative

$$_C\mathcal{D}^\nu E_\alpha(-t^\alpha) = \frac{1}{\Gamma(n-\nu)} \int_0^t \frac{\mathbf{D}^n E_\alpha(-\tau^\alpha)}{(t-\tau)^{\nu-n+1}} d\tau$$

being the parameter of the Mittag-Leffler function $\alpha \in \mathbb{R}$ and $\mathbf{D}^n = d^n/d\tau^n$. By introducing the series representation for the classical Mittag-Leffler function and permuting the order with the derivative operator, we can write

$$_C\mathcal{D}^\nu E_\alpha(-t^\alpha) = \frac{1}{\Gamma(n-\nu)} \int_0^t \frac{d\tau}{(t-\tau)^{\nu-n+1}} \sum_{k=0}^{\infty} \frac{(-1)^k}{\Gamma(\alpha k + 1)} \frac{d^n}{d\tau^n}(\tau^{\alpha k}).$$

Using the result of Exercise (1)

$$\frac{d^n}{d\tau^n}(\tau^{\alpha k}) = \frac{\Gamma(\alpha k + 1)}{\Gamma(\alpha k - n + 1)}\tau^{\alpha k - n}$$

replacing in the preceding and simplifying, we obtain

$$_C\mathcal{D}^\nu E_\alpha(-t^\alpha) = \frac{1}{\Gamma(n-\nu)} \int_0^t \frac{d\tau}{(t-\tau)^{\nu-n+1}} \sum_{k=0}^\infty \frac{(-1)^k}{\Gamma(\alpha k - n + 1)}\tau^{\alpha k - n}.$$

Changing the order of the summation with the integral, introducing the change of variable $\tau = tx$ and simplifying, we can write

$$_C\mathcal{D}^\nu E_\alpha(-t^\alpha) = \frac{1}{\Gamma(n-\nu)} \sum_{k=0}^\infty \frac{(-1)^k}{\Gamma(\alpha k - n + 1)} t^{\alpha k - \nu} \int_0^1 (1-x)^{n-\nu-1} x^{\alpha k - n + 1 - 1} dx.$$

Using the definition of beta function, as well as the relation between beta and gamma functions, we obtain

$$_C\mathcal{D}^\nu E_\alpha(-t^\alpha) = t^{-\nu} \sum_{k=0}^\infty \frac{(-t^\alpha)^k}{\Gamma(\alpha - \nu + 1)}$$

which, identifying with the Mittag-Leffler function with two parameters, provides

$$_C\mathcal{D}^\nu E_\alpha(-t^\alpha) = t^{-\nu} E_{\alpha, 1-\nu}(-t^\alpha).$$

In particular, in the case where $\alpha = \nu$, we obtain

$$_C\mathcal{D}^\nu E_\nu(-t^\nu) = t^{-\nu} E_{\nu, 1-\nu}(-t^\nu)$$

which is the desired result   ◇

17. We must evaluate the following derivative

$$\mathcal{D}^\nu E_\alpha(-t^\alpha) = \frac{1}{\Gamma(n-\nu)} \frac{d^n}{dt^n} \int_0^t \frac{E_\alpha(-\tau^\alpha)}{(t-\tau)^{\nu-n+1}} d\tau$$

being the parameter of the Mittag-Leffler function $\alpha \in \mathbb{R}$.

Introducing the series representation for the classical Mittag-Leffler function and permuting the order of integration with the summation, we can write

$$\mathcal{D}^\nu E_\alpha(-t^\alpha) = \frac{1}{\Gamma(n-\nu)} \frac{d^n}{dt^n} \sum_{k=0}^\infty \frac{(-1)^k}{\Gamma(\alpha k + 1)} \int_0^t \tau^{\alpha k}(t-\tau)^{n-\nu-1} d\tau.$$

Introducing the change of variable $\tau = tx$ and simplifying we get

$$\mathcal{D}^\nu E_\alpha(-t^\alpha) = \frac{1}{\Gamma(n-\nu)} \frac{d^n}{dt^n} \sum_{k=0}^\infty \frac{(-1)^k}{\Gamma(\alpha k+1)} t^{\alpha k+n-\nu} \int_0^1 x^{\alpha k+1-1}(1-x)^{n-\nu-1}\,dx.$$

Using the definition of the beta function, the relation between gamma and beta functions and simplifying, we can write

$$\mathcal{D}^\nu E_\alpha(-t^\alpha) = \frac{d^n}{dt^n} \sum_{k=0}^\infty \frac{(-1)^k}{\Gamma(\alpha k+n-\nu+1)} t^{\alpha k+n-\nu}.$$

Using the result of Exercise (1) and simplifying, we have

$$\mathcal{D}^\nu E_\alpha(-t^\alpha) = t^{-\nu} \sum_{k=0}^\infty \frac{(-t^\alpha)^k}{\Gamma(\alpha k-\nu+1)}$$

which, identifying with the Mittag-Leffler function with two parameters, provides

$$\mathcal{D}^\nu E_\alpha(-t^\alpha) = t^{-\nu} E_{\alpha,1-\nu}(-t^\alpha)$$

which is exactly the result as obtained with the Caputo derivative   ◇

18. We must calculate the following integral

$$\mathcal{J}^\nu E_{\alpha,1}^\gamma(-t^\alpha) = \frac{1}{\Gamma(\nu)} \int_0^t \frac{E_{\alpha,1}^\gamma(-\tau^\alpha)}{(t-\tau)^{1-\nu}}\,d\tau$$

with the parameters $\alpha, \gamma \in \mathbb{R}_+$.
Introducing the power series representation for the Mittag-Leffler function with three parameters and permuting the order of integration with the sum we have

$$\mathcal{J}^\nu E_{\alpha,1}^\gamma(-t^\alpha) = \frac{1}{\Gamma(\nu)} \sum_{k=0}^\infty \frac{(-1)^k}{\Gamma(\alpha k+1)} \frac{(\gamma)_k}{k!} \int_0^t \tau^{\alpha k}(t-\tau)^{\nu-1}\,d\tau.$$

In order to calculate the remaining integral, we introduce the change of variable $\tau = tx$, from where we can write, already simplifying

$$\mathcal{J}^\nu E_{\alpha,1}^\gamma(-t^\alpha) = t^\nu \sum_{k=0}^\infty \frac{(\gamma)_k}{\Gamma(\alpha k+\nu+1)} \frac{(-t^\alpha)^k}{k!}$$

which, by identifying with the Mittag-Leffler function with three parameters, provides

$$\mathcal{J}^\nu E_{\alpha,1}^\gamma(-t^\alpha) = t^\nu E_{\alpha,\nu+1}^\gamma(-t^\alpha).$$

In the particular case where $\gamma = 1$ we have $E_{\alpha,1}^1(\cdot) = E_\alpha(\cdot)$ and $E_{\alpha,\beta}^1(\cdot) = E_{\alpha,\beta}(\cdot)$ substitute in the previous one, allows to write

$$\mathcal{J}^\nu E_\alpha(-t^\alpha) = t^\nu E_{\alpha,\nu+1}(-t^\alpha)$$

which is exactly the result obtained in Exercise (4). Further, in the case where we consider $\nu \to -\nu$ in the previous one, we obtain the result according to Exercise (17), as we already mentioned $\mathcal{J}^{-\nu} = \mathcal{D}^\nu$ ◇

19. We must calculate the following integral

$$_H\mathcal{J}^\nu E_\alpha\left[\left(\ln\frac{t}{a}\right)^\alpha\right] = \frac{1}{\Gamma(\nu)}\int_a^t \left(\ln\frac{t}{\tau}\right)^{\nu-1} E_\alpha\left[\left(\ln\frac{t}{a}\right)^\alpha\right]\frac{d\tau}{\tau}.$$

Introducing the power series representation for the classical Mittag-Leffler function and permuting the order of integration with the summation, we can write

$$_H\mathcal{J}^\nu E_\alpha\left[\left(\ln\frac{t}{a}\right)^\alpha\right] = \frac{1}{\Gamma(\nu)}\sum_{k=0}^\infty \frac{1}{\Gamma(\alpha k+1)}\int_a^t \left(\ln\frac{t}{\tau}\right)^{\nu-1}\left(\ln\frac{\tau}{a}\right)^{\alpha k}\frac{d\tau}{\tau}.$$

By changing the variable $\ln\tau - \ln a = \xi$ in the preceding integral, we have

$$_H\mathcal{J}^\nu E_\alpha\left[\left(\ln\frac{t}{a}\right)^\alpha\right] = \frac{1}{\Gamma(\nu)}\sum_{k=0}^\infty \frac{1}{\Gamma(\alpha k+1)}\int_0^{\ln t-\ln a}\left(\ln\frac{t}{a}-\xi\right)^{\nu-1}\xi^{\alpha k}d\xi.$$

Another change of variable $\xi = (\ln t - \ln a)u$ leads us to the integral

$$_H\mathcal{J}^\nu E_\alpha\left[\left(\ln\frac{t}{a}\right)^\alpha\right] = \frac{1}{\Gamma(\nu)}\sum_{k=0}^\infty \frac{1}{\Gamma(\alpha k+1)}\left(\ln\frac{t}{a}\right)^{\alpha k+\nu}\int_0^1 (1-u)^{\nu-1}u^{\alpha k+1-1}du.$$

Using the definition of the beta function, the relation between gamma and beta functions and simplifying, we can write

$$_H\mathcal{J}^\nu E_\alpha\left[\left(\ln\frac{t}{a}\right)^\alpha\right] = \left(\ln\frac{t}{a}\right)^\nu\sum_{k=0}^\infty \frac{\left(\ln\frac{t}{a}\right)^{\alpha k}}{\Gamma(\alpha k+\nu+1)}$$

which, identifying with the power series representation of the Mittag-Leffler function with two parameters, allows us to write

$$_H\mathcal{J}^\nu E_\alpha\left[\left(\ln\frac{t}{a}\right)^\alpha\right] = \left(\ln\frac{t}{a}\right)^\nu E_{\alpha,\nu+1}\left[\left(\ln\frac{t}{a}\right)^\alpha\right]$$

which is the desired result. Note that this result, with the change $\ln t - \ln a \to t$ does not lead to the result obtained in the Riemann-Liouville fractional integral calculation of the classic Mittag-Leffler function, and, change that must be made so that one fractional integral is carried in the other   ◇

20. First, consider $\lambda = 0$. In this case we have $x(t) = E_0(0) = 1$ from where it follows

$$c\mathcal{D}^\nu x(t) = 0 = \lambda x(t).$$

Let $\lambda \neq 0$. We must evaluate the integral

$$c\mathcal{D}^\nu E_\nu(\lambda t^\nu) = \frac{1}{\Gamma(n-\nu)} \int_0^t \frac{D^n E_\nu(\lambda \tau^\nu)}{(t-\tau)^{\nu-n+1}} d\tau.$$

Introducing the power series representation for the Mittag-Leffler function in the definition of the Caputo fractional derivative, we can write

$$c\mathcal{D}^\nu E_\nu(\lambda t^\nu) = \frac{1}{\Gamma(n-\nu)} \int_0^t \frac{d\tau}{(t-\tau)^{\nu-n+1}} \sum_{k=1}^\infty \frac{\lambda^k}{\Gamma(\nu k+1)} D^n(\tau^{\nu k}).$$

Note that the sum begins at $k = 1$ and not at $k = 0$, since the Caputo fractional derivative of a constant is zero. Calculating the derivative, permuting the order of the sum with the integral and simplifying, we obtain

$$c\mathcal{D}^\nu E_\nu(\lambda t^\nu) = \frac{1}{\Gamma(n-\nu)} \sum_{k=1}^\infty \frac{\lambda^k}{\Gamma(\nu k-n+1)} \int_0^t \tau^{\nu k-n}(t-\tau)^{n-\nu-1} d\tau.$$

Consider the change of variable $\tau = xt$ in the resulting integral and rearrange, then

$$c\mathcal{D}^\nu E_\nu(\lambda t^\nu) = \frac{1}{\Gamma(n-\nu)} \sum_{k=1}^\infty \frac{\lambda^k}{\Gamma(\nu k-n+1)} t^{\nu k-\nu} \int_0^1 x^{\nu k-n+1-1}(1-x)^{n-\nu-1} dx.$$

The remaining integral can be identified with the definition of the beta function. Therefore, using the relation between beta and gamma functions and simplifying, we get

$$c\mathcal{D}^\nu E_\nu(\lambda t^\nu) = \sum_{k=1}^\infty \frac{\lambda^k}{\Gamma(\nu k-\nu+1)} t^{\nu k-\nu}.$$

Considering the change $k \to k+1$ we can write

$$c\mathcal{D}^\nu E_\nu(\lambda t^\nu) = \lambda \sum_{k=0}^\infty \frac{(\lambda t^\nu)^k}{\Gamma(\nu k+1)}$$

which, identifying with the Mittag-Leffler function, allows us to write

$$c\mathcal{D}^\nu E_\nu(\lambda t^\nu) = \lambda E_\nu(\lambda t^\nu)$$

which is the desired result   ◇

21. We must calculate the integral

$$_C\mathcal{D}^v(t+1)^m = \frac{1}{\Gamma(n-v)} \int_0^t \frac{D^n[(\tau+1)^m]}{(t-\tau)^{v-n+1}} d\tau.$$

Calculating the derivative and rearranging, we obtain

$$_C\mathcal{D}^v(t+1)^m = \frac{1}{\Gamma(n-v)} \frac{\Gamma(m+1)}{\Gamma(m-n+1)} \int_0^t (t-\tau)^{n-v-1}(1+\tau)^{m-n} d\tau.$$

Entering the variable change $\tau = tx$ we can write

$$_C\mathcal{D}^v(t+1)^m = \frac{1}{\Gamma(n-v)} \frac{\Gamma(m+1)}{\Gamma(m-n+1)} t^{n-v} \int_0^1 (1-x)^{n-v-1}(1+tx)^{m-n} dx.$$

In this case, compare with the hypergeometric function, we have $a = n - m, b = 1, c = n - v + 1$ and $x \to -t$, from where we can write, already simplifying

$$_C\mathcal{D}^v(t+1)^m = \frac{\Gamma(m+1)}{\Gamma(m-n+1)\Gamma(n-v+1)} t^{n-v} {}_2F_1(n-m, 1; n-v+1; -t)$$

where ${}_2F_1(a, b; c; x)$ is the classical hypergeometric function   ◇

22. First, the case $v = n$ integer, that is,

$$_a\mathcal{D}_t^v\left[_a\mathcal{J}_t^v f(t)\right] = D^n \int_a^t (t-\tau)^{n-1} f(\tau) d\tau$$

which, in this case, is reduced to the term $n = 1$, since all others are null, so

$$_a\mathcal{D}_t^1\left[_a\mathcal{J}_t^1 f(t)\right] = \frac{d}{dt} \int_a^t f(\tau) d\tau = f(t).$$

Let, now $n - 1 \leq v < n$. Using the result of Exercise (14) of Chap. 6, that is, the semigroup property, we can write to the fractional integral

$$_a\mathcal{J}_t^n f(t) = {}_a\mathcal{J}_t^{n-v}\left[_a\mathcal{J}_t^v f(t)\right]$$

from where it follows, for the v-order derivative

$$_a\mathcal{D}_t^v\left[_a\mathcal{J}_t^v f(t)\right] = D^n \left\{_a\mathcal{J}_t^{n-v}\left[_a\mathcal{J}_t^v f(t)\right]\right\}$$

or, in the following way

$$_a\mathcal{D}_t^v\left[_a\mathcal{J}_t^v f(t)\right] = D^n \left[_a\mathcal{J}_t^n f(t)\right] = f(t)$$

which is the desired result   ◇

23. We have, by definition, the Riemann-Liouville fractional derivative

$$_a\mathcal{D}_t^\nu f(t) = \mathsf{D}^n \left[_a\mathcal{J}_t^{n-\nu} f(t) \right]$$

that is, the integer-order derivative of a fractional integral.
Let $\nu = n - 1$. Thus, from the previous one, we can write

$$_a\mathcal{D}_t^{n-1} f(t) = \mathsf{D}^n \left[_a\mathcal{J}_t^1 f(t) \right] = f^{(n-1)}(t)$$

that is, the usual derivative of order $n - 1$. On the other hand, where $_a\mathcal{D}_t^0 f(t) = f(t)$ and considering $\nu = n \geq 1$, we have

$$_a\mathcal{D}_t^\nu f(t) = \mathsf{D}^n \left[_a\mathcal{D}_t^0 f(t) \right]$$

or, in the following way

$$_a\mathcal{D}_t^\nu f(t) = \mathsf{D}^n f(t) = f^{(n)}(t)$$

hence, the Riemann-Liouville fractional derivative of order $\nu = n$ matches the usual derivative of order n ◇

24. From the conditions in the parameters μ, ν, n and m and using the semigroup property, according to Exercise (14), we can write

$$_a\mathcal{D}_t^\mu \left[_a\mathcal{J}_t^\nu f(t) \right] = \mathsf{D}^m \left\{ _a\mathcal{J}_t^{m-\mu} \left[_a\mathcal{J}_t^\nu f(t) \right] \right\}$$

or, in the following way

$$_a\mathcal{D}_t^\mu \left[_a\mathcal{J}_t^\nu f(t) \right] = \mathsf{D}^m \left[_a\mathcal{J}_t^{\nu-\mu+m} f(t) \right].$$

Thus, the conditions in the parameters and the relationship $_a\mathcal{J}_t^{-\alpha} = {_a\mathcal{D}_t^\alpha}$ we have

$$_a\mathcal{D}_t^\mu \left[_a\mathcal{J}_t^\nu f(t) \right] = \mathsf{D}^n \left[_a\mathcal{D}_t^{\mu-\nu-n} f(t) \right].$$

Finally, using the relationship $\mathsf{D}^n {_a\mathcal{D}_t^{-n}} f(t) = f(t)$ we can write

$$_a\mathcal{D}_t^\mu \left[_a\mathcal{J}_t^\nu f(t) \right] = {_a\mathcal{D}_t^{\mu-\nu}} f(t)$$

which is the desired result   ◇

25. Since the fractional derivative in the Caputo sense is defined as the fractional integral of an integer-order derivative, it is immediate to conclude that this derivative is zero, since the usual derivative of a constant is zero,

$$_c\mathcal{D}_t^\nu C = 0.$$

On the other hand, for the Riemann-Liouville fractional derivative we get, by definition

$$\mathcal{D}_t^\nu C = \frac{1}{\Gamma(\nu)} D \int_{a\cdot}^t \frac{C}{(t-\tau)^\nu}\, d\tau,$$

with $0 < \nu \le 1$. Introducing the change of variable $t - \tau = x$ and integrating, we get

$$\mathcal{D}_t^\nu C = \frac{C}{\Gamma(2-\nu)} \frac{d}{dt}(t-a)^{1-\nu}.$$

Calculating the remaining derivative and simplifying, we obtain

$$\mathcal{D}_t^\nu C = \frac{C}{\Gamma(1-\nu)}(t-a)^{-\nu}$$

from which it follows that, the Riemann-Liouville fractional derivative of a constant function is not zero, except to consider initialization in $a = -\infty$.
It is important to note that in this particular case $a = -\infty$, the fractional derivatives of Riemann-Liouville and Caputo coincide, provided that suitable conditions for the function and its derivatives are satisfied   ◇

26. Consider the Riemann-Liouville fractional integral with $a = -\infty$,

$$-\infty \mathcal{J}_t^\nu f(t) = \frac{1}{\Gamma(\nu} \int_{-\infty}^t (t-\tau)^{\nu-1} f(\tau)\, d\tau$$

which can be written in terms of the convolution product, according to Eq. (5.3),

$$-\infty \mathcal{J}_t^\nu f(t) = \phi_\nu(t) \star f(t) \tag{5.15}$$

with $\phi_\nu(t)$ is the Gel'fand-Shilov function, given by

$$\phi_\nu(t) = \begin{cases} \dfrac{t^{\nu-1}}{\Gamma(\nu)}, & t > 0 \\ 0, & t \le 0. \end{cases}$$

Taking the Fourier transform on both sides of Eq. (5.15), we obtain

$$\mathscr{F}\left[-\infty \mathcal{J}_t^\nu f(t)\right] = F(\omega)\Phi(\omega)$$

where $F(\omega)$ and $\Phi(\omega)$ are the Fourier transforms of $f(t)$ and $\phi_\nu(t)$, respectively, both of parameter ω. Let us now calculate the Fourier transform of the Gel'fand-Shilov function. By introducing $s = i\omega$ with ω real, in the expression that provides the Laplace transform of the power function, we obtain

$$\int_0^\infty t^{\nu-1} e^{i\omega t}\, dt = \frac{\Gamma(\nu)}{(-i\omega)^\nu}$$

which converges for $0 < \nu < 1$, from where it follows

$$\Phi_\nu(\omega) = \mathscr{F}[\phi_\nu(t)] = (-i\omega)^{-\nu}.$$

Returning with this result in the expression for the convolution product, we have

$$\mathscr{F}\left[_{-\infty}\mathcal{J}_t^\nu f(t)\right] = (-i\omega)^{-\nu} F(\omega)$$

which is the desired result   ◇

27. Let $n - 1 < \nu < 1$. In analogy to the calculation of the Fourier transform of the Riemann-Liouville fractional derivative, we will write the Riemann-Liouville fractional derivative, for $a = -\infty$, in the form

$$_{-\infty}\mathcal{D}_t^\nu f(t) = \mathsf{D}^n \left[_{-\infty}\mathcal{J}_t^{n-\nu} f(t)\right]$$

or further, using the convolution product,

$$_{-\infty}\mathcal{D}_t^\nu f(t) = \mathsf{D}^n \left[\phi_{n-\nu}(t) \star f(t)\right].$$

Taking the Fourier transform on both sides of the preceding one and using the result of Exercise (26) we can write

$$\mathscr{F}\left[_{-\infty}\mathcal{D}_t^\nu f(t)\right] = (-i\omega)^{\nu-n} \mathscr{F}[\mathsf{D}^n f(t)].$$

The Fourier transform of the derivative of order n, assuming the conditions of existence, is given by the following expression

$$\mathscr{F}[\mathsf{D}^n f(t)] = (-i\omega)^n \mathscr{F}[f(t)]$$

which replaced in the previous one, allows us to write

$$\mathscr{F}\left[_{-\infty}\mathcal{D}_t^\nu f(t)\right] = (-i\omega)^{\nu-n}(-i\omega)^n F(\omega) = (-i\omega)^\nu F(\omega)$$

with $F(\omega) = \mathscr{F}[f(t)]$ ◇

28. The Riemann-Liouville fractional integral of order $\nu > 0$ and initialized in $a = 0$ is given by the following expression

$$_0\mathcal{J}_t^\nu f(t) = \frac{1}{\Gamma(\nu)} \int_0^t (t - \tau)^{\nu-1} f(\tau) \, d\tau.$$

By introducing in the preceding the variable change $\tau = tx$, we can write

$$_0\mathcal{J}_t^\nu f(t) = \frac{t^\nu}{\Gamma(\nu)} \int_0^1 (1 - x)^{\nu-1} f(tx) \, dx$$

or, in the following form

$$_0\mathcal{J}_t^\nu f(t) = \frac{t^\nu}{\Gamma(\nu)} \int_0^\infty g(x) f(tx)\, dx$$

where we define the function $g(t)$ such that

$$g(t) = \begin{cases} (1-t)^{\nu-1}, & 0 \le t < 1, \\ 0, & t \ge 1. \end{cases}$$

The Mellin transform of the function $g(t)$, denoted by $\mathcal{M}[f(t)] = G(s)$ is such that

$$G(s) = \int_0^1 (1-t)^{\nu-1} t^{s-1}\, dt$$

which, using the definition of the beta function, allows us to write

$$G(s) = B(s, \nu).$$

Using the Property 4.4.2 and the convolution theorem we have

$$\mathcal{M}\left[t^\mu \int_0^\infty x^\nu f(xt) g(x)\, dx \right] = F(s+\mu) G(1-s-\mu+\nu)$$

with $F(s)$ and $G(s)$ being the Mellin transform of $f(t)$ and $g(t)$, respectively. Thus, we can write formally for the Mellin transform of the Riemann-Liouville fractional integral

$$\mathcal{M}\left[_0\mathcal{J}_t^\nu f(t) \right] = \frac{1}{\Gamma(\nu)} F(s+\nu) B(\nu, 1-s-\nu).$$

Using the relation between gamma and beta functions and simplifying, we have

$$\mathcal{M}\left[_0\mathcal{J}_t^\nu f(t) \right] = \frac{\Gamma(1-s-\nu)}{\Gamma(1-s)} F(s+\nu)$$

which is the desired result. It is important to note that, in the case where the substitutions involving the limits $t = 0$ and $t = \infty$ do not provide contributions, the Mellin transform of the Riemann-Liouville fractional derivative coincides with the result of the Mellin transform of the Riemann-Liouville integral, with the identification imposed in the preceding expression $\nu \to -\nu$ and with the notation $_0\mathcal{J}_t^{-\nu} \equiv {_0\mathcal{D}_t^\nu}$ $\diamond$

29. We must calculate the integral

$$\mathscr{M}[{}_H\mathcal{J}_0^\nu f(t)] = \mathscr{M}\left\{\frac{1}{\Gamma(\nu)}\int_0^t \left(\ln\frac{t}{\tau}\right)^{\nu-1} f(\tau)\frac{d\tau}{\tau}\right\} \equiv F(s)G(s).$$

Through the product of convolution, we can identify the functions

$$F(s) = \mathscr{M}[f(t)] \quad \text{and} \quad G(s) = \frac{1}{\Gamma(\nu)}\mathscr{M}[(\ln t)^{\nu-1}].$$

Let's explicitly calculate the $G(s)$ function. Through the definition of the Mellin transform of parameter s, we can write

$$\frac{1}{\Gamma(\nu)}\mathscr{M}[(\ln t)^{\nu-1}] = \frac{1}{\Gamma(\nu)}\int_1^\infty (\ln t)^{\nu-1} t^{s-1}\, dt.$$

Note that we consider the lower limit equal to unity, since the integral has no sense in the interval $0 \le t < 1$ due to the logarithm in the integrand. Introducing the change of variable $\ln t = x$, we obtain

$$\frac{1}{\Gamma(\nu)}\mathscr{M}[(\ln t)^{\nu-1}] = \frac{1}{\Gamma(\nu)}\int_0^\infty x^{\nu-1} e^{x(s-1)} e^x\, dx$$

which by exchanging $-s = p > 0$ and simplifying, allows typing

$$\frac{1}{\Gamma(\nu)}\mathscr{M}[(\ln t)^{\nu-1}] = \frac{1}{\Gamma(\nu)}\int_0^\infty x^{\nu-1} e^{-px}\, dx.$$

Introducing the variable change $px = \xi$ and rearranging, we have

$$\frac{1}{\Gamma(\nu)}\mathscr{M}[(\ln t)^{\nu-1}] = \frac{p^{-\nu}}{\Gamma(\nu)}\int_0^\infty \xi^{\nu-1} e^{-\xi}\, d\xi$$

which, using the definition of the gamma function, provides, already returning in parameter s,

$$\frac{1}{\Gamma(\nu)}\mathscr{M}[(\ln t)^{\nu-1}] = \frac{p^{-\nu}}{\Gamma(\nu)}\Gamma(\nu) = (-s)^{-\nu} = G(s).$$

Returning with these results, $F(s)$ and $G(s)$, in the expression for the Mellin transform of the Hadamard fractional integral, we obtain

$$\mathscr{M}[{}_H\mathcal{J}_0^\nu f(t)] = (-s)^{-\nu} F(s)$$

which is the desired result. It should be noted that the calculation of the Mellin transform of the Hadamard fractional derivative is obtained with the same procedure, from where we can write

$$\mathcal{M}[{}_H\mathcal{D}_0^\nu f(t)] = (-s)^\nu F(s)$$

being also the parameter $s < 0$ and $F(s) = \mathcal{M}[f(t)]$ ◇

30. We must calculate the following integral

$$_0\mathcal{J}_t^\nu[t^{\mu/2}J_\mu(2\sqrt{t})] = \frac{1}{\Gamma(\nu)}\int_0^t (t-\xi)^{\nu-1}\xi^{\mu/2}J_\mu(2\sqrt{t})\,d\xi\cdot$$

Introducing the change of variable $\xi = tx$ and rearranging, we have

$$_0\mathcal{J}_t^\nu[t^{\mu/2}J_\mu(2\sqrt{t})] = \frac{t^{\nu+\mu/2}}{\Gamma(\nu)}\int_0^1 (1-x)^{\nu-1}x^{\mu/2}J_\mu(2\sqrt{tx})\,dx\cdot$$

Introducing the series representation for the Bessel function in the preceding one and changing the order of the integral with the sum, we can write

$$_0\mathcal{J}_t^\nu[t^{\mu/2}J_\mu(2\sqrt{t})] = \frac{t^{\nu+\mu}}{\Gamma(\nu)}\sum_{k=0}^\infty \frac{(-1)^k}{k!}\frac{t^k}{\Gamma(\mu+k+1)}\int_0^1 (1-x)^{\nu-1}x^{k+\mu+\nu+1-1}\,dx\cdot$$

The remaining integral is the representation of the beta function that, already expressed in terms of gamma functions, allows to write, simplifying,

$$_0\mathcal{J}_t^\nu[t^{\mu/2}J_\mu(2\sqrt{t})] = t^{\nu+\mu}\sum_{k=0}^\infty \frac{(-1)^k}{k!}\frac{t^k}{\Gamma(\mu+\nu+k+1)}$$

or, in the following way

$$_0\mathcal{J}_t^\nu[t^{\mu/2}J_\mu(2\sqrt{t})] = t^{\frac{\nu+\mu}{2}}\sum_{k=0}^\infty \frac{(-1)^k}{k!}\frac{(2\sqrt{t}/2)^{2k+\mu+\nu}}{\Gamma(\mu+\nu+k+1)}$$

which, identified with the series representation of the Bessel function, provides

$$_0\mathcal{J}_t^\nu[t^{\mu/2}J_\mu(2\sqrt{t})] = t^{\frac{\nu+\mu}{2}}J_{\mu+\nu}(2\sqrt{t})$$

which is the desired result   ◇

31. In the integral to be calculated, next,

$$_0\mathcal{J}_t^\nu\left[t^{\beta-1}E_{\alpha,\beta}^\gamma(at^\alpha)\right] = \frac{1}{\Gamma(\nu)}\int_0^t \tau^{\beta-1}E_{\alpha,\beta}^\gamma(a\tau^\alpha)(t-\tau)^{\nu-1}\,d\tau$$

we introduce the power series representation of the Mittag-Leffler function with three parameters and the change of variable $\tau = tx$, in order to obtain

$$_0\mathcal{J}_t^\nu \left[t^{\beta-1} E_{\alpha,\beta}^\gamma (at^\alpha) \right] = \frac{1}{\Gamma(\nu)} \int_0^1 (tx)^{\beta-1} \sum_{k=0}^\infty \frac{(\gamma)_k}{\Gamma(\alpha k + \beta)} \frac{a^k (tx)^{\alpha k}}{k!} t^{\nu-1} (1-x)^{\nu-1} t \, dx \cdot$$

We change the order of integration with summation and rearranging, we get

$$_0\mathcal{J}_t^\nu \left[t^{\beta-1} E_{\alpha,\beta}^\gamma (at^\alpha) \right] = \frac{t^{\beta+\nu-1}}{\Gamma(\nu)} \sum_{k=0}^\infty \frac{(\gamma)_k}{\Gamma(\alpha k + \beta)} \frac{(at^\alpha)^k}{k!} \int_0^1 x^{\beta+\alpha k-1} (1-x)^{\nu-1} \, dx \cdot$$

The remaining integral is the representation for the beta function. Using the relationship between beta and gamma functions and simplifying, we can write

$$_0\mathcal{J}_t^\nu \left[t^{\beta-1} E_{\alpha,\beta}^\gamma (at^\alpha) \right] = t^{\beta+\nu-1} \sum_{k=0}^\infty \frac{(\gamma)_k}{\Gamma(\alpha k + \beta + \nu)} \frac{(at^\alpha)^k}{k!}$$

or, using the Mittag-Leffer function with three parameters, in the following form

$$_0\mathcal{J}_t^\nu \left[t^{\beta-1} E_{\alpha,\beta}^\gamma (at^\alpha) \right] = t^{\beta+\nu-1} E_{\alpha,\beta+\nu}^\gamma (at^\alpha)$$

which is the desired result   ◊

32. Let $F(s) = \mathscr{L}[x(t)]$. Taking the Laplace transform of the equation we have

$$F(s) = \frac{\mu}{\Gamma(2-\alpha)} \mathscr{L}[t^{1-\alpha}] - \frac{\lambda}{\Gamma(1-\alpha)} \mathscr{L}\left[\int_0^t (t-\xi)^{-\alpha} x(\xi) \, d\xi \right] \cdot$$

Using the convolution theorem and the definition of the gamma function to calculate the integrals from the Laplace transform, we can write

$$F(s) = \frac{\mu}{\Gamma(2-\alpha)} \Gamma(2-\alpha) s^{\alpha-2} - \frac{\lambda}{\Gamma(1-\alpha)} \Gamma(1-\alpha) s^{\alpha-1} F(s)$$

where s is the parameter of the Laplace transform. Simplifying the precedent we obtain an algebraic equation for $F(s)$ whose solution, after simplification, is given by

$$F(s) = \mu \frac{s^{-1}}{s^{1-\alpha} + \lambda} \cdot$$

Let us now proceed with the inversion of the Laplace transform. Using the result of Exercise (30) of Chap. 4, we can write

$$x(t) = \mathscr{L}^{-1} \left[\mu \frac{s^{-1}}{s^{1-\alpha} + \lambda} \right] = \mu t^{1-\alpha} E_{1-\alpha, 2-\alpha}(-\lambda t^{1-\alpha})$$

where $E_\nu(\cdot)$ is a Mittag-Leffler function with two parameters. In order to obtain the solution in the requested form, we will use the relation involving the Mittag-Leffer functions with one and two parameters

$$E_\nu(z) = z E_{\nu,\nu+1}(z) + 1 \cdot$$

Identifying $\nu = 1 - \alpha$ and $z = -\lambda t^{1-\alpha}$ and rearranging, we have

$$x(t) = \frac{\mu}{\lambda} \left[1 - E_{1-\alpha}(-\lambda t^{1-\alpha}) \right]$$

which is the desired result   ◇

33. Using the Leibniz rule to derive an integral, we can write for the first parcel

$$\mathsf{D}[_0\mathcal{J}_t^\alpha f(t)] = \frac{\alpha - 1}{\Gamma(\alpha)} \int_0^t (t - \tau)^{\alpha-2} f(\tau) \, d\tau \cdot$$

On the other hand, using integration by parts in the second parcel, we have

$$_0\mathcal{J}_t^\alpha[\mathsf{D}f(t)] = \frac{1}{\Gamma(\alpha)} (t - \tau)^{\alpha-1} f(\tau) \Big|_{\tau=0}^{\tau=t} + \frac{\alpha - 1}{\Gamma(\alpha)} \int_0^t (t - \tau)^{\alpha-2} f(\tau) \, d\tau \cdot$$

By subtracting from each other and simplifying, we obtain

$$\mathsf{D}[_0\mathcal{J}_t^\alpha f(t)] - {}_0\mathcal{J}_t^\alpha[\mathsf{D}f(t)] = f(0) \frac{t^{\alpha-1}}{\Gamma(\alpha)}$$

which is the desired result   ◇

34. The Legendre function of first kind and degree α, given by

$$P_\alpha(x) = {}_2F_1[\alpha + 1, -\alpha; 1; \frac{1}{2}(1 - x)]$$

is valid for $|1 - x| < 2$, being $_2F_1(a, b; c; \cdot)$ a hypergeometric function. In order to express it in terms of a fractional derivative, we begin by considering the series expansion

$$x^\alpha(1 - x)^\alpha = \frac{1}{\Gamma(-\alpha)} \sum_{k=0}^\infty \frac{\Gamma(k - \alpha)}{k!} x^{k+\alpha}$$

valid for $|x| < 1$ and α an arbitrary parameter.
Using the expression for the derivative of order α

$$_0\mathcal{D}_x^\alpha x^{k+\alpha} = \frac{\Gamma(\alpha + k + 1)}{\Gamma(k + 1)} x^k$$

we can write to the derivative of order α of a power

$$_0\mathcal{D}_x^\alpha[x^\alpha(1-x)^\alpha] = \frac{1}{\Gamma(-\alpha)}\sum_{k=0}^{\infty}\frac{\Gamma(k-\alpha)}{k!}\frac{\Gamma(\alpha+k+1)}{\Gamma(k+1)}x^k$$

which, identified with the hypergeometric function, allows us to write

$$_0\mathcal{D}_x^\alpha[x^\alpha(1-x)^\alpha] = \Gamma(\alpha+1)_2F_1(\alpha+1,-\alpha;1;x)\cdot$$

Identifying with the expression that relates the Legendre functions and the hypergeometric function, we can write

$$P_\alpha(1-2x) = \frac{1}{\Gamma(\alpha+1)}\,_0\mathcal{D}_x^\alpha[x^\alpha(1-x)^\alpha]\cdot$$

Entering the variable $t = 1 - 2x$ and rearranging, we obtain

$$P_\alpha(t) = \frac{1}{2^\alpha\Gamma(\alpha+1)}\,_1\mathcal{D}_t^\alpha[(1-t^2)^\alpha$$

that, in the particular case where $\alpha = n$, a nonnegative integer, retrieves the classical Rodrigues formula for the orthogonal polynomials, here, the Legendre polynomials. Note that the lower end of the integral changes from zero to one $\diamond$

35. Introducing the power series for the zero order Bessel function, we have

$$_0\mathcal{J}_x^{1/2}\,J_0(\sqrt{x}) = {}_0\mathcal{J}_x^{1/2}\left\{\sum_{k=0}^{\infty}\frac{(-1)^k}{k!k!}\left(\frac{\sqrt{x}}{2}\right)^{2k}\right\}\cdot$$

By switching the order of the derivative to the summation, using the expression for the derivative of order $1/2$ of a power function and simplifying, we can write

$$_0\mathcal{J}_x^{1/2}\,J_0(\sqrt{x}) = \sum_{k=0}^{\infty}\frac{(-1)^k}{k!}\frac{2^{-2k}}{\Gamma(k+1/2)}x^{k-1/2}\cdot$$

Using the Legendre duplication formula

$$\frac{1}{\Gamma(k+1/2)} = \frac{2^{2k}\Gamma(k+1)}{\sqrt{\pi}\Gamma(2k+1)},$$

already rearranging, we can write for the derivative

$$_0\mathcal{J}_x^{1/2}\,J_0(\sqrt{x}) = \frac{1}{\sqrt{\pi x}}\sum_{k=0}^{\infty}\frac{(-1)^k}{(2k)!}(\sqrt{x})^{2k}$$

which, identified with the cosine series expansion, provides

$$_0\mathcal{J}_x^{1/2} J_0(\sqrt{x}) = \frac{\cos(\sqrt{x})}{\sqrt{\pi x}}$$

which is the desired result   ◇

36. Let $t > 0$. Using the definition of the Riemann-Liouville fractional derivative, with $\alpha = 1/2$ and $f(t) = \ln t$, we must calculate the following integral

$$_0\mathcal{D}_t^{1/2} \ln t = \frac{1}{\Gamma(1/2)} \frac{d}{dt} \int_0^t (t - \tau)^{-1/2} \ln \tau \, d\tau.$$

Entering the change of variable $t - \tau = u$, we obtain

$$_0\mathcal{D}_t^{1/2} \ln t = \frac{1}{\sqrt{\pi}} \frac{d}{dt} \int_0^t \frac{\ln(t - u)}{\sqrt{u}} \, du$$

that, from the change of variable $u = tx$ and using the property of the logarithms (log of the product is the sum of the logarithms, maintained the base) allows to write

$$_0\mathcal{D}_t^{1/2} \ln t = \sqrt{\frac{1}{\pi}} \frac{d}{dt} \left\{ \sqrt{t} \ln t \int_0^1 \frac{dx}{\sqrt{x}} + \sqrt{t} \int_0^1 \frac{\ln(1 - x)}{\sqrt{x}} \, dx \right\}.$$

The first integral is immediate and for the second integral we use the given result, as well as the property of the logarithms, so

$$_0\mathcal{D}_t^{1/2} \ln t = \sqrt{\frac{1}{\pi}} \frac{d}{dt} \left[2\sqrt{t}(\ln 4t - 2) \right]$$

and calculating the derivative and rearranging, as follows

$$_0\mathcal{D}_t^{1/2} \ln t = \frac{\ln 4t}{\sqrt{\pi t}}$$

which is the desired result   ◇

37. Introducing the change of variable $r \cos \theta = x$ we can write

$$\int_0^r \left(1 - \frac{x^2}{r^2} \right)^\alpha \phi(x) \, dx = r \, h(r).$$

From the substitution $1/r = \sqrt{z}$ and the notation $\frac{1}{\sqrt{z}} h\left(\frac{1}{\sqrt{z}} \right) = \psi(z)$ we get

$$\int_0^{1/\sqrt{z}} \left(\frac{1}{z} - x^2\right)^\alpha \phi(x)\,dx = z^{-\alpha}\psi(z).$$

Finally, by introducing the variable $x^2 = \tau$ and the parameter $1/z = t$ we drive the Poisson integral equation into an Abel equation, that is,

$$\int_0^{\sqrt{t}} (t - \tau)^\alpha f(\tau)\,d\tau = g(t)$$

where we introduce the notation $f(\tau) = \frac{\phi(\sqrt{\tau})}{\sqrt{\tau}}$ and $g(t) = 2t^\alpha\psi(\frac{1}{t})$.

Operating with the Riemann-Liouville fractional derivative, we obtain the solution

$$f(t) = \frac{1}{\Gamma(\alpha + 1)}\, {}_0\mathcal{D}_t^\alpha g(t).$$

Turning now, in the initial variables, we can write

$$\phi(\sqrt{\tau}) = \frac{2\sqrt{\tau}}{\Gamma(\alpha + 1)}\, {}_0\mathcal{D}_t^\alpha \left[t^{\alpha+1/2}h(\sqrt{t})\right]$$

with $h(\sqrt{t}) = h\left(\frac{1}{\sqrt{z}}\right) = h(r)$ ◇

38. Multiplying and dividing the first member by $\Gamma(1 - \alpha)$ and using the Riemann-Liouville fractional integral, we can write

$$\Gamma(1 - \alpha)\, {}_0\mathcal{D}_t^{\alpha-1} f(t) = 1$$

or, in the following form

$$f(t) = \frac{{}_0\mathcal{D}_t^{1-\alpha} 1}{\Gamma(1 - \alpha)}.$$

Using the exponents rule, we can write the precedent as

$$f(t) = \frac{1}{\Gamma(1 - \alpha)}\, {}_0\mathcal{D}_t^1\, {}_0\mathcal{D}_t^{-\alpha} 1$$

which, using the fractional integral, leads us to the integral

$$f(t) = \frac{1}{\Gamma(1 - \alpha)}\frac{1}{\Gamma(\alpha)}\, {}_0\mathcal{D}_t^1 \int_0^t (t - \tau)^{\alpha-1}\,d\tau.$$

Integrating the previous, we obtain

$$f(t) = \frac{1}{\Gamma(1 - \alpha)}\frac{1}{\Gamma(\alpha)}\frac{d}{dt}\left(\frac{t^\alpha}{\alpha}\right)$$

which, evaluate the derivative, allows to write

$$f(t) = \frac{1}{\Gamma(1 - \alpha)} \frac{1}{\Gamma(\alpha)} t^{\alpha-1}$$

which is the desired result

39. We must calculate the following integral

$$_H\mathcal{J}_t^\nu f(t) = \frac{1}{\Gamma(\nu)} \int_1^t \left(\ln \frac{t}{\tau}\right)^{\nu-1} \tau \frac{d\tau}{\tau}.$$

Introducing the change of variable $\tau = tx$ we can write

$$_H\mathcal{J}_t^\nu f(t) = \frac{t(-1)^{\nu-1}}{\Gamma(\nu)} \int_{1/t}^1 (\ln x)^{\nu-1} dx.$$

Now, the following change of variable $x = e^{-\xi}$ takes us, after simplifying, the integral

$$_H\mathcal{J}_t^\nu f(t) = \frac{t}{\Gamma(\nu)} \int_0^{\ln t} e^{-\xi} \xi^{\nu-1} d\xi$$

exactly the integral representation for the incomplete gamma function, as well as

$$_H\mathcal{J}_t^\nu f(t) = \frac{t}{\Gamma(\nu)} \gamma(\nu, \ln t)$$

which is the desired result

40. First, consider the derivative of order n, a positive integer, of the power function, since the Caputo fractional derivative requires that the function be differentiable. Thus, by substituting the expression

$$D^n t^\mu = \frac{\Gamma(\mu + 1)}{\Gamma(\mu - n + 1)} t^{\mu-n}$$

in the relation for the Caputo fractional derivative, we can write

$$_C\mathcal{D}_t^\nu f(t) = \frac{1}{\Gamma(n - \nu)} \int_0^t \frac{\Gamma(\mu + 1)}{\Gamma(\mu - n + 1)} \tau^{\mu-n} \frac{d\tau}{(t - \tau)^{\nu-n+1}}.$$

Introducing the change of variable $\tau = tx$ and rearranging, we obtain

$$_C\mathcal{D}_t^\nu f(t) = \frac{t^{\mu-\nu}}{\Gamma(n - \nu)} \frac{\Gamma(\mu + 1)}{\Gamma(\mu - n + 1)} \int_0^1 x^{\mu-n} (1 - x)^{-\nu+n-1} dx.$$

The remaining integral is the integral representation for the beta function,

$$_C\mathcal{D}_t^\nu f(t) = \frac{t^{\mu-\nu}}{\Gamma(n-\nu)} \frac{\Gamma(\mu+1)}{\Gamma(\mu-n+1)} B(\mu-n+1, n-\nu)$$

using the relation between beta and gamma functions and simplifying, we get

$$_C\mathcal{D}_t^\nu f(t) = \frac{\Gamma(\mu+1)}{\Gamma(\mu-\nu+1)} t^{\mu-\nu}$$

which is the desired result ◇

5.5.4 Proposed exercises

1. Let $b > t$ and $\text{Re}(\nu) > 0$. Show that the integral to the right of the power function is

$$[\mathcal{J}_{b-}^\nu (b-t)^{\beta-1}](t) = \frac{\Gamma(\beta)}{\Gamma(\beta+\nu)} (b-t)^{\nu+\beta-1}.$$

2. Let $a, \beta \in \mathbb{R}$. Show that $\mathcal{D}_x^\beta(\sin ax) = a^\beta \sin\left(ax + \frac{\pi\beta}{2}\right).$

3. Show that $\sqrt{2}\mathcal{D}^{1/2} \sin x = \sin x + \cos x \cdot$

4. Let $\text{Re}(\beta) > \text{Re}(\alpha) > 0$ and $0 < a < b < \infty$. Show that Hadamard integral $(_H\mathcal{J}_{b-}^\nu f)(t)$ of $f(t) = \left(\ln \frac{b}{t}\right)^{\beta-1}$ is given by

$$\left[_H\mathcal{J}_{b-}^\alpha \left(\ln \frac{b}{t}\right)^{\beta-1}\right](x) = \frac{\Gamma(\beta)}{\Gamma(\beta+\alpha)} \left(\ln \frac{b}{x}\right)^{\beta+\alpha-1}.$$

5. Let $\alpha, \beta \in \mathbb{C}$ such that $\text{Re}(\alpha) > 0$, $\text{Re}(\beta) > 0$ and $1 \le p \le \infty$. If $c > 0$, then, for $f \in X_c^p(\mathbb{R}^+)$, the semigroup property is valid

$$\left[_H\mathcal{J}_-^\alpha {}_H\mathcal{J}_-^\beta\right](x) = {}_H\mathcal{J}_-^{\alpha+\beta}(x)$$

being $_H\mathcal{J}_-^\alpha$ the Hadamard integral to the right.

6. Consider the space $L^p[a, b]$ with $p \in [1, \infty]$ and $0 \le a \le b < \infty$. Show that the Hadamard fractional derivative of order α to the right, with $0 < \alpha \le 1$ is the inverse operator to the left of the fractional integral, for $a \le t \le b$, that is,

$$(_H\mathcal{D}^\alpha {}_H\mathcal{J}^\alpha f(t) = f(t).$$

7. Let $\text{Re}(\mu) > 0$ and $n \in \mathbb{N}$ such that $n - 1 < \text{Re}(\mu) \le n$. Show that the Laplace transform of the fractional derivative in the sense of Riemann-Liouville is given by

$$\mathscr{L}[\mathcal{D}^{\mu} f(t)] = s^{\mu} \mathscr{L}[f(t)] - \sum_{k=0}^{n-1} s^{n-k-1} g^{(k)}(0)$$

being s the parameter of the Laplace transform and $g(t) = \mathcal{J}^{n-\mu} f(t)$, where $\mathcal{J}^{\alpha}$ is the Riemann-Liouville fractional integral operator of order α.

8. Use the Exercise (8) to show that

$$\int_0^t \int_0^{t_3} \int_0^{t_2} \int_0^{t_1} f(t_4) \, dt_4 \, dt_3 \, dt_2 \, dt_1 = \frac{1}{3!} \int_0^t (t - \xi)^3 d(\xi) \, d\xi \cdot$$

9. Using the Proposed exercise (8) show the result

$$\mathscr{L}^{-1}\left[\frac{F(s)}{s^n}\right] = \int_0^t \int_0^{t_1} \cdots \int_0^{t_{n-2}} \int_0^{t_{n-1}} f(t_n) \, dt_n \, dt_{n-1} \cdots dt_2 \, dt_1 \cdot$$

10. Use the Exercises (8) and (9) to show that

$$\int_0^t \int_0^{t_1} \cdots \int_0^{t_{n-2}} \int_0^{t_{n-1}} f(t_n) \, dt_n \, dt_{n-1} \cdots dt_2 \, dt_1 = \frac{1}{(n-1)!} \int_0^t (t - \xi)^{n-1} f(\xi) \, d\xi \cdot$$

11. Calculate the Caputo fractional derivative of order v with $n - 1 < v \le n$, of the Mittag-Leffler function with three parameters. Discuss, as a particular case, this derivative for the Mittag-Leffler function with two parameters.

12. Calculate the Riemann-Liouville fractional derivative of order v with $n - 1 < v \le n$, of the Mittag-Leffler function with three parameters. Compare the result with the Caputo fractional derivative.

13. Using the Proposed exercise (12), retrieve the result of the Riemann-Liouville derivative from the Mittag-Leffler function with two parameters and compare the result with the Caputo fractional derivative.

14. Let $v > 0$. Calculate the Riemann-Liouville fractional integral of order v of the Mittag-Leffler function with two parameters, denoted by $E_{\alpha,\beta}(-t^{\alpha})$ and discuss the case $\beta = v$.

15. Calculate the Hadamard fractional integral of order v of the function

$$E_{\alpha,1}^{\gamma}\left[\left(\ln \frac{t}{a}\right)^{\alpha}\right].$$

Discuss the particular case $\gamma = 1$.

16. Let $n - 1 < v \le n$. Show that the Caputo fractional derivative of the exponential function $f(t) = \exp(t)$ is

$$_C\mathcal{D}^{v} \exp(t) = t^{n-v} E_{1,n-v+1}(t)$$

where $E_{\alpha,\beta}(\cdot)$ is a Mittag-Leffler function with two parameters. Discuss the case $\nu = n$.

17. Let $\mathrm{Re}(\alpha) > 0$, $\lambda \in \mathbb{R}$ and $\mathrm{Re}(\nu) > 0$. Show that the Riemann-Liouville fractional integral of the function $f(t) = e^{\lambda t} t^{\alpha-1}$ is given by

$$_0\mathcal{J}_t^\nu[f(t)] = \frac{\Gamma(\alpha)}{\Gamma(\alpha+\nu)} t^{\alpha+\nu-1} \, _1F_1(\alpha; \alpha+\nu; \lambda t)$$

where $_1F_1(a; c; z)$ is a confluent hypergeometric function.

18. Let $\alpha, \beta, \gamma, \nu > 0$ and $a \in \mathbb{R}$. Show that the Riemann-Liouville fractional derivative of the function $f(t) = t^{\beta-1} E_{\alpha,\beta}^\gamma(at^\alpha)$ is given by

$$_0\mathcal{D}_t^\nu \left[t^{\beta-1} E_{\alpha,\beta}^\gamma(at^\alpha) \right] = t^{\beta-\nu-1} E_{\alpha,\beta-\nu}^\gamma(at^\alpha).$$

Compare with the Exercise (31) and verify that one result follows the other with the changes: in the order $\nu \to -\nu$ and in the operator $\mathcal{D}^{-\nu} \to \mathcal{J}^\nu$ and vice-versa.

19. Let $t > 0$, $\alpha > 0$ and $\lambda \in \mathbb{R}$. Show that the solution of the differential equation [20]

$$_0\mathcal{D}_t^\alpha x(t) - \lambda x(t) = f(t),$$

in terms of the Laplace convolution product, is given by

$$x(t) = \int_0^t (t-\tau)^{\alpha-1} E_{\alpha,\alpha}[\lambda(t-\tau)^\alpha] f(\tau) \, d\tau$$

where $E_{\alpha,\alpha}(\cdot)$ is a Mittag-Leffler function with two parameters.

20. In analogy to Exercise (33) in Chap. 6, show using the Laplace transform methodology that the solution of the non-homogeneous fractional differential equation

$$_0\mathcal{D}_t^\alpha x(t) - \lambda x(t) = h(t)$$

for $t > 0$ and satisfying the conditions

$$_0\mathcal{D}_0^{\alpha-k} x(t)\big|_{t=0} = b_k$$

with $k = 1, 2, \ldots, n$ where $n - 1 < \alpha < n$ is given by [21]

$$x(t) = \sum_{k=1}^n b_k t^{\alpha-k} E_{\alpha,\alpha-k+1}(\lambda t^\alpha) + \int_0^t (t-\tau)^{\alpha-1} E_{\alpha,\alpha}[\lambda(t-\tau)^\alpha] h(\tau) \, d\tau$$

where $E_{\alpha,\alpha}(\cdot)$ is a Mittag-Leffler function with two parameters.

21. Let $J_0(\sqrt{x})$ a zero order Bessel function. Show that

$$_0\mathcal{J}_x^{1/2}\, J_0(\sqrt{x}) = \frac{2\sin(\sqrt{x})}{\sqrt{\pi}}.$$

22. Let $J_0(\sqrt{x})$ a zero order Bessel function. Show that

$$_0\mathcal{D}_x^{1/2}\, \sin(\sqrt{x}) = \frac{\sqrt{\pi}}{2}\, J_0(\sqrt{x}).$$

23. Let $a > 0,\, b > 0$ and $c > 0$. Use the integral representation for hypergeometric function, $_2F_1(a, b; c; x)$ in order to express it as a fractional integral

$$_2F_1(a, b; c; x) = \frac{\Gamma(c)}{\Gamma(b)} x^{1-c}\, _0\mathcal{J}_x^{c-b}\left[x^{b-1}(1-x)^{-a}\right].$$

24. Use the result of Exercise (36) to calculate the fractional integral of the order $1/2$, in the Riemann-Liouville sense, initialized at zero, of the function $f(t) = \ln t\cdot$
25. Show that the Caputo fractional derivative of the order $1/2$, initialized at zero, of the function function $f(t) = \ln t$, is not defined.
26. Let $\alpha > 0$. Show that the Laplace transform of the Riemann-Liouville fractional integral of the order α of the function $\sin t$ is such that

$$\mathscr{L}\left[_0\mathcal{J}_t^\alpha \sin t\right] = \frac{1}{s^\alpha(s^2 + 1)}.$$

27. Let $\alpha > 0$. Show that the Laplace transform of the Riemann-Liouville fractional integral of the order α of the function $\cos t$ is such that

$$\mathscr{L}\left[_0\mathcal{J}_t^\alpha \sin t\right] = \frac{s}{s^\alpha(s^2 + 1)}.$$

28. Solve the Exercise (38) and use the Laplace transform to obtain the solution as follows

$$f(t) = \frac{\sin \pi \alpha}{\pi} t^{\alpha-1}.$$

29. Let $\mu > 0,\, t > 0$ and $\nu > 0$ the order of fractional integral. Calculate the Hadamard fractional integral, initialized at $a = 1$, of the function $f(t) = t^\mu$, in order to show that

$$_\mathsf{H}\mathcal{J}_t^\nu f(t) = \frac{t^\mu}{\Gamma(\nu)}\frac{\gamma(\nu, \mu \ln t)}{\mu^\nu}$$

where $\gamma(\cdot, \cdot)$ is an incomplete gamma function.
30. Let $\mu > -1$ and $\nu > 0$. Calculate the Riemann-Liouville fractional derivative of order ν of the function $f(t) = t$. Compare with the result of Exercise (40).

31. Let $\alpha > 0$ be the order of the integral equation and $f(t)$ a continuous function. Show that the solution of the integral equation with the Riemann-Liouville operator

$$_0\mathcal{J}_t^\alpha f(t) = 1 + f(t)$$

is given by $f(t) = -E_\alpha(t^\alpha)$ where $E_\alpha(\cdot)$ is the classical Mittag-Leffler function.

32. Let $\mu > 0$ and $\nu > 0$ the order of the Riemann-Liouville fractional integral. Show that

$$_0\mathcal{J}_t^\nu[t^{\mu-1} \sin t] = \frac{t^{\mu+\nu-1}}{2i} \frac{\Gamma(\mu)}{\Gamma(\mu+\nu)} [\,_1F_1(\mu; \mu+\nu; ix) - \,_1F_1(\mu; \mu+\nu; -ix)]$$

where $_1F_1(a; a; \cdot)$ is the confluent hypergeometric function. Discuss the case $\mu = 1$.

33. Let $0 < \alpha \leq 1$ be the order of the integrodifferential equation and $f(t)$ a continuous function such that $f'(0) = 0$. Show that the solution of the integrodifferential equation, being $_C\mathcal{D}_t^\alpha$ the Caputo fractional derivative,

$$_C\mathcal{D}_t^\alpha f(t) = 1 + f(t)$$

is given by $f(t) = t^\alpha E_{\alpha,\alpha+1}(t^\alpha)$ where $E_{\alpha,\beta}(\cdot)$ is a Mittag-Leffler function with two parameters. Discuss the case $\alpha = 1$.

34. Let $_0\mathcal{J}_t^\alpha f(t)$ be the Riemann-Liouville fractional integral of order $\alpha > 0$. Show that

$$_0\mathcal{J}_t^\alpha[t\, f(t)] = t\, _0\mathcal{J}_t^\alpha f(t) - \alpha\, _0\mathcal{J}_t^{\alpha+1} f(t).$$

35. Let $\alpha > 0$ be the order of the Caputo fractional derivative, denoted by $_C\mathcal{D}_t^\alpha$ and the Riemann-Liouville fractional integral, by $_0\mathcal{J}_t^\alpha$. Set n, a positive integer, such that: $n = [\alpha] + 1$, if $\alpha \neq \mathbb{N}$ and $n = \alpha$ if $\alpha \in \mathbb{N}$, being $[\alpha]$ the integer part of α. Show that

$$_0\mathcal{J}_t^\alpha \,_C\mathcal{D}_t^\alpha f(t) = f(t) - \sum_{k=0}^{n-1} f^{(k)}(0) \frac{t^k}{k!}$$

being $f^{(k)}(0)$ the k-th derivative of $f(t)$ in $t = 0$. Discuss the case $0 < \alpha \leq 1$.

36. Show that the Hadamard derivative is a linear operator.

37. Since the Riemann-Liouville fractional derivative is the inverse operator to the left of the fractional integral, show that the solution of the integrodifferential equation

$$t = \int_0^t (t-\tau)^{-1/2} f'(\tau)\, d\tau$$

is given by $f(t) = \frac{4}{3\pi}\sqrt{x^3}$.

38. Nigmatullin [22] obtained the solution of the one-dimensional diffusion equation, fractional in the temporal part and obtained for the Green function the expression

$$G(x, t) = \frac{1}{\pi} \int_0^\infty t^{\alpha-1} E_{\alpha,\alpha}(-\lambda^2 k^2 t^\alpha) \cos(kx) \, dk$$

where λ is a constant and $0 < \alpha \leq 1$ the order of derivative. Let $x > 0$. Show that

$$G(x, t) = \frac{t^{\alpha-1}}{x} H_{1,1}^{1,0} \left[\frac{x^2}{\lambda^2 t^\alpha} \middle| \begin{matrix} (\alpha, \alpha) \\ (1, 2) \end{matrix} \right]$$

being $H_{a,b}^{c,d}(\cdot)$ a Fox's H-function.

39. Let $_0\mathcal{D}_t^\alpha$ be the Riemann-Liouville fractional derivative of order $\alpha > 0$. Show that

$$_0\mathcal{D}_t^\alpha[t^s \, e^t] = \frac{\Gamma(s+1)}{\Gamma(s-\alpha+1)} t^{s-\alpha} \, {}_1F_1(s+1; s-\alpha+1; t)$$

where $_1F_1(a; b; x)$ is a confluent hypergeometric function and $s > -1$. Discuss the particular case $s = 0$.

40. Let $c > 0$, $t > 0$ and $v > 0$. Use the Laplace transform to solve [23]

$$x(t) - x_0 f(t) = -c^v \, {}_0\mathcal{J}_t^v x(t)$$

where $f(t)$ is an integrable function and x_0 a constant, that is, show that

$$x(t) = c \, x_0 \int_0^t H_{1,2}^{1,1} \left[c^v (t - \tau)^v \middle| \begin{matrix} (-1/v, 1) \\ (-1/v, 1), (0, v) \end{matrix} \right] f(\tau) \, d\tau$$

· being $H_{a,b}^{c,d}(\cdot)$ a Fox's H-function.

41. Let $f(x) = e^{\mu x}$ with $\mu > 0$. Considering $n > 0$ with $n \notin \mathbb{N}$, show the relation

$$\mathcal{D}_a^n f(x) = \frac{e^{\mu a}}{\Gamma(1-n)} (x - a)^{-n} \, {}_1F_1(1; 1 - n; \mu(x - a))$$

with $_1F_1(b; b; \cdot)$ a confluent hypergeometric function and a is the lower index. (i) Discuss the case $a = -\infty$ and compare with the classical result

$$\mathcal{D}_a^n f(x) = \mu^n \, e^{\mu a}$$

that holds for $n \in \mathbb{N}$. (ii) As in the previous item, discuss the case $a = 0$.

42. Consider the noninteger order fractional differential equation, with the Caputo fractional derivative,

$$\mathcal{D}^\alpha[(1 - x^2)\mathcal{D}^\beta y(x)] = 1$$

with $0 < \alpha \leq 1$, $0 < \beta \leq 1$ and $-1 < x < 1$. Show that: (i) the solution $y(x)$ can be written as follows

$$y(x) = \frac{x^{\alpha+\beta}}{2\Gamma(\alpha + \beta + 1)} \Omega(\alpha, \beta, x)$$

where we have introduced the notation

$$\Omega(\alpha, \beta, x) \equiv {_2F_1}(1, \alpha + 1; \alpha + \beta + 1; x) + {_2F_1}(1, \alpha + 1; \alpha + \beta + 1; -x)$$

with ${_2F_1}(1, \alpha + 1; \alpha + \beta + 1; \pm x)$ being classical hypergeometric functions. (ii) for $\alpha = 1 = \beta$ the solution is given by

$$y(x) = -\frac{1}{2} \ln(1 - x^2).$$

References

1. Capelas de Oliveira, E., Tenreiro Machado, J.A.: A review of definitions for fractional derivatives and integrals. Math. Prob. Ing. **2014**, ID 238459 (2014)
2. Almeida, R.: A Caputo fractional derivative of a function with respect to another function. Commun. Nonlinear Sci. Numer. Simulat. **44**, 460–481 (2017)
3. Sousa, J.V.C., Capelas de Oliveira, E.: On the ψ-Hilfer fractional derivative. Commun. Nonlinear Sci. Numer. Simulat. **60**, 72–91 (2018)
4. Sonin, N.Y.: On differentiation with arbitrary index. Moscou Mat. Sb. **6**, 1–38 (1869)
5. Grünwald, A.K.: Derivationen und deren Anwendung. Z. Angew. Math. Phys. **12**, 441–480 (1867)
6. Letnikov, A.V.: Theory of differentiation with an arbitrary index. Matem. Sbornik. **3**, 1–66 (1868)
7. Prudnikov, A.P., Brychkov, Y.A., Marichev, O.I.: Integral and Series (Elementary Functions). Gordon and Breach Science Publishers, London (1986)
8. Tricomi, F.G., Erdélyi, A.: The asymptotic expansion of a ratio of gamma functions. Pacific J. Math. **1**, 133–142 (1951)
9. Love, E.R.: Changing the order of integration. J. Austral. Math. Soc. **9**, 421–432 (1970)
10. Miller, K.S., Ross, B.: An Introduction to the Fractional Calculus and Fractional Differential Equations. Wiley, New York (1993)
11. Samko, S.G., Kilbas, A.A., Marichev, O.E.: Fractional Integrals and Derivatives: Theory and Applications. Gordon and Breach Science Publishers, Switzerland (1993)
12. Caputo, M.: Linear models of dissipation whose Q is almost frequency independent—II. Geophys. J. Roy. Astron. Soc. **13**, 529–539 (1967). Reprinted in Fract. Cal. Appl. Anal. **11**, 4–14 (2008)
13. Caputo, M., Mainardi, F.: Linear models of dissipation in anelastic solids. Riv. Nuovo Cimento **1**, 161–198 (1971). Reprinted in Fract. Cal. Appl. Anal. **10**, 309-324 (2007)
14. Dzhrbashian, M. M., Nersesyan, A. B.: Fractional derivatives and the Cauchy problem for fractional differential equations. Izv. Akad. Nauk. Armyan **3**, 3–29 (1968) (in Russian)
15. Davis, H.T.: The Theory of Linear Operators. The Principia Press, Bloomington (1936)
16. Hilfer, R. (ed.): Applications of Fractional Calculus in Physics. World Scientific, Singapore (2000)

17. Rubanraj, S., Jernith, S.S.: Fractional calculus of logarithmic functions. Int. J. Math. Comp. Research. **4**, 1296–1303 (2016)

18. Debnath, L., Bhatta, D.: Integral Transform and Their Applications, 2nd edn. Chapman & Hall/CRC, Boca Raton (2007)

19. Ortigueira, M.D.: Fractional Calculus for Scientists and Engineers. Springer, New York (2011)

20. Kilbas, A. A., Srivastava, H. M., Trujillo, J. J.: The Theory and Applications of Fractional Differential Equations. North-Holland Mathematics Studies, vol. 204. Elsevier, Amsterdam (2006)

21. Podlubny, I.: Fractional Differential Equations. Mathematical in Sciences and Engineering, vol. 198. Academic Press, San Diego (1999)

22. Nigmatullin, R.R.: The realization of the generalized transfer equation in a medium with fractal geometry. Phys. Sta. Sol. B. **133**, 425–430 (1986)

23. Mathai, A.M., Saxena, R.K., Haubold, H.J.: The H-Function: Theory and Applications. Springer, New York (2010)

Chapter 6
Applications and Add-ons

> We present here examples in which the classical special
> functions play a fundamental role, as well as examples envolving
> fractional derivative which give rise to special function proper to
> fractional calculus. Differential, integral and integrodifferential
> equations of fractional order will be discussed in specific
> exercises. As an add-on we address issues that, in our view,
> would not fit in the main text, because they require more lengthy
> calculation, for example, fractional integration.

1. *Simple pendulum.* A simple pendulum consists of a mass m fixed at one end of a
   rod of length ℓ whose mass is negligible with respect to mass m. The other end
   of the rod is fixed at one point so that the system (rod and mass) can move freely
   in the plane under the influence of gravity. Use Newton's second law to obtain
   the differential equation that describes the movement of the spring, as well as
   the respective solution. Discuss the case for small oscillations.

   We begin by writing the respective differential equation, an ordinary one, of
   second order, but not linear. The forces acting on the mass m are the weight,
   given by mg, due to gravity, where g is the gravitational constant, and the stress,
   denoted by T, due to the rod. Using Newton's second law, we can write for
   the tangential and centripetal components of the force, respectively, the motion
   equations

$$-mg \sin \theta = m \frac{d^2}{dt^2}(\ell \theta)$$

   and

$$T - mg \cos \theta = m\omega^2 \ell$$

© Springer Nature Switzerland AG 2019
E. Capelas de Oliveira, *Solved Exercises in Fractional
Calculus*, Studies in Systems, Decision and Control 240,
https://doi.org/10.1007/978-3-030-20524-9_6

where $\theta = \theta(t)$ is the angle that the rod forms with the vertical and ω is the angular velocity. The differential equation associated with the tangential component can be rearranged in the following form

$$\frac{d^2}{dt^2}\theta + \frac{g}{\ell}\sin\theta = 0$$

which is a nonlinear, second order ordinary differential equation. Let us briefly discuss this differential equation, which describes the exact movement of the simple pendulum and for which we shall seek its solution. First, in the case of oscillations with small amplitudes (small oscillations), we consider the approximation $\sin\theta \simeq \theta$ which leads us to a nonlinear, second order ordinary differential equation, whose solution is given in terms of the sine and cosine functions, or even in terms of complex exponentials. Now, in our case (general) we begin by multiplying the differential equation by the factor $m\ell^2\dfrac{d\theta}{dt}$, a factor $(m\ell^2)$ multiplying the velocity $\left(\dfrac{d\theta}{dt}\right)$ which, after rearranging, can be written in the following form

$$\frac{d}{dt}\left[\frac{m\ell^2}{2}\left(\frac{d\theta}{dt}\right)^2 - mg\ell\cos\theta\right] = 0 \qquad (6.1)$$

and which clearly shows that the quantity that lies between the brackets must be a constant, since the derivative of order one of a constant is zero. We mention that this is the well-known conservation law of mechanical energy, that is, the sum of kinetic energy, the term proportional to the velocity square, and the potential energy, the term proportional to the length of the rod. In order to determine this constant, we must impose a condition which, in this case, we assume that the pendulum is at rest for $\theta = \theta_0$, that is, the initial value is given. With this, Eq. (6.1) is given by

$$\frac{d}{dt}\theta = \left[\frac{2g}{\ell}(\cos\theta - \cos\theta_0)\right]^{1/2}.$$

Using the trigonometric relation involving the double arc $\cos 2x = 1 - 2\sin^2 x$, we can write, already separating the variables and integrating

$$2\sqrt{\frac{g}{\ell}}\int dt = \int \frac{d\theta}{[k^2 - \sin^2(\theta/2)]^{1/2}}$$

where we introduce the notation $k = \sin(\theta_0/2)$. Once $\theta \leq \theta_0$, we make a change of variable, that is, we introduce the variable ϕ such that $\sin(\theta/2) = k\sin\phi$ from where it follows, after simplification,

$$\sqrt{\frac{g}{\ell}} \int dt = \int \frac{d\phi}{\sqrt{1 - k^2 \sin^2(\phi)}}.$$

The integral in the second member is the called first kind elliptic integral, denoted by the expression [1]

$$F(k, \phi) = \int_0^{\phi} \frac{d\xi}{\sqrt{1 - k^2 \sin^2(\xi)}}.$$

In the case where the time t, required for the angle to change from zero to θ_0, is equal to one quarter of the oscillation period, T, we have ϕ going from zero to $\pi/2$, from where it follows

$$F\left(k, \frac{\pi}{2}\right) \equiv F(k) = \int_0^{\pi/2} \frac{d\xi}{\sqrt{1 - k^2 \sin^2(\xi)}},$$

a first kind complete elliptic integral. Further, the called complete first kind elliptic integral is defined by

$$E(k, \phi) = \int_0^{\phi} \sqrt{1 - k^2 \sin^2(\xi)}\, d\xi,$$

while the complete second kind elliptic integral is defined by

$$E\left(k, \frac{\pi}{2}\right) \equiv E(k) = \int_0^{\pi/2} \sqrt{1 - k^2 \sin^2(\xi)}\, d\xi$$

which has an application in calculating the arc length of an ellipse, for example. Let's go back to the simple pendulum problem. To calculate the resulting integral, we begin by expanding the integrand, $F(k)$, with the help of Newton's binomial, that is,

$$\frac{1}{\sqrt{1 - k^2 \sin^2 \xi}} = 1 + \frac{k^2}{2} \sin^2 \xi + \frac{3}{8} k^4 \sin^4 \xi + \cdots$$

$$= \sum_{n=0}^{\infty} \frac{(2n - 1)!!}{(2n)!!} k^{2n} \sin^{2n} \xi = \sum_{n=0}^{\infty} \frac{(1/2)_n}{n!} k^{2n} \sin^{2n} \xi$$

where we introduce the double factorial notation, as well as the Pochhammer symbol, according to Chap. 2. Returning with this expression in the integral to $F(k)$ and permuting the order of the integration, we obtain

$$F(k) = \sum_{n=0}^{\infty} \frac{(1/2)_n}{n!} k^{2n} \int_0^{\pi/2} \sin^{2n} \xi \, d\xi.$$

The resulting integral can be considered as a particular case of the expression

$$\int_0^{\pi/2} \cos^p dx = \int_0^{\pi/2} \sin^p x \, dx = \frac{\sqrt{\pi}\Gamma\left(\frac{p+1}{2}\right)}{2\Gamma\left(\frac{p+2}{2}\right)}$$

which is valid for $p > -1$, being $\Gamma(\cdot)$ the gamma function, according to Chap. 2, and which can be written, in terms of the Pochhammer symbol [2], in the form

$$\int_0^{\pi/2} \sin^p x \, dx = \begin{cases} \dfrac{\pi}{2} \dfrac{\left(\frac{1}{2}\right)_{p/2}}{\left(\frac{p}{2}\right)!} & p = 0, 2, 4, \dots \\[3mm] \dfrac{\left(\frac{p-1}{2}\right)!}{\left(\frac{3}{2}\right)_{\frac{p-1}{2}}} & p = 1, 3, 5, \dots \end{cases}$$

Substituting this result, for p an even number, in the expression for $F(k)$ we get

$$F(k) = \frac{\pi}{2} \sum_{n=0}^{\infty} \frac{\left(\frac{1}{2}\right)_n \left(\frac{1}{2}\right)_n}{n!(1)_n} k^{2n}$$

which, compared to the series representation of the hypergeometric function, according to Chap. 2, allows to write

$$F(k) = \frac{\pi}{2} \, {}_2F_1\left(\frac{1}{2}, \frac{1}{2}; 1; k^2\right),$$

being $_2F_1(a, b; c; x)$ the hypergeometric function. Reincorporating the factor $\sqrt{\ell/g}$ we can write, definitively, the period of oscillation of a simple pendulum, in the form

$$T = 2\pi \sqrt{\frac{\ell}{g}} \, {}_2F_1\left[\frac{1}{2}, \frac{1}{2}; 1; \sin^2(\theta_0/2)\right]$$

which is the exact solution of the motion equation. For $\theta_0 \simeq 0$ (small oscillations) we recover the classical result $T = 2\pi\sqrt{\ell/g}$.   ◇

2. *Fractional radial diffusion equation.* The fractional radial diffusion equation [3], obtained from the Weyl fractional derivatives in the spatial variable, and Hilfer-Katugampola, in the time variable, was discussed using similarity and the Mellin transform. The solution was presented in terms of the redefinition of the initial variables, r, radial coordinate and t, temporal coordinate, that is, in terms of the

similarity variables η and $F(\eta)$. The solution is given in terms of an integral in the complex plane, namely

$$F(\eta) = \frac{1}{2\pi i} \int_{\mathfrak{L}} \frac{\Gamma\left(\dfrac{s+\alpha-k-\ell}{1+\alpha}\right) \Gamma\left(\dfrac{s-\ell-1}{2}\right) \Gamma\left(\dfrac{s-\ell}{2}\right)}{\Gamma\left(\dfrac{s-\ell-1}{1+\alpha}\right) \Gamma\left(\dfrac{\gamma s+\alpha+1}{1+\alpha}\right)} z^{-s} \, ds$$

(6.2)

where $\mathfrak{L}$ is a contour that separates the poles to the left of $\mathrm{Re}(s) = b > 0$ from those to the right of $\mathrm{Re}(s) = b > 0$ for the gamma functions found in numerator of the integrand and we define $z = \frac{\eta}{2}(-\rho^{\gamma}/c)^{1/(1+\alpha)}$. Further, with respect to the parameters, they are such that $c, k \in \mathbb{R}$, with c due to the physics of the problem and k associated with the geometry ($k = 0$, flat; $k = 1$, cylindrical; $k = 2$, spherical and $k > 2$, hyperspherical); $\rho > 0$, $0 < \alpha < 1$ and $0 < \gamma < 1$, due to the fractional derivatives; ℓ arbitrary and $\eta = rt^{-\rho\gamma/(1+\alpha)}$, arising from the transformation of similarity. Write $F(\eta)$ as a Fox's H-function and justify whether you can write it in terms of a Meijer's G-function.

Comparing with the kernel of Fox's function, as in Chap. 2, we identified $m = 3$, $n = 0$, $p = 2$ and $q = 3$, from where it follows, for the only two products that are not 'empty', that is, $\displaystyle\prod_{j=1}^{3} \Gamma(b_j + \beta_j s)$ and $\displaystyle\prod_{j=1}^{2} \Gamma(a_j + \alpha_j s)$.

Explaining the parameters we have

$$a_1 = -\frac{\ell+1}{1+\alpha} \qquad\qquad \alpha_1 = \frac{1}{1+\alpha},$$

$$a_2 = 1 \qquad\qquad \alpha_2 = \frac{\gamma}{1+\alpha},$$

$$b_1 = \frac{\alpha-k-\ell}{1+\alpha} \qquad\qquad \beta_1 = \frac{1}{1+\alpha},$$

$$b_2 = -\frac{\ell+1}{2} \qquad\qquad \beta_2 = \frac{1}{2},$$

$$b_3 = -\frac{\ell}{2} \qquad\qquad \beta_3 = \frac{1}{2}.$$

Returning directly to the expression for Fox's H-function, we can write

$$F(\eta) = H_{2,3}^{3,0}\left[z \, \middle|\, \begin{array}{l} \left(-\frac{\ell+1}{1+\alpha}, \frac{1}{1+\alpha}\right), \left(1, \frac{\gamma}{1+\alpha}\right) \\ \left(\frac{\alpha-k-\ell}{1+\alpha}, \frac{1}{1+\alpha}\right), \left(-\frac{\ell+1}{2}, \frac{1}{2}\right), \left(-\frac{\ell}{2}, \frac{1}{2}\right) \end{array} \right]$$

with $z = \frac{\eta}{2}(-\rho^{\gamma}/c)^{1/(1+\alpha)}$. Finally, since, in this case, all the parameters α_1, α_2, β_1, β_2, β_3 are different from unity, it is not possible to express the solution in terms of the Meijer's G-function. Just to reinforce, in order for us to have a Meijer's function, all these parameters should be equal to unity. ◇

3. *Radial diffusion equation in a fractal medium.* As a particular case of the former, let us consider the fractional radial diffusion equation in a fractal medium. This equation is parameterized by the called fractal Hausdorff dimension and the anomalous diffusion exponent. For this, we must consider the parameters such that: $\ell = -d_f$ (fractal Hausdorff dimension); $\gamma = (1+\alpha)/d_\omega$ with d_ω (exponent of anomalous diffusion); $k = d_s - 1$ with d_s (fractal spectral dimension), the relation $d_s = (1+\alpha)d_f/d_\omega$, and $c = -1$. Considering $\omega(d_f) = 2\pi^{d_f/2}/\Gamma(d_f/2)$, the limits $\rho \to 1$ and $\alpha \to 1$, we obtain the expression for $p(r,t)$, the called probability density function, given by

$$
p(r,t) = \frac{d_\omega/\omega(d_f)2^{d_f}t^{\frac{d_f}{d_\omega}}}{\Gamma\left(1 + \frac{d_f}{2} - \frac{d_f}{d_\omega}\right)\Gamma\left(\frac{d_f}{2}\right)} H_{1,2}^{2,0}\left[\frac{r^{d_\omega}}{2^{d_\omega}t}\left|\begin{array}{c}\left(1 - \frac{d_f}{d_\omega}, 1\right) \\ \left(1 - \frac{d_f}{d_\omega}, \frac{d_\omega}{2}\right), \left(0, \frac{d_\omega}{2}\right)\end{array}\right.\right].
$$
(6.3)

This result was obtained in a different way by Duan et al. [4]. Express the Fox's H-function in terms of an integral in the complex plane, explaining the shape of the integration contour.

This case is exactly the inverse procedure of the previous one, that is, the expression for H is known and we want to write the integral in the complex plane, while in the previous one we had the integral in the complex plane and we wanted to explicit the expression to H. Then, comparing with the definition of Fox's H-function, we directly identify the parameters $m = 2, n = 0, p = 1$, and $q = 2$. Even more, we can write for the only two products other than zero

$$
\prod_{j=1}^{2}\Gamma(b_j + \beta_j s) \text{ and } \prod_{j=1}^{1}\Gamma(a_j + \alpha_j s).
$$

Explaining the parameters we have

$$
\begin{aligned}
a_1 &= 1 - \frac{d_f}{d_\omega} & \alpha_1 &= 1, \\
b_1 &= 1 - \frac{d_f}{d_\omega} & \beta_1 &= \frac{d_\omega}{2}, \\
b_2 &= 0 & \beta_2 &= \frac{d_\omega}{2}.
\end{aligned}
$$

With these parameters, we return in the definition in order to write the contour integral

$$
\frac{1}{2\pi i}\int_{\mathcal{L}} \frac{\Gamma\left(1 - \frac{d_f}{d_\omega} + \frac{d_\omega}{2}s\right)\Gamma\left(\frac{d_\omega}{2}s\right)}{\Gamma\left(1 - \frac{d_f}{d_\omega} + s\right)}z^{-s}\,ds.
$$
(6.4)

The contour $\mathcal{L}$ separates the poles from the gamma function $\Gamma(d_\omega s/2)$, leaving them to the left of $\mathrm{Re}(s) = b > 0$, from the poles of the gamma function $\Gamma(1 - d_f/d_\omega + d_\omega s/2)$, leaving them to the right of $\mathrm{Re}(s) = b > 0$.   ◇

4. *A limiting case of the anomalous diffusion equation.* Use Eq. (6.3) to discuss the limiting case of the anomalous diffusion exponent $d_\omega \to 2$ and the fractal Hausdorff dimension $d_f \to 1$ to obtain the solution of the classical diffusion equation. The solution of the classical diffusion equation was also obtained, as a particular case, by Lenzi et al. [5], through a different methodology. Substituting $d_\omega = 2$ and $d_f = 1$ in Eq. (6.3) we get

$$p(r, t) = \frac{1}{2t^{1/2}\Gamma(1/2)} H_{1,2}^{2,0}\left[\frac{r^2}{4t} \, \middle| \, \begin{matrix} (\frac{1}{2}, 1) \\ (\frac{1}{2}, 1), (0, 1) \end{matrix}\right]$$

or by using the result $\Gamma(1/2) = \sqrt{\pi}$ we can write in terms of the Meijer's G-function,

$$p(r, t) = \frac{1}{2\sqrt{t\pi}} G_{1,2}^{2,0}\left[\frac{r^2}{4t} \, \middle| \, \begin{matrix} \frac{1}{2} \\ \frac{1}{2}, 0 \end{matrix}\right].$$

This Meijer's G-function is exactly the exponential, then

$$p(r, t) = \frac{1}{2\sqrt{t\pi}} \exp(-r^2/4t)$$

which is the solution of the classical diffusion equation.   ◇

5. *Wave equation in a fractional space.* The time-dependent wave equation (non-homogeneous partial differential equation) in a fractional space was solved by means of the Fourier transform where a retarded Green's function was obtained [6]. As an application, the authors discussed the movement of a charged particle. Assume that the charged particle moves in a fractal medium along the z-axis, with fractal dimensionality $D = 2 + \alpha$ and $0 < \alpha < 1$. For given charge and current densities, the potential, $\phi(z, t)$, was obtained as the integral

$$\phi(z, t) = \Omega(\alpha) \frac{(ct + z)^{-\frac{1+\alpha}{2}}}{(ct - z)^{\frac{\alpha-1}{2}}} \int_0^1 (1 - \xi)^{-\frac{1+\alpha}{2}} (1 - u\xi)^{-\frac{1+\alpha}{2}} \, d\xi$$

where $u = \dfrac{c + v}{c - v} \dfrac{ct - z}{ct + z}$ and we introduce the constant $\Omega(\alpha) = \dfrac{q/(c - v)}{2c\Gamma((\alpha + 1)/2)\pi^{(\alpha+1)/2}}$.

Express the integral as a hypergeometric function and discuss the limiting case $\alpha \to 0$.

Using the integral representation for the hypergeometric function, given by Eq. (2.10), we identify the parameters as $b = 1$, $c - b - 1 = -(1 + \alpha)/2$ and $a = (1 + \alpha)/2$, with $c = (3 - \alpha)/2$. Turning in the expression for the potential, we have

$$\phi(z,t) = \Omega(\alpha)\frac{(ct+z)^{-\frac{1+\alpha}{2}}}{(ct-z)^{\frac{\alpha-1}{2}}}\frac{\Gamma(\frac{1-\alpha}{2})}{\Gamma(\frac{3-\alpha}{2})}{}_2F_1\left(\frac{\alpha+1}{2},1;\frac{3-\alpha}{2};u\right).$$

In the limit case $\alpha \to 0$ we obtain

$$\phi_0(z,t) = \Omega(0)\frac{(ct+z)^{-\frac{1}{2}}}{(ct-z)^{\frac{-1}{2}}}\frac{\Gamma(\frac{1}{2})}{\Gamma(\frac{3}{2})}{}_2F_1\left(\frac{1}{2},1;\frac{3}{2};u\right)$$

or, in the following form, substituting $\Omega(0)$,

$$\phi_0(z,t) = \frac{q}{\pi c(c-v)}\sqrt{\frac{ct-z}{ct+z}}\,{}_2F_1\left(\frac{1}{2},1;\frac{3}{2};u\right).$$

Just to emphasize, because we are not entering the physical context of the problem, in this case, it is possible to express the hypergeometric function in terms of elementary functions. In the case where $0 < u < 1$ the solution can be given in terms of a logarithm, since it is worth the relation $\ln\left(\frac{1+z}{1-z}\right) = 2z \cdot {}_2F_1\left(\frac{1}{2},1;\frac{3}{2};z^2\right)$, while in the case for which we have $u < 0$, the solution can be given in terms of an arc tangent function, since it is worth the relation $\arctan z = z \cdot {}_2F_1\left(\frac{1}{2},1;\frac{3}{2};z^2\right)$.   ◇

6. *Inverse Laplace transform without complex plane.* There is a way to get the inverse Laplace transform without using the complex plane. Recently, this methodology was discussed in the particular case of a Mittag-Leffler function with three parameters and several integrals were obtained. In this specific case the following results were found

$$\int_0^\infty \frac{\tau^3 + 3\tau}{\tau^4 + 6\tau^2 + 25}\sin(\tau t)\,d\tau = \frac{\pi}{2}e^{-2t}\cos t, \qquad t > 0$$

and

$$\int_0^\infty \frac{\tau^3 + 5\tau}{\tau^4 + 10\tau^2 + 9}\sin(\tau t)\,d\tau = \frac{\pi}{2}e^{-2t}\cosh t, \qquad t > 0.$$

Let $E_\alpha(\cdot)$ be the classical Mittag-Leffler function, show that $E_4(t^4)$ can be express as follows [7]

$$E_4(t^4) = \frac{e^{2t}}{2}\int_0^\infty\left[\frac{\tau^3 + 3\tau}{\tau^4 + 6\tau^2 + 25} + \frac{\tau^3 + 5\tau}{\tau^4 + 10\tau^2 + 9}\right]\sin(\tau t)\,d\tau$$

for $t > 0$.

Using the result of Exercise (8) in Chap. 3, we have

$$E_{2\alpha}(x^2) = \frac{1}{2}\left[E_\alpha(x) + E_\alpha(-x)\right].$$

Considering in this equality $\alpha = 2$ and $x = t^2$, we can write

$$E_4(t^4) = \frac{1}{2}\left[E_2(t^2) + E_2(-t^2)\right].$$

The two classical Mittag-Leffler functions found in the second member can be expressed in terms of trigonometric functions, as the Proposed exercise (2), and the hyperbolic functions, according to the Proposed exercise (3), both of Chap. 3, that is, are valid the relations

$$E_2(-z^2) = \cos z \qquad \text{and} \qquad E_2(z^2) = \cosh z$$

then we have for $E_4(t^4)$,

$$E_4(t^4) = \frac{1}{2}\left(\cos t + \cosh t\right).$$

Finally, replacing the two results, involving the trigonometric cosine and the hyperbolic cosine, obtained in the mentioned work and rearranging, we can write

$$E_4(t^4) = \frac{e^{2t}}{2}\int_0^\infty\left[\frac{\tau^3 + 3\tau}{\tau^4 + 6\tau^2 + 25} + \frac{\tau^3 + 5\tau}{\tau^4 + 10\tau^2 + 9}\right]\sin(\tau t)\,d\tau,$$

an integral representation for this particular Mittag-Leffler function, $E_4(t^4)$. ⋄

7. *Relaxation equations and oscillations.* In the classic study of ordinary differential equations we are faced with the two simplest equations, representative of the relaxation and oscillation processes,

$$\frac{d}{dt}x(t) \equiv x'(t) = -x(t) + q(t)$$

and

$$\frac{d^2}{dt^2}x(t) \equiv x'(t) = -x(t) + q(t),$$

respectively. The first one being classified as first order and the second of order two and $q(t)$ is the non-homogeneity term.

In the case of analogous equations, in the fractional sense, we can write them in a unique way so that we transfer this classification, in relation to the order, to the parameter α, such that $m - 1 < \alpha \leq m$ with $m \in \mathbb{N}^*$. We have mentioned that an analogous procedure can be done by taking $m - 1 \leq \alpha < m$, introducing, in this case, the fractional derivative on the right. Here, as already mentioned, we will work only with the derivative on the left. Then, using a fractional derivative in the Caputo sense, we write

$$\frac{d^\alpha}{dt^\alpha} x(t) = -x(t) + q(t) \tag{6.5}$$

where α is the order of derivative with $m - 1 < \alpha \le m$ and $m \in \mathbb{N}^*$. Then, for $m = 1$ we recover the classical case of a relaxation process, while for $m = 2$, we obtain the analogue of a process involving oscillations.

In summary, we can work with Eq. (6.5) and, depending on the value of the parameter α, we have a respective value of m, as well as adding the initial conditions we have a problem (equation + conditions) in the fractional sense. Further, another derivative could be considered, however, here we discuss a specific case with the formulation of the derivative as proposed by Caputo. Let's solve Eq. (6.5).

Using the Laplace transform methodology, remembering that the operator is linear, we can write, from Eq. (6.5),

$$s^\alpha \mathscr{L}[x(t)] - \sum_{k=0}^{m-1} s^{\alpha-k-1} \frac{d^k}{dt^k} x(t) \bigg|_{t=0} = -\mathscr{L}[x(t)] + Q(s)$$

where $Q(s) = \mathscr{L}[q(t)]$ denotes the Laplace transform with parameter s. Solving the algebraic equation for $\mathscr{L}[x(t)]$ we obtain

$$\mathscr{L}[x(t)] = \sum_{k=1}^{m-1} \frac{s^{\alpha-k-1}}{s^\alpha + 1} x^{(k)}(0) + \frac{Q(s)}{s^\alpha + 1} \tag{6.6}$$

where $x^{(k)}(0)$ denotes the k-th derivative of $x(t)$ evaluated at $t = 0$.

Finally, taking the inverse Laplace transform of Eq. (6.6), which is also a linear operator, and using the result of Exercise (30) from Chap. 4, we can write

$$x(t) = \sum_{k=0}^{m-1} \mathcal{J}^k E_\alpha(-t^\alpha) x^{(k)}(0) + \mathscr{L}^{-1}\left[\frac{Q(s)}{s^\alpha + 1}\right].$$

In order to further simplify, let us use the Laplace convolution product by writing the inverse Laplace transform in terms of that product, using the result of the Exercise (31) from Chap. 4, hence

$$x(t) = \sum_{k=0}^{m-1} t^k E_{\alpha,k+1}(-t^\alpha) x^{(k)}(0) - q(t) \star E_\alpha'(-t^\alpha)$$

where the line denotes derivative with respect to the variable t and $\star$ is the Laplace convolution product. Finally, the function $q(t)$, nonhomogeneous term, and the order of the derivative, that is, the value of the integer m are known, the problem, composed of a differential equation and initial conditions, will be solved. $\diamond$

8. *Relation between the Wright and Mittag-Leffler functions with two parameters.*
   Let us show that the Wright and Mittag-Leffler functions of two parameters are
   related by means of the Laplace transform.
   Let $x \in \mathbb{R}$ and $\alpha, \beta \in \mathbb{C}$ with $\mathrm{Re}(\alpha > 0)$ and $\mathrm{Re}(\beta > 0)$. Consider the Wright
   function, denoted by $W(x; \alpha, \beta)$, and we will show that

$$\mathscr{L}\left[W(x; \alpha, \beta)\right](s) = \frac{1}{s} E_{\alpha, \beta}\left(\frac{1}{s}\right).$$

Introducing the power series representation for the Wright function, we obtain
for the Laplace transform of the Wright function

$$\mathscr{L}\left[W(x; \alpha, \beta)\right](s) = \int_0^\infty e^{-sx} \sum_{k=0}^\infty \frac{1}{\Gamma(\alpha k + \beta)} \frac{x^k}{k!}\, dx.$$

By changing the order of integration with the summation, we can write

$$\mathscr{L}\left[W(x; \alpha, \beta)\right](s) = \sum_{k=0}^\infty \frac{1}{k! \Gamma(\alpha k + \beta)} \int_0^\infty e^{-sx} x^k\, dx.$$

Introducing the change of variable $\xi = sx$ and rearranging, we have

$$\mathscr{L}\left[W(x; \alpha, \beta)\right](s) = \sum_{k=0}^\infty \frac{1}{k! \Gamma(\alpha k + \beta)} \frac{1}{s^{k+1}} \int_0^\infty e^{-\xi} \xi^k\, d\xi.$$

The remaining integral is nothing more than the integral representation for the
gamma function. Then, using the relation $\Gamma(k + 1) = k!$ and simplifying, we
obtain

$$\mathscr{L}\left[W(x; \alpha, \beta)\right](s) = \frac{1}{s} \sum_{k=0}^\infty \frac{(s^{-1})^k}{\Gamma(\alpha k + \beta)}.$$

The summation on the second member is exactly the power series for the Mittag-
Leffler function with two parameters, then

$$\mathscr{L}\left[W(x; \alpha, \beta)\right](s) = \frac{1}{s} E_{\alpha, \beta}\left(\frac{1}{s}\right),$$

which is the desired result.   ◇

9. *Abel and the tautochronous problem.* The problem of the tautochronous or
   isochronic curve is characterized in determining a curve in which the time spent
   by an object to slide, without friction, in uniform gravity until its minimum point,
   is independent of the starting point. This problem is given in terms of an integral

and is probably the first application of the fractional calculus. The solution can be obtained in two ways, using the usual calculation and via fractional calculus. The solution proposed by Abel is based on the *energy conservation principle* which states that the total amount of energy in an isolated system remains constant, or the sum between gravitational potential energy and kinetic energy is constant. In this problem we are assuming that the particle moves without friction and therefore its kinetic energy is exactly equal to the difference between the potential energy at its initial point and the potential energy at the point where it is located. Let us denote by m the mass of the particle, $v(t)$ its velocity at time t, y_0 the height at which it was abandoned, and $y(t)$ the height at instant t. From Mechanics it is known that the kinetic and potential energies are given by the expressions

$$\frac{1}{2}m\,v^2 \qquad \text{and} \qquad m\,g\,y.$$

respectively.

Since the particle is restricted to moving on the curve, its velocity can be written in the form $v = ds/dt$, where s is the distance measured along the curve. According to the energy conservation principle, we have

$$\frac{1}{2}m\left(\frac{ds}{dt}\right)^2 = mg(y_0 - y)$$

which, by specifying dt and rearranging, provides

$$dt = \pm\frac{1}{\sqrt{2g}}(y - y_0)^{-1/2}\left(\frac{ds}{dy}\right)dy.$$

We have two signs and we must decide for one of them. Since the $s(y)$ function describes the remaining distance on the curve in terms of the remaining height y and as the distance and height decrease as time passes, we must consider the negative sign. Integrating this equation on both sides of y_0 to zero, we obtain

$$\tau = t(y_0) = \int_{y_0}^0 dt = -\frac{1}{\sqrt{2g}}\int_{y_0}^0 (y - y_0)^{-1/2}\left(\frac{ds}{dy}\right)dy,$$

or, in the following form,

$$\tau = \frac{1}{\sqrt{2g}}\int_0^{y_0} (y - y_0)^{-1/2}\left(\frac{ds}{dy}\right)dy \qquad (6.7)$$

in which τ is the time of descent. This equation is an integrodifferential equation, since it contains the dependent variable as a derivative within the integration signal and can be solved in several ways, among which we mention the classic solution that uses the Laplace transform and the solution as proposed by

Abel using the fractional calculus. Here we discuss only the second way of approaching it. The solution proposed by Abel is based on the observation that the integral given in the second member of Eq. (6.7), except for the multiplicative factor $1/\Gamma(1/2)$ is exactly the definition of fractional integration, in the case of order $1/2$. Applying the derivative, according to Riemann-Liouville derivative of order $1/2$, on both sides of the equation we have

$$\frac{\mathrm{d}^{1/2}}{\mathrm{d}y^{1/2}}\tau = \frac{\sqrt{\pi}}{\sqrt{2g}}\frac{\mathrm{d}^{1/2}}{\mathrm{d}y^{1/2}}\left[\frac{1}{\Gamma(1/2)}\int_0^{y_0}(y-y_0)^{-1/2}\left(\frac{\mathrm{d}s}{\mathrm{d}y}\right)\mathrm{d}y\right].$$

Since the Riemann-Liouville fractional derivative operator is the left inverse operator of the fractional integral operator and that the Riemann-Liouville half-derivative is equal to $(\sqrt{\pi\,y})^{-1}$ we can write

$$\frac{\tau}{\sqrt{\pi\,y}} = \frac{\sqrt{\pi}}{\sqrt{2g}}\left(\frac{\mathrm{d}s}{\mathrm{d}y}\right),$$

or by separating the variables, as follows

$$\mathrm{d}s = \frac{\tau\sqrt{2g}}{\pi}y^{-1/2}\mathrm{d}y$$

whose integration provides the solution of the tautochronous problem,

$$s(y) = \frac{2\tau\sqrt{2g}}{\pi}y^{1/2}$$

which, as we can see, is independent of the starting point, y_0.   ◇

10. *Radioactive decay and population growth.* The decay of the mass of a radioactive material is proportional to the remaining mass of the same. Analogous model is the population growth, both described by a first order ordinary differential equation, whose solution is an exponential, with negative argument and with positive argument, respectively, being the proportionality constant dependent on the material, in the first case, and dependent on the initial population, in the second case. In short, what distinguishes the two models is the signal of the proportionality constant. Here we go consider only the fractional version associated with the models for a more accurate description since the solution will depend on the order of the fractional derivative, that is, a parameter to be chosen adequately.

Consider the fractional differential equation

$$\frac{\mathrm{d}^{\alpha}m(t)}{\mathrm{d}t^{\alpha}} = -k\,m(t),$$

with $0 < \alpha \leq 1$ and k a constant of proportionality.[1] The dependent variable $m(t)$ can denote both the mass of the radioactive material and the population, both for a given instant of time. In order to solve this equation, we use the Laplace transform methodology, from which follows an algebraic equation

$$s^\alpha F(s) - s^{\alpha-1} m(0) = -k F(s)$$

being $F(s) \equiv \mathscr{L}[m(t)](s) = \int_0^\infty m(t) e^{-st} dt$, whose solution is given by

$$F(s) = m_0 \frac{s^{\alpha-1}}{s^\alpha + k}.$$

In order to recover the solution of the equation in the independent variable, t, the respective inverse Laplace transform methodology is applied, as well as using the result of the Exercises (13) and (14) from Chap. 3, we have

$$m(t) = m_0 E_\alpha(-kt^\alpha), \tag{6.8}$$

where m_0 is the initial mass and $E_\alpha(z)$ is the classical Mittag-Leffler function. In an analogous way we can obtain the solution of the problem of population growth, in its fractional version, with solution

$$P(t) = P_0 E_\alpha(kt^\alpha), \tag{6.9}$$

where P_0 denotes the initial population. Note the difference only in the signal of the proportionality constant. Anyway, in the limit $\lim_{\alpha \to 1}$ both cases lead us to the classical solutions, in the case of radioactive decay $m(t) = m_0 E_1(-kt^1) = m_0 e^{-kt}$, and in the case of population growth $P(t) = P_0 E_1(kt^1) = P_0 e^{kt}$, which are the solutions of the respective differential equation with derivative of order one.   ◇

11. *Fractional harmonic oscillator.* Here we discuss the classic problem of the harmonic oscillator in the fractional version. This is a clear case in which modeling by a non-integer order equation gives a more accurate description of reality in contrast to the respective integer order equation obtained as the limiting case of the parameter associated with the order of the equation. The differential equation describing the harmonic oscillator without damping, also known as the mass-spring system, in its fractional version, has the solution given by the damped harmonic oscillator which, as is well-known, more accurately describes reality, since in all real systems we have frictions [9]. As is known from Mechanics,

---

[1]It is important to note that, in all equations involving a constant that multiply the dependent variable, this constant must be related with the order of the differential equation. In this particular case we can also consider $k \to k^\alpha$ in the solution. For a general discussion one can see [8].

Newton's second law of motion, applied to time-repeating systems, provides the ordinary differential equation

$$m\frac{d^2}{dt^2}x(t) + \mu\frac{d}{dt}x(t) + k\,x(t) = g(t)$$

which describes the displacement (elongation), $x(t)$, of a body of mass m, in time t, from the equilibrium position, subject to a Hooke force, $-kx(t)$, to a damping force $-\mu\frac{d}{dt}x(t)$ and to an external force $g(t)$, where μ and k are positive constants. In the particular case in which there are no frictions or external forces acting on the system, we have the following problem, composed of an ordinary differential equation of second order and with constant coefficients

$$m\frac{d^2}{dt^2}x(t) + k\,x(t) = 0 \tag{6.10}$$

and the initial conditions $x(0) = x_0$ and $x'(0) = 0$. The solution to this problem is

$$x(t) = x_0\cos(\omega_0 t) \tag{6.11}$$

where we introduce the frequency ω_0, such that $\omega_0^2 = k/m$.

Another way to formulate this problem is by means of an integral equation, that is, the equation whose function we wish to obtain, is under the integral signal. In order to rewrite Eq. (6.10) as an integral equation, we consider

$$x''(t) = -\frac{k}{m}x(t) = -\omega_0^2 x(t) \quad\Rightarrow\quad \int_0^t x''(u)du = -\omega_0^2\int_0^t x(u)du,$$

then

$$x'(t) = x'(0) - \omega_0^2\int_0^t x(u)du.$$

Integrating the previous one and using Goursat's theorem

$$\int_0^t\int_0^v x(u)dudv = \int_0^t\int_t^v x(u)dvdu = \int_0^t x(u)du\int_t^u dv = \int_0^t (u-t)x(u)du.$$

we obtain, by simplifying, the following integral equation

$$x(t) = x(0) + tx'(0) + \omega_0^2\int_0^t (t-u)x(u)du. \tag{6.12}$$

Note that the unknown function, $x(t)$, is found as integrating. Moreover, the two equations Eqs. (6.10) and (6.12) are equivalent. Finally, the integral equation already carries with it the initial conditions, unlike the differential equation.

As we have already mentioned, we will compare integer-order modeling with the respective fractional modeling, in particular, we consider the fractional generalization of the ordinary differential equation, associated with the simple harmonic oscillator problem [10], obtained from the substitution of the derivative of order two of Eq. (6.10), by the fractional derivative of α, as formulated by Caputo, that is,

$$\frac{d^\alpha}{dt^\alpha} x(t) + \omega^\alpha x(t) = 0.$$

with $1 < \alpha \le 2$, and we introduce the notation $\omega^\alpha = \omega_0^2 = k/m$. In complete analogy to the differential equation, we proceed with the integral equation, that is, we introduce an integral of order α in Eq. (6.12) in order to obtain a fractional integral equation

$$x(t) = x(0) + tx'(0) - \frac{\omega^\alpha}{\Gamma(\alpha)} \int_0^t (t - u)^{\alpha-1} x(u) \cdot du. \tag{6.13}$$

In order to solve Eq. (6.13), we opted for the Laplace transform methodology. Then, taking the Laplace transform on both members, we have,

$$F(s) = \frac{1}{s} x(0) + \frac{1}{s^2} x'(0) - \omega^\alpha \int_0^\infty \left[\frac{t^{\alpha-1}}{\Gamma(\alpha)} \star x(t) \right] e^{-st} dt$$

in which $\mathscr{L}[x(t)](s) \equiv F(s) = \int_0^\infty z(t) e^{-st} dt$ is the Laplace transform of $x(t)$, with parameter s and in the second member we use the definition of convolution product, denoted by $\star$. Since the Laplace transform of the convolution product is equal to the product of the Laplace transforms and that the Laplace transform of $t^{\alpha-1}/\Gamma(\alpha)$ is $s^{-\alpha}$ we can write, solving for $F(s)$,

$$F(s) = x(0) \frac{s^{-1}}{1 + \omega^\alpha s^{-\alpha}} + x'(0) \frac{s^{-2}}{1 + \omega^\alpha s^{-\alpha}}. \tag{6.14}$$

In order to recover the solution, $x(t)$, we apply the inverse Laplace transform on both sides of the Eq. (6.14), then,

$$x(t) = x(0) E_\alpha(-\omega^\alpha t^\alpha) + x'(0) E_{\alpha,2}(-\omega^\alpha t^\alpha),$$

where $E_\alpha(z)$ and $E_{\alpha,\beta}(z)$ are Mittag-Leffler functions with one and two parameters, respectively. Using the initial conditions $x(0) = x_0$ and $x'(0) = 0$, we can write

$$x(t) = x_0 E_\alpha(-\omega^\alpha t^\alpha), \tag{6.15}$$

which is the solution of the fractional harmonic oscillator problem. On the other hand, considering $\alpha = 2$ in the previous equation we have

$$x(t) = x_0 \, E_2(-(\omega t)^2) = x_0 \, \cos(\omega t),$$

that is, the solution of the harmonic oscillator of the integer order is recovered. Let's go back to the harmonic oscillator case. Consider the mass-spring system, consisting of a spring of elastic constant, k, a mass, m, and a damper with the damping coefficient μ. The displacement, $x(\tau)$ of the mass from the equilibrium position is described by the ordinary differential equation

$$\frac{d^2}{d\tau^2}x(\tau) + 2\sigma\frac{d}{d\tau}x(\tau) + x(\tau) = 0 \tag{6.16}$$

where, to simplify the notation, we introduce $2\sigma = \mu/m\omega$, a dimensionless measure of the damping, and the initial conditions are given by $x(0) = 1$, in the displacement, and $\left.\dfrac{d}{d\tau}x(\tau)\right|_{\tau=0} = 0$, in the velocity. We introduce the called natural frequency of the undamped system, ω, defined by $\omega^2 = k/m$, normalize the displacement in $x = a$ and delimit the time dimension, τ, defined by $\tau = \omega t$. The solution of Eq. (6.16) satisfying the initial conditions is given by

$$x(\tau) = \left(\frac{d}{d\tau} + 2\sigma\right) e^{-\sigma\tau}\frac{\text{sen}\,\omega\tau}{\omega}, \tag{6.17}$$

with $\omega = \sqrt{1 - \sigma^2}$.

In complete analogy to the fractional harmonic oscillator, we now introduce two parameters α and β, associating them with the order of the fractional differential equation,

$$\frac{d^\alpha}{d\tau^\alpha}x(\tau) + 2\sigma\frac{d^\beta}{d\tau^\beta}x(\tau) + x(\tau) = 0 \tag{6.18}$$

such that the parameters satisfy the inequalities $1 < \alpha \le 2$ and $0 < \beta \le 1$ and the derivative considered in the Caputo sense.

In order to solve this system, the fractional differential equation, satisfying the initial conditions, we consider the Laplace transform methodology, from where we can write

$$s^\alpha F(s) - s^{\alpha-1}x(0) - s^{\alpha-2}\left.\frac{d}{d\tau}x(\tau)\right|_{\tau=0} + 2\sigma\left[s^\beta F(s) - s^{\beta-1}x(0)\right] + F(s) = 0$$

where $\mathscr{L}[x(t)](s) \equiv F(s) = \displaystyle\int_0^\infty e^{-s\tau}x(\tau)\,d\tau$ with $\text{Re}(s) > 0$ is the Laplace transform of $x(\tau)$. Using the initial conditions and rearranging, we obtain an algebraic equation whose solution is given by

$$F(s) = \frac{s^{\alpha-1} + 2\sigma s^{\beta-1}}{s^\alpha + 2\sigma s^\beta + 1} . \tag{6.19}$$

In order to proceed with the inversion of the Laplace transform, we introduce the notation $\mathscr{L}^{-1}[F(s)] = x(\tau)$, follows, using Eq. (6.19),

$$x(\tau) = \mathscr{L}^{-1}\left[\frac{s^{\alpha-1}}{s^\alpha + 2\sigma s^\beta + 1}\right] + 2\sigma \mathscr{L}^{-1}\left[\frac{s^{\beta-1}}{s^\alpha + 2\sigma s^\beta + 1}\right].$$

To calculate these two inverse Laplace transforms, we use Exercise (33) of Chap. 3. Then, we can write

$$x(\tau) = \sum_{r=0}^{\infty} (-2\sigma)^r \tau^{(\alpha-\beta)r} E_{\alpha,1+(\alpha-\beta)r}^{r+1}(-\tau^\alpha)$$

$$+ 2\sigma\tau^{\alpha-\beta} \sum_{r=0}^{\infty} (-2\sigma)^r \tau^{(\alpha-\beta)r} E_{\alpha,1+(\alpha-\beta)r+\alpha-\beta}^{r+1}(-\tau^\alpha) \tag{6.20}$$

which is the solution of the fractional mass-spring problem.

We will conclude this application, recovering the result of the limit case. We consider $\alpha = 2$ and $\beta = 1$. Introducing these values in Eq. (6.20) and using relation involving the Mittag-Leffler function with three parameters [11]

$$\frac{d}{dz}\left[z^{\gamma-1} E_{2,\gamma}^{r+1}(-z^2)\right] = z^{\gamma-2} E_{2,\gamma-1}^{r+1}(-z^2)$$

we can write for the solution of the harmonic oscillator of integer order

$$x(\tau) = \left(\frac{d}{d\tau} + 2\sigma\right) \sum_{r=0}^{\infty} (-2\sigma)^r \tau^{r+1} E_{2,r+2}^{r+1}(-\tau^2) \tag{6.21}$$

which can be shown to match Eq. (6.17).

Let us also recover the free case, that is, where we have no damping. From Eq. (6.21) and considering $\sigma = 0$ we get

$$x(\tau) = \frac{d}{d\tau}\left[\tau E_{2,2}(-\tau^2)\right]$$

where $E_{\mu,\nu}(\cdot)$ is a Mittag-Leffler function with two parameters. Since the relationship $z E_{2,2}(-z^2) = \text{sen } z$ is known, finally, get

$$x(\tau) = \cos \tau$$

which is exactly the solution found when solving the differential equation associated with the harmonic oscillator problem, in particular, here, with the parameters $x_0 = 1$ and $\omega = 1$, that is, Eq. (6.11)   ◇

12. *RLC electric circuit.* The *RLC* circuit is the analogous of the mass-spring problem of mechanics. Let's discuss the called *RLC* circuit in a fractional version, through an integrodifferential equation, that is to say, besides the unknown function being under the signal of integral, it also bears a derivative. The integrodifferential equation describing the system, containing a capacitor, whose capacitance we denote by C, and an inductor, whose inductance is L, connected in parallel, and to this set is connected with a resistor with resistance R, in series. In order to simplificate, we consider a Heaviside-type source denoted by $\varepsilon(t)$ [12]. The integrodifferential equation associated with the current passing through the capacitor, denoted by $i_C(t)$, is given by

$$R\frac{d^\alpha}{dt^\alpha}i_C(t) - \frac{1}{C}i_C(t) + \frac{R/LC}{\Gamma(\alpha)}\int_0^t (t-\xi)^{\alpha-1} i_C(\xi)\, d\xi = \frac{d}{dt}\varepsilon(t) \qquad (6.22)$$

with $0 < \alpha \le 1$. For $\alpha = 1$ we recover the classical result.

Let us use the Laplace transform methodology. To this end, we define

$$\mathscr{L}[i_C(t)] \equiv F(s) = \int_0^\infty e^{-st} i_C(t)\, dt$$

with $\mathrm{Re}(s) > 0$ and we use the Laplace transform of the convolution product, from which we obtain an algebraic equation whose solution is given by

$$F(s) = \frac{1}{R}\frac{s^\alpha}{s^{2\alpha} + as^\alpha + b}$$

where we introduce the parameters $a \equiv 1/RC$ and $b \equiv 1/LC$, both positive. In order to recover the solution of the integrodifferential equation, we must proceed with the respective inversion of the Laplace transform. Then, the current through the capacitor is given by the following expression

$$i_C(t) = \frac{1}{R}\mathscr{L}^{-1}\left[\frac{s^\alpha}{s^{2\alpha} + as^\alpha + b}\right].$$

Using Exercise (33) from Chap. 4, we can write to the current in the capacitor

$$i_C(t) = \frac{t^{\alpha-1}}{R}\sum_{r=0}^\infty (-a)^r t^{\alpha r} E_{2\alpha,\alpha+\alpha r}^{r+1}(-bt^{2\alpha})\theta(t) \qquad (6.23)$$

where $E_{\mu,\nu}^\rho(\cdot)$ is the Mittag-Leffler function with three parameters and $\theta(t)$ is the Heaviside function. To explicitly calculate this sum, we use the theorem involving the functions of Mittag-Leffler [12], then

$$i_C(t) = \frac{t^{\alpha-1}}{R} \frac{\mu E_{\alpha,\alpha}(\mu\, t^\alpha) - \nu E_{\alpha,\alpha}(\nu\, t^\alpha)}{\mu - \nu} \theta(t) \qquad (6.24)$$

where the pair (μ, ν) is the solution of the algebraic system composed of the following equations $\mu + \nu = -1/RC$ and $\mu\nu \equiv 1/LC$.

Taking the parameter $\alpha = 1$ we recover the classical case, that is,

$$i_C(t) = \frac{1}{R} \frac{\mu e^{\mu t} - \nu e^{\nu t}}{\mu - \nu} \theta(t) \qquad (6.25)$$

where μ and ν are given above. Note that, in the case $\mu = \nu$, we must use, in both Eqs. (6.23) and (6.24), the l'Hôpital rule.   ◇

13. *Mellin transform of second kind modified Bessel function.* First, we will introduce, from the Proposed exercise (26) of Chap. 2, the called first kind modified Bessel function and then, using this modified Bessel function, introduce the second kind modified Bessel function of order μ, denoted by $K_\mu(x)$, for which we will show that the Mellin transform is given by the expression

$$\mathcal{M}[K_\mu(x)] = 2^{p-2}\Gamma\left(\frac{p+\mu}{2}\right)\Gamma\left(\frac{p-\mu}{2}\right) \qquad (6.26)$$

being p the parameter of the Mellin transform and μ is the order.

Using Proposed exercise (26) from Chap. 2

$$J_\mu(x) = \sum_{k=0}^{\infty} \frac{(-1)^k}{\Gamma(\mu+k+1)k!}\left(\frac{x}{2}\right)^{\mu+2k}$$

we introduce the change of variable $x = iz$ that results in

$$J_\mu(iz) = \sum_{k=0}^{\infty} \frac{i^\mu}{\Gamma(\mu+k+1)k!}\left(\frac{z}{2}\right)^{\mu+2k}$$

or, in the following form

$$e^{-i\frac{\pi}{2}\mu} J\left(x\, e^{i\frac{\pi}{2}\mu}\right) = \sum_{k=0}^{\infty} \frac{1}{\Gamma(\mu+k+1)k!}\left(\frac{z}{2}\right)^{\mu+2k}.$$

We define the first kind modified Bessel function, denoted by $I_\mu(z)$, by the expression

$$I_\mu(z) = \sum_{k=0}^{\infty} \frac{1}{\Gamma(\mu+k+1)k!}\left(\frac{z}{2}\right)^{\mu+2k} \qquad (6.27)$$

while the second kind modified Bessel function, denoted by $K_\mu(z)$, is given by

$$K_\mu(z) = \frac{\pi}{2} \frac{I_{-\mu}(z) - I_\mu(z)}{\sin \mu \pi}.$$

It is suitable to make a remark, for it is easy to note, how much $\mu \in \mathbb{Z}$, the quotient that defines the second kind modified Bessel function becomes indeterminate. In this case we must consider the limit $I_n(z) = \lim_{\mu \to n} I_\mu(z)$ with $\mu \in \mathbb{Z}$. In order to show the result of Eq. (6.26), we use the inverse Mellin transform, which allows us to write

$$K_\mu(z) = \frac{1}{2\pi i} \int_{c-i\infty}^{c+i\infty} 2^{p-2} \Gamma\left(\frac{p+\mu}{2}\right) \Gamma\left(\frac{p-\mu}{2}\right) z^{-p} \, dp$$

with $c > \mu$. To calculate this integral in the complex plane, we use the residue theorem. We begin by admitting that $\mu \notin \mathbb{Z}$, so the integrating has simple poles at $p = \mu - 2k$, with $k = 0, 1, 2, \ldots$ Let's close contour of integration in order to leave the singular points to the left of the vertical line $\text{Re}(p) = c$, thus the residues are the only contributions to the integral, as in Fig. 6.1.

Let's explicitly calculate one of the residues, since the other, besides providing the same result, has the calculation in a similar way, suffice to consider $\mu \to -\mu$. So let's calculate the limit

$$\Lambda_1 = \lim_{p \to -\mu-2k} (p + \mu + 2k) \left\{ 2^{p-2} \Gamma\left(\frac{p+\mu}{2}\right) \Gamma\left(\frac{p-\mu}{2}\right) z^{-p} \right\}.$$

Since the product limit is the product of the limits, we can write

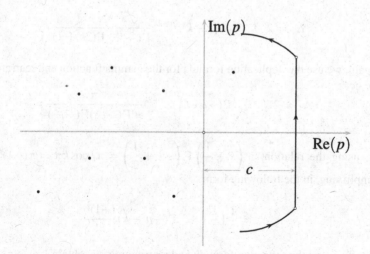

Fig. 6.1 Contour for second kind modified Bessel function

$$\Lambda_1 = \lim_{p \to -\mu-2k} \left\{ 2^{p-2} \Gamma\left(\frac{p-\mu}{2}\right) z^{-p} \right\} \cdot \lim_{p \to -\mu-2k} (p+\mu+2k)\Gamma\left(\frac{p+\mu}{2}\right).$$

For the limit on the left, replace directly, as we have no indetermination, while on the right, first, we must raise indetermination. To calculate the limit on the right, we first introduce the duplication formula for the gamma function, so we can write the product $(p+\mu+2k)\Gamma(p+\mu)$, in the place of the sum $(p+\mu)/2$, from where we get

$$\Lambda_1 = 2^{-\mu-2k-2} \Gamma(-\mu-k) z^{\mu+2k} \cdot \lim_{p \to -\mu-2k} (p+\mu+2k) \left[\frac{\sqrt{\pi}\Gamma(p+\mu)}{2^{p+\mu-1}\Gamma(\frac{p+\mu+1}{2})} \right],$$

or, in the following form

$$\Lambda_1 = 2^{-\mu-2k-2} \Gamma(-\mu-k) z^{\mu+2k} \frac{\sqrt{\pi}2^{1+2k}}{\Gamma(\frac{1-2k}{2})} \cdot \lim_{p \to -\mu-2k} (p+\mu+2k)\Gamma(p+\mu) \cdot$$

For the remaining limit we write separately

$$\lim_{p \to -\mu-2k} (p+\mu+2k)\Gamma(p+\mu) = \frac{(p+\mu+2k)(p+\mu+2k-1)\cdots(p+\mu)\Gamma(p+\mu)}{(p+\mu+2k-1)\cdots(p+\mu)}$$

$$= \lim_{p \to -\mu-2k} \frac{\Gamma(p+\mu+2k+1)}{(p+\mu+2k-1)\cdots(p+\mu)} = \frac{1}{(2k)!}.$$

Returning in the expression to Λ_1 and simplifying, we get

$$\Lambda_1 = 2^{-\mu-1} \Gamma(-\mu-k) z^{\mu+2k} \frac{\sqrt{\pi}}{\Gamma(\frac{1-2k}{2})} \frac{1}{\Gamma(2k+1)}.$$

Again, we use the duplication formula for the gamma function and rearranging, we get

$$\Lambda_1 = 2^{-\mu-2k-1} \Gamma(-\mu-k) z^{\mu+2k} \frac{\pi}{k!\Gamma(\frac{1-2k}{2})\Gamma(\frac{1+2k}{2})},$$

or, using the relation, $\Gamma\left(k+\frac{1}{2}\right)\Gamma\left(-k+\frac{1}{2}\right) = \pi/\cos k\pi = \pi(-1)^k$ and simplifying, in the following form

$$\Lambda_1 = \frac{1}{2}\left(\frac{z}{2}\right)^{\mu+2k} \Gamma(-\mu-k) \frac{(-1)^k}{k!}.$$

Finally, using the reflection formula and rearranging, we obtain

$$\Lambda_1 = -\frac{\pi/2}{\sin\mu\pi}\left(\frac{z}{2}\right)^{\mu+2k} \frac{1}{\Gamma(\mu+k+1)}.$$

In complete analogy to these calculations, the evaluation of the Λ_2, i.e.,

$$\Lambda_2 = \lim_{p \to \mu - 2k} (p - \mu + 2k) \left\{ 2^{p-2} \Gamma \left(\frac{p - \mu}{2} \right) \Gamma \left(\frac{p + \mu}{2} \right) z^{-p} \right\}$$

provides the expression

$$\Lambda_2 = \frac{\pi/2}{\sin \mu \pi} \left(\frac{z}{2} \right)^{-\mu+2k} \frac{1}{\Gamma(-\mu + k + 1)}.$$

Using the two expressions for the limits and remembering that we must add, since in the residue theorem we have the sum of the residues from which follows the expression for the second kind modified Bessel function,

$$K_\mu(z) = \frac{\pi/2}{\sin \mu \pi} \sum_{k=0}^{\infty} \frac{1}{k!} \left[\frac{1}{\Gamma(1 - \mu + k)} \left(\frac{z}{2} \right)^{-\mu+2k} - \frac{1}{\Gamma(1 + \mu + k)} \left(\frac{z}{2} \right)^{\mu+2k} \right].$$

Using the series representation for the first kind modified Bessel function, we can write

$$K_\mu(z) = \frac{\pi}{2} \frac{I_{-\mu}(z) - I_\mu(z)}{\sin \pi \mu}$$

which is exactly the expression that defines the second kind modified Bessel function, thus, returning with the inverse Mellin transform, we obtain

$$\mathcal{M}[K_\mu(x)] = 2^{p-2} \Gamma \left(\frac{p + \mu}{2} \right) \Gamma \left(\frac{p - \mu}{2} \right)$$

which is the desired result   ◇

14. *Semigroup property.* Semigroup property, also known as the law of exponents, plays an important role, for example, whether an operator acting on the left (on the right) in another operator reproduces the same result as acting on the right (left) or even to know whether a differential (integral) operator acting on an integral (differential) operator reproduces the own function in which that product operates, that is, if one is the inverse of the other. Before discussing this result, it is necessary to discuss the inversion in the order of integration, that is, in a product of two integrals which we must integrate first.

We begin with the particular result involving the exchange of the order of integrals whose proof is found in [13]. Let $f(x, y)$ be a continuous function on the variables x and y and the parameters λ, μ and ν such that $0 < \lambda \leq 1, 0 < \mu \leq 1$ and $0 < \nu \leq 1$ then

$$\int_0^1 dx \left\{ \int_0^{1-x} x^{\lambda-1} y^{\mu-1} (1 - x - y)^{\nu-1} f(x, y) \, dy \right\} =$$

$$= \int_0^1 dy \left\{ \int_0^{1-y} x^{\lambda-1} y^{\mu-1} (1 - x - y)^{\nu-1} f(x, y) \, dx \right\}.$$

From this expression, we get the called Dirichlet formula that extends the previous result. To do this, we introduce the real variables ξ and η defined by

$$\xi = a + (b - a)x \quad \text{and} \quad \eta = b + (a - b)y$$

with $a \neq b$, in the previous equality and we rearrange in order to obtain

$$\int_a^b d\xi \left\{ \int_\xi^b (\xi - a)^{\lambda-1} (b - \eta)^{\mu-1} (\eta - \xi)^{\nu-1} F(\xi, \eta) \, d\eta \right\} =$$

$$= \int_a^b d\eta \left\{ \int_a^\eta (\xi - a)^{\lambda-1} (b - \eta)^{\mu-1} (\eta - \xi)^{\nu-1} F(\xi, \eta) \, d\xi \right\} \qquad (6.28)$$

assuming that the function $F(\xi, \eta)$ is a continuous function on the variables ξ and η.

The semigroup property associated with the fractional integral, on the left and right, in the case where the orders are not integers, states that

$$_a\mathcal{J}_\pm^\alpha \, _a\mathcal{J}_\pm^\beta f(x) = \, _a\mathcal{J}_\pm^{\alpha+\beta} f(x)$$

for real constants $\alpha, \beta \geq 0$. In this expression $_a\mathcal{J}_+^\alpha f(x)$ is defined by the expression

$$_a\mathcal{J}_+^\alpha f(x) = \frac{1}{\Gamma(\alpha)} \int_a^x (x - \xi)^{\alpha-1} f(\xi) \, d\xi$$

with $x > a$, the fractional integral on the left, while $_a\mathcal{J}_-^\alpha f(x)$, defined by the expression

$$_a\mathcal{J}_-^\alpha f(x) = \frac{1}{\Gamma(\alpha)} \int_x^b (\xi - x)^{\alpha-1} f(\xi) \, d\xi$$

with $x < b$, the fractional integral on the right. As we have already mentioned, in the case $a = 0$ the integral is said to be a fractional integral in the Riemann sense while, for $a = -\infty$, it is said to be a fractional integral in the Liouville sense. If we do not specify the extreme a, the integral is called a fractional integral in the Riemann-Liouville sense, or simply Riemann-Liouville integral. In order to explain the calculations, we will show the property only for the fractional integral on the left, that is,

$$_a\mathcal{J}_+^\alpha \, _a\mathcal{J}_+^\beta f(x) = \, _a\mathcal{J}_+^{\alpha+\beta} f(x)$$

for real constants $\alpha, \beta \geq 0$.

Let's start with the first member, writing it in terms of the respective expressions for the fractional integrals, then

$$
{}_a\mathcal{I}_+^\alpha \, {}_a\mathcal{I}_+^\beta \, f(x) = \frac{1}{\Gamma(\alpha)} \int_a^x (x - \xi)^{\alpha - 1} f(\xi) \, d\xi \, \frac{1}{\Gamma(\beta)} \int_a^x (x - \eta)^{\beta - 1} f(\eta) \, d\eta.
$$

(6.29)

Using the Dirichlet's formula Eq. (6.28) in the particular case $\lambda = 1$ and $F(\xi, \eta)$ is a function of only the variable η, given by $F(\xi, \eta) = f(\eta)g(\xi)$ with $g(\xi) = 1$, we get

$$
\int_a^x d\xi (x - \xi)^{\alpha - 1} \int_a^\xi (\xi - \eta)^{\beta - 1} f(\eta) \, d\eta = \int_a^x f(\eta) \, d\eta \int_\eta^x (x - \xi)^{\alpha - 1} (\xi - \eta)^{\beta - 1} \, d\xi.
$$

Returning with the previous expression in Eq. (6.29), we can write

$$
{}_a\mathcal{I}_+^\alpha \, {}_a\mathcal{I}_+^\beta \, f(x) = \frac{1}{\Gamma(\alpha)\Gamma(\beta)} \int_a^x f(\eta) \, d\eta \int_\eta^x d\xi \, (x - \xi)^{\alpha - 1} (\xi - \eta)^{\beta - 1}
$$

$$
= \frac{1}{\Gamma(\alpha)\Gamma(\beta)} \int_a^x f(\eta) \, d\eta \, \mathcal{K}(x, \eta)
$$

where $\mathcal{K}(x, \eta)$, interpreted as the kernel of a convolution integral, is given by

$$
\mathcal{K}(x, \eta) = \int_\eta^x d\xi \, (x - \xi)^{\alpha - 1} (\xi - \eta)^{\beta - 1}.
$$

In order to evaluate the integral that appears in the kernel, we first introduce the variable, t, defined by

$$
t = \frac{\xi - \eta}{x - \eta}
$$

and, then, we can write

$$
\mathcal{K}(x, \eta) = (x - \eta)^{\alpha + \beta - 1} \int_0^1 dt \, (1 - t)^{\alpha - 1} t^{\beta - 1},
$$

which can be identified with the beta function, then we have for the kernel

$$
\mathcal{K}(x, \eta) = (x - \eta)^{\alpha + \beta - 1} \frac{\Gamma(\alpha)\Gamma(\beta)}{\Gamma(\alpha + \beta)},
$$

where we use the relation between gamma and beta functions.

Finally, the expression for the product of two fractional integrals is given by

$$
{}_aJ^{\alpha}_+ \, {}_aJ^{\beta}_+ f(x) = \frac{1}{\Gamma(\alpha)\Gamma(\beta)} \int_a^x f(\eta)\, d\eta \left\{ (x-\eta)^{\alpha+\beta-1} \frac{\Gamma(\alpha)\Gamma(\beta)}{\Gamma(\alpha+\beta)} \right\}
$$

$$
= \frac{1}{\Gamma(\alpha+\beta)} \int_a^x f(\eta)\, d\eta (x-\eta)^{\alpha+\beta-1}.
$$

From the preceding expression and from the definition of fractional integral we get

$$
{}_aJ^{\alpha}_+ \, {}_aJ^{\beta}_+ f(x) = \frac{1}{\Gamma(\alpha+\beta)} \int_a^x (x-\eta)^{\alpha+\beta-1} f(\eta)\, d\eta \equiv {}_aJ^{\alpha+\beta}_+ f(x)
$$

which is the desired result.   ◇

15. *Heat transfer.* The equation describing heat transfer, also known as Newton's law, emerges, for example, in the diffusion problem of a solute, which in this context is known as Fick's law. Let us consider a body without internal heat source (not heat) placed in a certain environment. A concrete example is a cup of coffee, that is, the temperature of the coffee decreases until it reaches the temperature of the environment. It is observed that, by heat exchange between the body and the environment, the temperature of the body, denoted by $T(t)$, over time tends to equal the ambient temperature, denoted by T, admitted constant, fact that is expressed by the limit $\lim_{t\to\infty} T(t) = T$.

Newton's law on heat transfer states that: As $T(t)$ approaches T, the velocity at which $T(t)$ tends to T, decreases gradually. Assuming that this velocity $dT(t)/dt$ is proportional to $T(t) - T$, which represents the difference between temperatures, we have

$$
\frac{d}{dt} T(t) = -k[T(t) - T] \tag{6.30}
$$

where $k > 0$ is a constant of proportionality.

From this differential equation, we will consider the fractional analogue, that is, a differential equation of order α, with $0 < \alpha \le 1$ and assuming that the initial condition is such that $T(0) = T_0$, for $t = 0$ we have a constant temperature T_0, that is, we must solve the following fractional differential equation

$$
\frac{d^{\alpha}}{dt^{\alpha}} T(t) = -k[T(t) - T] \tag{6.31}
$$

with $0 < \alpha \le 1$ and k a positive constant, satisfying the condition $T(0) = T_0$. To solve this problem, we introduce the Laplace transform methodology. Denoting the Laplace transform of $T(t)$ by $\mathscr{L}[T(t)] = F(s)$, such that

$$
\mathscr{L}[T(t)] := \int_0^{\infty} e^{-st} T(t)\, dt
$$

with $\mathrm{Re}(s) > 0$ and taking the Laplace transform on both members of the preceding differential equation, we obtain the following algebraic equation in the variable $F(s)$

$$s^\alpha F(s) - s^{\alpha-1} T(0) = -k\left[F(s) - \frac{T}{s}\right].$$

Using the initial condition $T(0) = T_0$ and solving the algebraic equation, we have

$$F(s) = T_0 \frac{s^{\alpha-1}}{s^{\alpha+k}} + kT \frac{s^{-1}}{s^\alpha + k}. \tag{6.32}$$

In order to recover the solution of the initial differential equation, we proceed with the calculation of the inverse Laplace transform, that is,

$$T(t) \equiv \mathscr{L}^{-1}[F(s)] = \frac{1}{2\pi i} \int_{\gamma-i\infty}^{\gamma+i\infty} F(s)\, e^{st}\, ds.$$

Taking the inverse Laplace transform from both members of the Eq. (6.32) and using linearity, we can write

$$T(t) = T_0 \mathscr{L}^{-1}\left[\frac{s^{\alpha-1}}{s^\alpha + k}\right] + kT \mathscr{L}^{-1}\left[\frac{s^{-1}}{s^\alpha + k}\right].$$

To calculate these two inverse Laplace transforms, we use the result obtained in the Exercise (13) of Chap. 4,

$$\mathscr{L}^{-1}\left[\frac{s^{\alpha-\beta}}{s^\alpha + \lambda}\right] = t^{\beta-1} E_{\alpha,\beta}(-\lambda t^\alpha),$$

that allow us to write the following expression,

$$T(t) = T_0 E_\alpha(-kt^\alpha) + kTt^\alpha E_{\alpha,\alpha+1}(-kt^\alpha) \tag{6.33}$$

being $E_\alpha(\cdot)$ and $E_{\alpha,\beta}(\cdot)$ Mittag-Leffler functions with one and two parameters, respectively. Note that the constant T_0 emerges naturally in the solution. Let's discuss the limit case $\alpha = 1$. Taking this limit in the Eq. (6.33), we have

$$T(t) = T_0 E_1(-kt) + kTt E_{1,2}(-kt)$$

or even using the relation involving the Mittag-Leffler functions, that is,

$$E_\alpha(z) = z E_{\alpha,\alpha+1}(z) + 1$$

and remembering that $E_1(z) = \exp(z)$ we obtain

$$T(t) = (T_0 - T)\,e^{-kt} + T$$

which is exactly the solution for the classical problem, that is, the expression that describes the heat transfer when we use the integer order calculus.   ◇

16. *Inverse Laplace transform and the Mittag-Leffler function.* As previously discussed, the Laplace transform of the classical Mittag-Leffler function $f(t) = E_\alpha(-\lambda t^\alpha)$, with $\mathrm{Re}(s) > 0$ and $\lambda > 0$, is given by

$$F(s) = \frac{s^{\alpha-1}}{s^\alpha + \lambda}$$

being $s = \sigma + i\tau$, with $\sigma, \tau \in \mathbb{R}$, the parameter of Laplace transform. This result was obtained by replacing the series expression of the classical Mittag-Leffler function and integrating via gamma function to finally use the geometric series. On the other hand, we know that the inverse Laplace transform recovers the classical Mittag-Leffler function, that is,

$$\mathscr{L}^{-1}\left[\frac{s^{\alpha-1}}{s^\alpha + \lambda}\right] = E_\alpha(-\lambda t^\alpha)$$

for $\mathrm{Re}(\alpha) > 0$ and $\lambda > 0$.

Let's work a little bit with the function $F(s) = s^{\alpha-1}/(s^\alpha + \lambda)$, in the particular case in which the parameter, associated with the order of the function, is such that $0 < \alpha \leq 2$. From now on, we highlight two distinct values $\alpha = 1$ and $\alpha = 2$, the only two values of integer α in the range we are considering for α.

In the first case, $\alpha = 1$, the function has only one simple pole, that is, $s = -\lambda$ is the root of the denominator, which leads us directly to an exponential function,

$$\mathscr{L}^{-1}\left[\frac{1}{s + \lambda}\right] = e^{-\lambda t}.$$

On the other hand, for $\alpha = 2$, the function has two simple poles, that is, the points $s = \pm i\sqrt{\lambda}$ (note that $\lambda > 0$), cancel the denominator which also leads us to a cosine (trigonometric) function, that is,

$$\mathscr{L}^{-1}\left[\frac{s}{s^2 + \lambda}\right] = \cos(\lambda t).$$

For any other value of α in the range we are considering, singularities are not simple poles. With this, it is convenient to divide the interval into two others, $0 < \alpha \leq 1$ and $1 < \alpha \leq 2$. In the first case, the integrand of the inverse Laplace transform has a branch point in $s = 0$ while in the second case, $s = 0$ is a zero of the function.

Since in the numerator we have a power function s^μ which is only defined by $s^\mu = |s|^\mu e^{i\,\arg(s)} = r^\mu e^{i\theta}$, being the module given by $r = |s| \geq 0$ and the argument such that $\pi < \arg(s) = \theta < \pi$, that is, with the cut line along the negative real half axis.

So, let's separate $f(t)$ into two parcels, $g_\alpha(t)$ and $h_\alpha(t)$, such that $f(t) \equiv g_\alpha(t) + h_\alpha(t)$. Through a continuous deformation, we conduct the Bromwich contour, in a Hankel contour (same as in Chap. 4), denoted by $\mathsf{Ha}(\varepsilon)$, begining at $-\infty$, along the lower side of the negative real half axis, bypassing, in the positive direction, the circumference $|s| = |\lambda|^{\frac{1}{\alpha}} = r$ and ending at $-\infty$, along the upper side of the negative real half axis. So we can write

$$f(t) = g_\alpha(t) + h_\alpha(t)$$

being $g_\alpha(t)$ defined by the following integral in the complex plane

$$g_\alpha(t) = \frac{1}{2\pi i} \int_{\mathsf{Ha}(\varepsilon)} e^{st} \frac{s^{\alpha-1}}{s^\alpha + \lambda} \, ds$$

where $\mathsf{Ha}(\varepsilon)$ denotes a *loop* consisting of the circle $|s| = \varepsilon$ with $\varepsilon \to 0$ and the two edges (bottom and top) of the cut line on the negative real half axis.

On the other hand, $h_\alpha(t)$ is defined by

$$h_\alpha(t) = \sum_j e^{\mathfrak{s}\,t} \mathrm{Res} \left[\frac{s^{\alpha-1}}{s^\alpha + \lambda} \right]_{s=\mathfrak{s}}$$

where, on the second member, we have the sum of the residues and $\mathfrak{s}$, with $j = 0, 1, 2 \ldots$ are the poles of the function $F(s)$. Calculating the residues, all simple poles, we have

$$\mathrm{Res} \left[\frac{s^{\alpha-1}}{s^\alpha + \lambda} \right]_{s=\mathfrak{s}} = \lim_{s \to \mathfrak{s}} (s - \mathfrak{s}) \frac{s^{\alpha-1}}{s^\alpha + \lambda} = \lim_{s \to \mathfrak{s}} \frac{1}{\alpha s^{\alpha-1}} \cdot s^{\alpha-1} = \frac{1}{\alpha}$$

from where it follows, then,

$$h_\alpha(t) = \frac{1}{\alpha} \sum_j e^{\mathfrak{s}\,t} .$$

Now the poles are such that $\mathfrak{s} = \lambda^{\frac{1}{\alpha}} \exp[(2j + 1)\frac{\pi}{\alpha}i]$ with $j = 0, 1, 2 \ldots$ however, those that actually contribute to the residue are only those that are situated on the Riemann main sheet, that is, with such an argument that $-\pi < \arg(\mathfrak{s}) < \pi$. In the first case, where $0 < \alpha < 1$ there are no poles, since for every integer j we have $|\arg(\mathfrak{s})| = |2j + 1|\frac{\pi}{\alpha} > \pi$ from where it follows $h_\alpha(t) = 0$, then $f(t) = g_\alpha(t)$.

In the second case, $1 < \alpha < 2$ there are two poles that contribute, that is, $j = 0$ which implies $s_0 = \exp(i\pi/\alpha)$ and $j = -1$ which imply $s_{-1} = \exp(-i\pi/\alpha)$, both located on the left half-plane. Note that, $|\arg(s_0)| = |\arg(s_{-1})| = \frac{\pi}{\alpha} < \pi$, because $\alpha > 1$, from where we can write

$$h_\alpha(t) = \frac{1}{\alpha}\left[e^{t\exp(i\pi/\alpha)} + e^{t\exp(-i\pi/\alpha)}\right].$$

Using the Euler relation, rearranging and simplifying, we have

$$h_\alpha(t) = \frac{2}{\alpha}e^{t\cos(\pi/\alpha)}\cos[t\sin(\pi/\alpha)].$$

Note that in the limiting case $\alpha = 2$ we have $h_2(t) = \cos t$. In addition, we have oscillations with angular frequency $\omega(\alpha) = \sin(\pi/\alpha)$ whose amplitude decays exponentially with reason $\xi(\alpha) = |\cos(\pi/\alpha)|$. In the first case, the function $g_\alpha(t)$ is completely monotonous while, in the second case, we must insert a minus sign, that is, $-g_\alpha(t)$, if we want the function to be completely monotonous. Finally, in both cases, $g_\alpha(t)$ tends to zero as $t \to \infty$, above the cut line in the first case and below the cut line in the second case. For more details, see [14].   ◇

17. *Spectral function. Titchmarsh's formula.* The spectral function is another way of explaining the function, albeit in terms of an integral. The expression that provides the spectral function is known as Titchmarsh's formula. Consider the functions $f(t)$ and $F(s)$ such that $F(s)$ is the Laplace transform of $f(t)$, provided it exists, while $f(t)$ is the inverse Laplace transform of $F(s)$, too, provided it is defined. The Titchmarsh's formula relates these two functions and is given by

$$f(t) = -\frac{1}{\pi}\int_0^\infty e^{-rt}\text{Im}\{F(re^{i\pi})\}\,dr$$

with $r = \text{Re}(s)$, being s the parameter of Laplace transform and $\text{Im}(\cdot)$ is the immaginary part of $F(s)$, evaluated in $s = re^{i\pi}$, that is, along the radius $re^{i\pi}$ with $r \geq 0$, a cut branch of the $F(s)$. The expression

$$-\frac{1}{\pi}\text{Im}\{F(re^{i\pi})\} = K(r) \qquad\qquad (6.34)$$

is known as a spectral function and is always positive.

As a particular case, we will consider a Mittag-Leffler type function, as discussed in the Proposed exercise (14) of Chap. 4. Let $\mathscr{E}_{\alpha,\beta}(t,\lambda)$ be the function given by

$$\mathscr{E}_{\alpha,\beta}(t,\lambda) = t^{\beta-1}E_{\alpha,\beta}(-\lambda t^\alpha)$$

where $E_{\alpha,\beta}(\cdot)$ is the Mittag-Leffler function with two parameters, with the imposition on the parameter $0 < \alpha \leq \beta < 1$ and λ a positive real constant. Let's show that

$$t^{\beta-1}E_{\alpha,\beta}(-\lambda t^\alpha) = \int_0^\infty e^{-rt}K_{\alpha,\beta}(r,\lambda)\,dr$$

with $0 < \alpha \le \beta < 1$, being the spectral function, $K_{\alpha,\beta}(r, \lambda)$, given by

$$K_{\alpha,\beta}(r, \lambda) = \frac{1}{\pi} \frac{\lambda \sin[(\beta - \alpha)\pi] + r^{\alpha} \sin \beta\pi}{r^{2\alpha} + 2\lambda r^{\alpha} \cos \alpha\pi + \lambda^2} r^{\alpha-\beta} \ge 0.$$

Then, using Eq. (6.34), we must show that

$$-\mathrm{Im}\left\{\frac{s^{\alpha-\beta}}{s^{\alpha} + \lambda}\right\}_{s=re^{i\pi}} = \pi K_{\alpha,\beta}(r, \lambda).$$

We start with the quotient

$$\Omega \equiv \frac{s^{\alpha-\beta}}{s^{\alpha} + \lambda} = \frac{(re^{i\pi})^{\alpha-\beta}}{(re^{i\pi})^{\alpha} + \lambda} = \frac{\cos(\alpha - \beta)\pi + i \sin(\alpha - \beta)\pi}{r^{\alpha}(\cos \alpha\pi + i \sin \alpha\pi) + \lambda} r^{\alpha-\beta}$$

where in the last passage we used the Euler formula. By multiplying numerator and denominator by the conjugate of the denominator, we can write

$$\Omega = \frac{\cos(\alpha - \beta)\pi + i \sin(\alpha - \beta)\pi}{(r^{\alpha} \cos \alpha\pi + \lambda) - ir^{\alpha} \sin \alpha\pi} \cdot \frac{(r^{\alpha} \cos \alpha\pi + \lambda) + ir^{\alpha} \sin \alpha\pi}{(r^{\alpha} \cos \alpha\pi + \lambda) - ir^{\alpha} \sin \alpha\pi} r^{\alpha-\beta}.$$

Taking only the imaginary part and changing the signal, we have

$$-\mathrm{Im}\left\{\frac{s^{\alpha-\beta}}{s^{\alpha} + \lambda}\right\}_{s=re^{i\pi}} = -\frac{r^{\alpha} \sin(\alpha - \beta - \alpha)\pi + \lambda \sin(\alpha - \beta)\pi}{r^{2\alpha} + 2\lambda r^{\alpha} \cos \alpha\pi + \lambda^2} r^{\alpha-\beta}$$

from where, by simplifying, we obtain for the spectral function [15, 16]

$$K_{\alpha,\beta}(r, \lambda) = \frac{1}{\pi} \frac{\lambda \sin[(\beta - \alpha)\pi] + r^{\alpha} \sin \beta\pi}{r^{2\alpha} + 2\lambda r^{\alpha} \cos \alpha\pi + \lambda^2} r^{\alpha-\beta}$$

which is the desired result.   ◇

18. *Spectral function for the classical Mittag-Leffler function.* Using the result of the precedent, obtain the spectral function associated with the classical Mittag-Leffler function. For this, it suffices that we consider $\beta = 1$, because

$$E_{\alpha,1}(-\lambda t^{\alpha}) = E_{\alpha}(-\lambda t^{\alpha})$$

where $E_{\alpha}(\cdot)$ is the classical Mittag-Leffler function. Then, taking $\beta = 1$ in the expression that provides the spectral function and simplifying, we can write

$$K_{\alpha}(r, \lambda) = \frac{1}{\pi} \frac{\lambda \sin \alpha\pi}{r^{2\alpha} + 2\lambda r^{\alpha} \cos \alpha\pi + \lambda^2} r^{\alpha-1}.$$

which is the desired result.   ◇

19. *Derivatives of integer order and the Grünwald-Letnikov derivative.* Using the
    definition of first derivative, obtain an expression for the n derivative with
    $n \in \mathbb{N}$ and extend it to the order $\alpha \in \mathbb{R}$ to introduce the derivative according
    to the Grünwald-Letnikov formulation. To do this, consider $t \in \mathbb{R}$, the interval
    $a \leq t \leq b$, with $b > a$ and a function $f(t)$ that admits derivative up to order n ,
    with $n \in \mathbb{N}$.

From the definition of derivative (order one) we have

$$\frac{\mathrm{d}}{\mathrm{d}t} f(t) = \lim_{h \to 0} \frac{f(t) - f(t-h)}{h}.$$

Note that, unlike the way in which we have in the numerator the difference of an
increased function, $f(t+h)$, and the function, $f(t)$, here we have the function
difference, $f(t)$, and the function with the increment $t - h$. These two ways
differentiate the left and right derivatives. From the previous one, we have for
the derivative of order two

$$\begin{aligned}
\frac{\mathrm{d}^2}{\mathrm{d}t^2} f(t) &= \lim_{h \to 0} \frac{f'(t) - f'(t-h)}{h} \\
&= \lim_{h \to 0} \frac{1}{h} \left\{ \frac{f(t) - f(t-h)}{h} - \frac{f(t-h) - f(t-2h)}{h} \right\} \\
&= \frac{1}{h^2} \{ f(t) - 2f(t-h) + f(t-2h) \} \cdot
\end{aligned}$$

Using the finite induction process, we can write for the derivative of order n

$$\frac{\mathrm{d}^n}{\mathrm{d}t^n} f(t) = \lim_{h \to 0} \frac{1}{h^n} \sum_{j=0}^{n} (-1)^j \frac{n!}{(n-j)! j!} f(t - jh)$$

where the factorial quotient is the binomial coefficient and $hn = t - a$.

Let now be the parameter $n = \alpha \in \mathbb{R}$. So, instead of the factorial, we must use
the gamma function, since this is the generalization of the factorial, we can
then write the fractional derivative, according to the formulation proposed by
Grünwald-Letnikov that, in this case, unlike the derivatives that were introduced
previously, through the integrals, is given in terms of a limit

$$\frac{\mathrm{d}^\alpha}{\mathrm{d}t^\alpha} f(t) = \lim_{h \to 0} \frac{1}{h^\alpha} \sum_{j=0}^{n} (-1)^j \frac{\Gamma(\alpha+1)}{\Gamma(\alpha - j + 1) j!} f(t - jh) \tag{6.35}$$

so that we have $nh = t - a$. Another way of denoting such a derivative is $_a D_t^\alpha$
where a is the lower bound, $\alpha \in \mathbb{R}$ is the order of the derivative, and t is the
independent variable. Further, if $\alpha = m$, with $m \in \mathbb{N}$, we have the derivative of
order m, while for $\alpha = -m$, with $m \in \mathbb{N}$, the integral of order m.

As we have seen, the derivative is a nonlocal operator, analogous to integration, that is, it depends on the value of the f function over the entire range of a to t. This can be seen from the factor $f(t - jh)$ in the sum of Eq. (6.35), showing that since j ranges from zero to $(t - a)/h$, the f argument can vary below zero. Finally, we write Eq. (6.35) as follows

$$_aD_t^\alpha f(t) = \sum_{k=0}^m \frac{(t-a)^{-\alpha+k}}{\Gamma(-\alpha+k+1)} f^{(k)}(a) + \frac{1}{\Gamma(-\alpha+m+1)} \int_a^t (t-\tau)^{m-\alpha} f^{(m+1)}(\tau)\, d\tau$$

provided that the derivatives $f^{(k)}(t)$ with $k = 1, 2, 3, \ldots, m+1$ be continuous in the closed interval $[a, t]$ and m an integer satisfying the condition $m + 1 > \alpha$. The lowest possible value of m is such that the double inequality $m < \alpha < m + 1$ is satisfied [17].   ◇

20. *Grünwald-Letnikov derivative. A particular case.* Consider the function $f(t) = (t - a)^2$. Calculate the derivative of Grünwald-Letnikov, from the previous exercise and with the order $\alpha = m = 2$, $\alpha = m = -2$ and $\alpha = 1/2$. Note that $a = 0$. We star with $\alpha = m = 2$. Substituting in the expression of the previous exercise and noting that only the part involving the series contributes, we can write

$$_aD_t^2(t-a)^2 = \sum_{k=0}^2 \frac{(t-a)^{-2+k}}{\Gamma(-2+k+1)} f^{(k)}(a) = \sum_{k=0}^2 \frac{(t-a)^{-2+k}}{\Gamma(k-1)} f^{(k)}(a).$$

From the previous expression, only the term $k = 2$ contributes, so

$$_aD_t^2(t-a)^2 = f^{(2)}(a) = 2.$$

This value is nothing more than the second derivative of the function $f(t) = (t - a)^2$ calculated at the point $t = a$. Now let us consider $\alpha = m = -2$ which has interpretation of the repeated integral and which we call the order. In order to maintain the rating, both in relation to Chap. 5 and the previous exercise, we highlight equality

$$_aD_t^{-2}(t-a)^2 = {_a}\mathscr{I}_t^2(t^2).$$

Only the integral part contributes, so we can write

$$_aD_t^{-2}(t-a)^2 = \frac{1}{\Gamma(1)} \int_a^t f^{(-2+1)}(\tau)\, d\tau = \int_a^t f^{(-1)}(\tau)\, d\tau$$

that is, the integral of an integral. As shown in Chap. 5, this integral can be given by the expression

$$_aD_t^{-2}(t-a)^2 = \int_a^t f^{(-1)}(\tau)\, d\tau = \int_a^t (t-\xi)(\xi-a)^2\, d\xi.$$

This is an immediate integral, from where it follows

$$_aD_t^{-2}(t-a)^2 = {_a}\mathscr{I}_t^2(t^2) = \frac{(t-a)^4}{12}$$

which is the desired result.   ◇

21. *Maxwell model of fractional order*. In the study of fractional viscoelastic models, we find the constitutive equation, in the relaxation state, given by [18]

$$\sigma(t) + \tau^{\alpha-\beta}\frac{d^{\alpha-\beta}}{dt^{\alpha-\beta}}\sigma(t) = E\tau^{\alpha}\frac{d^{\alpha}}{dt^{\alpha}}\varepsilon(t),$$

with $\alpha, \beta \in \mathbb{R}$ such that $0 < \alpha \leq \beta < 1$.

In this equation, $\sigma(t)$ denotes the voltage, $\varepsilon(t)$ the tension, E is the Yang module and $\tau = \eta/E$ with η the shear. Solve the equation by admitting the stress a Heaviside (step) function. Then, first, the fractional derivative of the Heaviside function,

$$\frac{d^{\alpha}}{dt^{\alpha}}\varepsilon(t) = \frac{d^{\alpha}}{dt^{\alpha}}1 = \frac{t^{\alpha}}{\Gamma(1-\alpha)}$$

from where, returning in the constitutive equation, we can write

$$\sigma(t) + \tau^{\alpha-\beta}\frac{d^{\alpha-\beta}}{dt^{\alpha-\beta}}\sigma(t) = E\tau^{\alpha}\frac{t^{\alpha}}{\Gamma(1-\alpha)}.$$

Let us use the Laplace transform methodology to solve the differential equation. We introduce the notation

$$\int_0^{\infty} e^{-st}\sigma(t)\,dt = F(s)$$

from where it follows to the differential equation

$$F(s) + \tau^{\alpha-\beta}\int_0^{\infty} e^{-st}\frac{d^{\alpha-\beta}}{dt^{\alpha-\beta}}\sigma(t)\,dt = \frac{E\tau^{\alpha}}{\Gamma(1-\alpha)}\int_0^{\infty} e^{-st}t^{-\alpha}\,dt.$$

By imposing the condition $\sigma(0) = 0$, introducing the change of variable $st = u$ in the second member and using the definition of the gamma function, we obtain the following algebraic equation in the variable $F(s)$

$$F(s) + \tau^{\alpha-\beta}s^{\alpha-\beta}F(s) = E\tau^{\alpha}s^{\alpha-1}$$

whose solution is given by

$$F(s) = E\tau^{\beta}\frac{s^{\alpha-1}}{s^{\alpha-\beta} + \left(\frac{1}{\tau}\right)^{\alpha-\beta}}.$$

In order to recover the solution of the differential equation, we use the inverse Laplace transform, that is, we can write

$$\sigma(t) = E\tau^\beta \mathscr{L}^{-1}\left[\frac{s^{\alpha-1}}{s^{\alpha-\beta}+\left(\frac{1}{\tau}\right)^{\alpha-\beta}}\right]$$

which provides, using the Exercise (13) of Chap. 4,

$$\sigma(t) = E\left(\frac{t}{\tau}\right)^{-\beta} E_{\alpha-\beta,1-\beta}\left[\left(-\frac{t}{\tau}\right)^{\alpha-\beta}\right]$$

which is the desired result   ◇

22. *Viscoelastic object of fractional order. Periodic voltage.* Let us consider the constitutive equation

$$\sigma(t) = E^{1-\alpha}\eta^\alpha\frac{d^\alpha}{dt^\alpha}\varepsilon(t)$$

with $0 < \alpha < 1$ and coefficients as in the previous exercise.

Using the Caputo derivative, and assuming a sinusoidal voltage $\varepsilon = \varepsilon_0 \sin(\omega t)$ with ε_0 and ω positive constants, we can write for the voltage [18]

$$\sigma(t) = E^{1-\alpha}\eta^\alpha\left\{\frac{\varepsilon_0\omega}{\Gamma(1-\alpha)}\int_0^t (t-\xi)^{-\alpha}\cos(\omega\xi)\,d\xi\right\}.$$

In order to perform the integration we will use the Laplace transform methodology, with parameter s, then

$$\sigma(s) = E^{1-\alpha}\eta^\alpha\frac{\varepsilon_0\omega}{\Gamma(1-\alpha)}\mathscr{L}\left[\frac{\varepsilon_0\omega}{\Gamma(1-\alpha)}\int_0^t (t-\xi)^{-\alpha}\cos(\omega\xi)\,d\xi\right]$$

with $\sigma(s) = \int_0^\infty e^{-st}\sigma(t)\,dt.$

We have identified this integral as the convolution product, from which we can write

$$\sigma(s) = E^{1-\alpha}\eta^\alpha\frac{\varepsilon_0\omega}{\Gamma(1-\alpha)}F(s)G(s)$$

where $F(s) = \mathscr{L}[t^{-\alpha}]$ and $G(s) = \mathscr{L}[\cos\omega t]$. From known relations we can write, respectively

$$F(s) = \Gamma(1-\alpha)\,s^\alpha \quad \text{and} \quad G(s) = \frac{1}{s^2+\omega^2}$$

from where it follows to the Laplace transform of the voltage

$$\sigma(s) = E^{1-\alpha}\eta^{\alpha}\varepsilon_0\omega\frac{s^{\alpha}}{s^2+\omega^2}.$$

To recover the solution of the initial differential equation, we consider the inverse Laplace transform, that is,

$$\sigma(t) = \varepsilon\omega E^{1-\alpha}\eta^{\alpha}\mathscr{L}^{-1}\left[\frac{s^{2-(2-\alpha)}}{s^2+\omega^2}\right]$$

which, with the Exercise (13) of Chap. 4, provides

$$\sigma(t) = \varepsilon\omega E^{1-\alpha}\eta^{\alpha}E_{2,2-\alpha}(-\omega^2 t^2)$$

where $E_{\alpha,\beta}(\cdot)$ is a Mittag-Leffler function with two parameters.   ◇

23. *Iterated integrals.* As we discussed in Chap. 4, using the notation $\mathscr{J}^{\mu}$, to denote the integral of order μ, let us here, through power-type functions, discuss integration as a derivative of negative order. Then, in the case where the order of the derivative is a negative number, considered as an integration, we can write for the first three

$$\frac{d^{-1}}{dx^{-1}}f(x) = \int_0^x f(\xi)\,d\xi$$
$$\frac{d^{-2}}{dx^{-2}}f(x) = \int_0^x \int_0^{\xi} f(\xi)\,d\xi\,d\xi_1 = \int_0^x (x-\xi)f(\xi)\,d\xi$$
$$\frac{d^{-3}}{dx^{-3}}f(x) = \int_0^x \int_0^{\xi} \int_0^{\xi_1} f(\xi)\,d\xi\,d\xi_1\,d\xi_2 = \frac{1}{2!}\int_0^x (x-\xi)^2 f(\xi)\,d\xi$$

where the last equalities are obtained after integration by parts.

Before we write an expression for the derivative of $-\mu$, we retrieve the case of the derivative of order μ from the power-type function. Consider a power-type function,

$$f(x) = x^n$$

with $n = 0, 1, 2, \ldots$ Calculating the derivative of order $k = 0, 1, 2, \ldots$ we can write

$$\frac{d^k}{dx^k}\left(x^n\right) = \frac{n!}{(n-k)!}x^{n-k}. \tag{6.36}$$

Thus, with $k = \mu \in \mathbb{R}$ we have the expression

$$\frac{d^{\mu}}{dx^{\mu}}\left(x^n\right) = \frac{n!}{\Gamma(n-\mu+1)}x^{n-\mu}$$

that is, the derivative of order μ of a power function.

In analogy to the generalization of the derivative of order μ, we have

$$\frac{d^{-\mu}}{dx^{-\mu}}f(x) = \frac{1}{\Gamma(\mu)} \int_0^x (x - \xi)^{\mu-1} f(\xi) \, d\xi \qquad (6.37)$$

which is the generalization of the Cauchy formula for repeated integrals [19], that is, the integral of order μ or the derivative of order $-\mu$. Note that Eq. (6.37) can be interpreted in two distinct ways, namely: we first consider a derivative and then an integral, using the expression given by Eq. (6.37), right hand, or first an integral, using Eq. (6.37) and then a derivative, called the left hand definition. These two ways of calculating a derivative in general do not provide the same result, as will be clear from an example. Finally, note that Eq. (6.37) makes clear the nonlocal character of operations involving derivatives (and integrals) of non-integer orders.

Suppose we want to calculate the derivative of order 3/2 of a given function, for example $f(x)$. One possibility is to calculate, using Eq. (6.37) with $\mu = 1/2$ and then calculate the derivative of order two. Another way is to calculate the derivative of order two and then use Eq. (6.37) with $\mu = 1/2$. Explicitly, $f(x) = x$ is allowed in order to retrieve the result as proposed by l'Hôpital and answered by Leibniz.

First, we calculate the integral, using Eq. (6.37) with $\mu = 1/2$, that is,

$$\frac{d^{-1/2}}{dx^{-1/2}}(x) = \frac{1}{\Gamma(1/2)} \int_0^x (x - \xi)^{1/2-1}(\xi) \, d\xi$$

whose integration provides

$$\frac{d^{-1/2}}{dx^{-1/2}}(x) = \frac{4}{3\sqrt{\pi}} x^{3/2}$$

whose second derivative leads to the desired result

$$\frac{d^2}{dx^2} \frac{d^{-1/2}}{dx^{-1/2}}(x) = \frac{1}{\sqrt{\pi x}}, \qquad (6.38)$$

which is, after a full-order integration (order one), exactly the result questioned by l'Hôpital and answered by Leibniz.

In the second way, the second derivative which results in zero is first calculated and then using Eq. (6.37) is also zero. In this case, different results, which justifies the order being fundamental. Further, the result obtained in Eq. (6.38) is exactly the same as would be obtained with the use of Eq. (6.36) with the order of the derivative equal to 3/2. It is concluded that this operation is not commutative, that is, derive first to integrate is not always the same as integrating first and then deriving. ◇

24. *Fractional derivative and fractional integral.* Let us consider two functions, the exponential and the power. We show that the order of operators, derivative and integral, does not change the result. We start, for example, with the exponential

function $f(x) = e^{ax}$ with $a \in \mathbb{R}$ and $x > 0$. For $n \in \mathbb{N}$ it is well-known the result

$$\mathsf{D}^n f(x) = a^n f(x)$$

where we introduce the notation $\mathsf{D} = \dfrac{d}{dx}$. On the other hand, also known, we have a similar expression for the iterated integration, that is, with the notation D^{-n}, then

$$\mathsf{D}^{-n} f(x) = \frac{f(x)}{a^n}.$$

We will calculate, explicitly, still considering the exponential function, the effect of first operate D^n and after D^{-n}, that is,

$$\mathsf{D}^{-n}\,\mathsf{D}^n f(x) = \mathsf{D}^n \left\{ a^n f(x) \right\} = a^n \frac{f(x)}{a^n} = f(x)$$

it recovers its own function, that is, one operator is the inverse of the other. The same result is obtained if we operate in reverse order. This result can be extended to a non-integer order, $n = \mu \in \mathbb{R}$, then

$$\mathsf{D}^{-\mu}\,\mathsf{D}^\mu f(x) = \mathsf{D}^\mu\,\mathsf{D}^{-\mu} f(x) = f(x).$$

Let us now consider a power function $f(x) = x^\nu$ with $x > 0$ and $n \in \mathbb{R}$ and compute the derivative of $\mu \in \mathbb{R}$, that is, by extending to non-integers. We have, for the derivative

$$\mathsf{D}^\mu x^\nu = \frac{\Gamma(\nu + 1)}{\Gamma(\nu - \mu + 1)} x^{\nu - \mu}$$

which, in the case of a integer-order derivative with integer exponent, the gamma functions become the factorial. Exchanging $\mu \to -\mu$, that is, instead of the derivative we now have an integration, we can write

$$\mathsf{D}^{-\mu} x^\nu = \frac{\Gamma(\nu + 1)}{\Gamma(\nu + \mu + 1)} x^{\nu + \mu}.$$

In analogy to the case of the exponential, except that we reverse the order of the operators, we will recover the function through the inverse

$$\mathsf{D}^\mu \mathsf{D}^{-\mu} x^\nu = \mathsf{D}^\mu \left[\frac{\Gamma(\nu + 1)}{\Gamma(\nu + \mu + 1)} x^{\nu + \mu} \right] = \frac{\Gamma(\nu + 1)}{\Gamma(\nu + \mu + 1)} \mathsf{D}^\mu x^{\nu + \mu} = x^\nu.$$

For these two functions we obtain the same expression in order to recover the original function from the operation differentiation (integration) and then the integration (differentiation). Looking from another point of view, through the Fourier transform, we have that the exponential admits a Fourier transform while the power only admits if the interval is finite. In short, the differentiation of non-integer order, in analogy to integration, is not unique and also not local, that is, it depends on previous terms   ◇

25. *Grünwald-Letnikov derivative*. Let us introduce another way of presenting the fractional derivative, here, through a limit process, in analogy to the integer derivative $n \in \mathbb{N}$. Let $x \in \mathbb{R}$ and $n \in \mathbb{N}$. The n-th derivative of $f(x)$ is given by expression

$$\mathbf{D}^n f(x) = \lim_{h \to 0} \frac{1}{h^n} \sum_{k=0}^{n} (-1)^k \binom{n}{k} f(x - kh).$$

In order to extend to a $\mu \in \mathbb{R}$ order, in addition to the binomial coefficient becoming a gamma function quotient, we must take into account the upper end of the sum, which will be written $[x/h]$, where the bracket denotes the integer part of the quotient, then

$$^{\mathsf{GL}}_{\ 0}\mathbf{D}^\mu_x f(x) = \lim_{h \to 0} \frac{1}{h^\mu} \sum_{k=0}^{[x/h]} (-1)^k \frac{\Gamma(\mu+1)}{k!\Gamma(\mu-k+1)} f(x - kh). \tag{6.39}$$

This expression is known by Grünwald-Letnikov derivative and is quite useful in numeric problems, since the sum can be "cut" in a particular term, according to the problem in question. This notation will be used only in the definition, in order not to overload indexes, and means GL, denoting Grünwald-Letnikov; 0 that the expansion is considering $x_0 = 0$; x to mention that the derivative is relative to the independent variable x and μ denotes the order of the derivative. Let us now obtain the integral of order μ, from the previous expression, with simplified notation. To do this, we enter $\mu \to -\mu$ from where it follows

$$\mathbf{D}^{-\mu} f(x) = \lim_{h \to 0} \frac{1}{h^{-\mu}} \sum_{k=0}^{[x/h]} (-1)^k \frac{\Gamma(-\mu+1)}{k!\Gamma(-\mu-k+1)} f(x - kh). \tag{6.40}$$

We begin by manipulating the quotient of gamma functions, using the reflection formula, we can write

$$\frac{\Gamma(-\mu+1)}{k!\Gamma(-\mu-k+1)} = \frac{1}{k!} \frac{\pi}{\Gamma(\mu)\sin\pi\mu} \frac{\Gamma(\mu+k)\sin(\mu+k)\pi}{\pi}$$

which, by simplifying, provides

$$\frac{\Gamma(-\mu+1)}{k!\Gamma(-\mu-k+1)} = \frac{(-1)^k}{k!} \frac{\Gamma(\mu+k)}{\Gamma(\mu)}.$$

Substituting this result into Eq. (6.40), we obtain

$$D^{-\mu} f(x) = \lim_{h \to 0} h^{\mu} \sum_{k=0}^{[x/h]} \frac{\Gamma(\mu + k)}{k! \Gamma(\mu)} f(x - kh)$$

which is the expression for the iterated integral of Grünwald-Letnikov of order μ.

Finally, we can show that the law of exponents is valid for the Grünwald-Letnikov derivative (integral), as well as a generalization of the Leibniz rule for the product of two functions [20].   ◇

26. *Memory effect.* As we have already mentioned, the fractional derivative is a nonlocal operator, which leads to the called memory effect, which, as we shall see below, carries with it what happened in the past, that is, there is a dependence of times before the time considered initial. In order to introduce what we call the memory effect name, let us consider a particular initial value problem, composed by a fractional ordinary differential equation

$$\mathbb{D}^{\mu} y(x) = f(x)$$

with $x \in \mathbb{R}$ and $0 < \mu \leq 1$, satisfying the initial condition $y(0) = 0$. The derivative operator $\mathbb{D}$ is such that $\mathbb{D} \equiv d/dx$. The general case can be found in [9].

The solution of the initial value problem can be obtained through the Laplace transform methodology and the use of the convolution theorem, then

$$y(x) = \frac{1}{\Gamma(\mu)} \int_0^x (x - \xi)^{\mu-1} f(\xi) \, d\xi. \tag{6.41}$$

From Eq. (6.41), we consider two distinct values x_1 and x_2 such that $x_1 < x_2$. One can then write, separating in two intervals

$$y(x_2) = \frac{1}{\Gamma(\mu)} \int_0^{x_2} (x_2 - \xi)^{\mu-1} f(\xi) \, d\xi$$

and

$$y(x_1) = \frac{1}{\Gamma(\mu)} \int_0^{x_1} (x_1 - \xi)^{\mu-1} f(\xi) \, d\xi,$$

whose difference, $y(x_2) - y(x_1)$, provides

$$y(x_2) - y(x_1) = \frac{1}{\Gamma(\mu)} \int_0^{x_2} (x_2 - \xi)^{\mu-1} f(\xi) \, d\xi - \frac{1}{\Gamma(\mu)} \int_0^{x_1} (x_1 - \xi)^{\mu-1} f(\xi) \, d\xi$$

which, by rearranging, can be written in the form

$$y(x_2) - y(x_1) = \frac{1}{\Gamma(\mu)} \int_0^{x_1} [(x_2 - \xi)^{\mu-1} - (x_1 - \xi)^{\mu-1}] f(\xi) \, d\xi +$$

$$\frac{1}{\Gamma(\mu)} \int_{x_1}^{x_2} (x_2 - \xi)^{\mu-1} f(\xi) \, d\xi \qquad (6.42)$$

since $x_1 < x_2$.

From Eq. (6.42), we can conclude that: the first integral involves values prior to x_1, while the second integral only values between x_1 and x_2. For all values of $\mu \neq 1$, the two parcels contribute, whereas in the case $\mu = 1$, only the second part contribute. In this case, there is no dependence on the value of the first integral from which we conclude that integer-order equations model systems without memory. On the other hand, in the case where $0 < \mu < 1$ the first parcel also contributes, that is to say, there is dependence of the two parcels, from which it is concluded that equations of non integer order model systems with memory, that is, has a memory effect relative to the integral in the range of zero to x_1, before (in the past) x_1.   ◇

27. *Hilfer derivative.* Although the book, as mentioned at the begining, is devoted to some particular definitions of derivatives, the called Hilfer fractional derivative operator [21] will be presented here, as it contains as limit cases the Riemann-Liouville and Caputo fractional derivatives. Let us then, after presenting the definition, retrieve the two limit cases, as well as mention that this derivative can be given in terms of the Riemann-Liouville derivative or the Caputo derivative. Finally, show that the Hilfer derivative retrieves the ordinary derivative when the order is a positive integer. For more properties we suggest the Ref. [20, 21]   ◇

Definition 6.1 (*Hilfer fractional derivative operator*) The Hilfer fractional derivative operator, of order α, with $0 < \alpha < 1$ and type β, with $0 \leq \beta \leq 1$, of a function $f(t)$, well-behaved, is given by

$$\mathcal{D}_{a\pm}^{\alpha,\beta} f(t) = \pm \mathcal{J}_{a\pm}^{\beta(1-\alpha)} \mathcal{D} \mathcal{J}_{a\pm}^{(1-\beta)(1-\alpha)} f(t)$$

with $\mathcal{D} = \dfrac{d}{dt}$ being the first order derivative and the integrals such that

$$\mathcal{J}_{a+}^{\alpha} f(t) = \frac{1}{\Gamma(\alpha)} \int_a^t f(\tau)(t-\tau)^{\alpha-1} \, d\tau, \qquad t \geq a$$

and

$$\mathcal{J}_{b-}^{\alpha} f(t) = \frac{1}{\Gamma(\alpha)} \int_t^b f(\tau)(\tau-t)^{\alpha-1} \, d\tau, \qquad t \leq b$$

the Riemann-Liouville fractional integrals on the left and right, respectively. Note that this fractional derivative is defined from an integral to the left of an integer-order derivative that is to the left of another integral. The integrals may eventually be integers, but the derivative is always of first order. Moreover, we do not bother to denote by a particular letter, because beyond the order it has another parameter that is the type. So, with two superscripts, we already know whether to treat the Hilfer derivative operator, at least in this introductory book.

Limit cases. Riemann-Liouville and Caputo. As we have already mentioned, the Riemann-Liouville and Caputo fractional derivatives can be recovered as limit cases of the Hilfer fractional derivative. So let's consider, first, the type $\beta = 0$. Replacing in the definition, we have

$$\mathcal{D}_{a\pm}^{\alpha,0} f(t) = \mathcal{J}_{a\pm}^{0} \, \mathcal{D} \, \mathcal{J}_{a\pm}^{1-\alpha} f(t) = \mathcal{D} \, \mathcal{J}_{a\pm}^{1-\alpha} f(t)$$

where the second equality comes from the fact that $\mathcal{J}_{a\pm}^{0}$ is the identity. Finally, since the integral of order μ is equal to the derivative of order $-\mu$ we can write

$$\mathcal{D}_{a\pm}^{\alpha,0} f(t) = \mathcal{D} \, \mathcal{D}_{a\pm}^{\alpha-1} f(t) = \mathcal{D}_{a\pm}^{1+\alpha-1} f(t) = \mathcal{D}_{a\pm}^{\alpha} f(t)$$

which is exactly the Riemann-Liouville fractional derivative.

On the other hand, consider now $\mu = 1$, then

$$\mathcal{D}_{a\pm}^{\alpha,1} f(t) = \mathcal{J}_{a\pm}^{1-\alpha} \, \mathcal{D} \, \mathcal{J}_{a\pm}^{0} f(t)$$

which, using $\mathcal{J}_{a\pm}^{0}$ again equal to the identity, allow us to write

$$\mathcal{D}_{a\pm}^{\alpha,1} f(t) = \mathcal{J}_{a\pm}^{1-\alpha} \, \mathcal{D} f(t).$$

In analogy to the above, the integral of order μ is equal to the derivative of order $-\mu$, and simplifying, we can write

$$\mathcal{D}_{a\pm}^{\alpha,1} f(t) = \mathcal{D}_{a\pm}^{\alpha-1} \, \mathcal{D} f(t) = {}_{c}\mathcal{D}_{\pm}^{\alpha} f(t)$$

which is exactly the Caputo fractional derivative.

Hilfer derivative as Riemann-Liouville or Caputo derivatives. In order that we may express the Hilfer fractional derivative either as a Riemann-Liouville fractional derivative or as a Caputo fractional derivative, we must introduce the following parameter $\gamma = \beta + \alpha - \alpha\beta$ so that $0 < \gamma < 1$. Thus, from the definition of Hilfer fractional derivative, we have

$$\mathcal{D}_{a\pm}^{\alpha,\beta} f(t) = \pm \mathcal{J}_{a\pm}^{\beta(1-\alpha)} \, \mathcal{D} \, \mathcal{D}_{a\pm}^{(-1+\beta)(1-\alpha)} f(t)$$

from where it follows, simplifying

$$\mathcal{D}_{a\pm}^{\alpha,\beta} f(t) = \pm \mathcal{J}_{a\pm}^{\beta(1-\alpha)} \, \mathcal{D}_{a\pm}^{1-1+\alpha+\beta-\beta\alpha} f(t) = \mathcal{J}_{a\pm}^{\beta(1-\alpha)} \, \mathcal{D}_{a\pm}^{\gamma} f(t)$$

which is exactly a Riemann-Liouville fractional derivative.

In analogy to the Hilfer fractional derivative, expressed as a Riemann-Liouville fractional derivative, we can express the Hilfer derivative in terms of a Caputo fractional derivative, provided that the constraint in the parameters $0 < \alpha\beta - \beta + 1 < 1$, then

$$\mathcal{D}_{a\pm}^{\alpha,\beta} f(t) = \pm \mathcal{I}_{a\pm}^{1-(\alpha\beta-\beta+1)} \, \mathcal{D} \, \mathcal{I}_{a\pm}^{(1-\beta)(1-\alpha)} f(t)$$

or by already introducing the suitable notation, in the form

$$\mathcal{D}_{a\pm}^{\alpha,\beta} f(t) = \pm {_C}\mathcal{D}_{\pm}^{\alpha\beta-\beta+1} \, \mathcal{I}_{a\pm}^{(1-\beta)(1-\alpha)} f(t)$$

which is exactly the Caputo fractional derivative
Hilfer derivative of integer order. In order to show that the Hilfer fractional derivative retrieves the ordinary derivative, when the order is a positive integer, we must show that $\mathcal{D}_{a+}^{n,\beta} \mathcal{I}_{a+}^{n} f(t) = f(t)$, independently of the type and $n \in \mathbb{N}$.

Then, from the definition of the Hilfer fractional derivative, we have

$$\mathcal{D}_{a+}^{n,\beta} \mathcal{I}_{a+}^{n} f(t) = \left[\mathcal{I}_{a\pm}^{\beta(1-n)} \, \mathcal{D} \, \left(\mathcal{I}_{a\pm}^{(1-\beta)(1-n)} \right) \right] \mathcal{I}_{a+}^{n} f(t).$$

The derivative of order μ is equal to the integral of order $-\mu$, so we can write

$$\mathcal{D}_{a+}^{n,\beta} \mathcal{I}_{a+}^{n} f(t) = \mathcal{I}_{a+}^{\beta(1-n)} \cdot \left(\mathcal{I}_{a\pm}^{(1-\beta)(1-n)+n-1} \right) f(t)$$

which, by simplifying, provides

$$\mathcal{D}_{a+}^{n,\beta} \mathcal{I}_{a+}^{n} f(t) = \mathcal{I}_{a+}^{0} f(t) = f(t)$$

which is the desired result. $\diamond$

28. *Violation of the Leibniz rule.* Let us present a theorem by asserting that if the Leibniz rule (derivative of the two-function product) is satisfied, then the derivative can not be considered a fractional derivative [22]. To prove this theorem, we begin with another result, due to Hadamard.

Theorem 6.1 (Hadamard theorem) *If $f \in C^1(U)$ ($C^m(U)$ is the space of functions m times continuously differentiable in $U \subset \mathbb{R}$) in a neighborhood U of the point x_0, then f can be represented in the form*

$$f(x) = f(x_0) + (x - x_0)g(x)$$

with $g \in C^1(U)$.

To prove this theorem, we introduce a function $F(t) = f(x_0 + (x - x_0)t)$, so that $F(0) = f(x_0)$ and $F(1) = f(x)$. By the fundamental theorem of calculus we have

$$\int_0^1 \frac{d}{dt} F(t) \, dt = F(1) - F(0) = f(x) - f(x_0).$$

On the other hand, using the chain rule, we can write

$$\int_0^1 \frac{d}{dt} F(t)\, dt = (x - x_0) \int_0^1 f'(x_0 + (x - x_0)t)\, dt \cdot$$

Denoting $g(x)$ by $g(x) = \int_0^1 f'(x_0 + (x - x_0)t)\, dt$ follows the result.

Leibniz rule. If E is a linear operator that satisfies the Leibniz rule,

$$E[f(x)g(x)] = g(x)Ef(x) + f(x)Eg(x)$$

then $E1 = 0$.

The demonstration is immediate. Since E, by hypothesis, is a linear operator and is worth the Leibniz rule, we can write

$$E1 = E(1 \cdot 1) = 1E1 + 1E1 = 2E1$$

from where it follows, $E1 = 0$ □

Theorem 6.2 (Order of the derivative equal to one) *If E is a linear operator that satisfies the Leibniz rule and can be applied to functions $C^2(U)$, being $U \subset \mathbb{R}$ a neighborhood of the point x_0, then this operator is a multiple of the derivative of first order,*

$$E = a(x)\frac{d}{dx}$$

where $a(x)$ is a function in $\mathbb{R}$.

In order to show this result, we start with the Hadamard theorem. If $f \in C^2(U)$ we have

$$f(x) = f(x_0) + (x - x_0)g(x)$$

with $g \in C^2(U)$. Again, we can write

$$g(x) = g(x_0) + (x - x_0)h(x)$$

with $h \in C^2(U)$. Then, rearranging the two preceding equations, we obtain

$$f(x) = f(x_0) + (x - x_0)g(x_0) + (x - x_0)^2 h(x),$$

from where it follows, for the derivative of order one,

$$f'(x) = g(x_0) + 2(x - x_0)h(x) + (x - x_0)^2 h'(x)$$

and therefore, $f'(x_0) = g(x_0)$. With this result, we obtain

$$f(x) = f(x_0) + (x - x_0)f'(x_0) + (x - x_0)^2 h(x).$$

Let us apply the operator E to the previous equation and which, by hypothesis, besides being a linear operator, the Leibniz rule is valid, therefore

$$E[f(x)] = E[f(x_0)] + E[(x - x_0)f'(x_0)] + E[(x - x_0)^2 h(x)].$$

Since $E[f(x_0)] = 0 = E[f'(x_0) = g(x_0)]$, because they are constant, we have

$$E[f(x)] = f'(x_0)E[x - x_0] + 2h(x)(x - x_0)E[x - x_0] + (x - x_0)^2 E[h(x)]$$

or, in the following form

$$E[f(x)] = f'(x_0)E[x] + 2h(x)(x - x_0)E[x] + (x - x_0)^2 E[h(x)].$$

We introduce the notation $E[x] = a(x)$ which, replacing in the previous one, leads us to

$$E[f(x)] = f'(x_0)a(x) + 2h(x)(x - x_0)a(x) + (x - x_0)^2 E[h(x)]$$

which, for $x = x_0$, provides $E[f(x_0)] = f'(x_0)a(x_0)$. Since x_0 is arbitrary, we obtain

$$E = a(x)\frac{\mathrm{d}}{\mathrm{d}x}$$

which proves the theorem. □

29. *Langevin fractional equation.* A natural way to approach anomalous diffusion is through a fractional order Langevin equation. It can be shown that the long-term behavior of the quadratic mean displacement, for the systems described by a Langevin integrodifferential equation, depend on the properties of the correlation function and the memory kernel [19]. Let us first discuss the generalized Langevin equation and then present its fractional version.

We address results associated with the classical generalized Langevin equation, in particular the Laplace, which is used to obtain the solution in terms of the relaxation function. We discussed the classic generalized Langevin equation and we present its solution in terms of a relaxation function that characterizes the particular physical process, obtained through the Laplace transform.

It is important to mention some theoretical facts about the Langevin equation. The classical approach, known as the Ornstein-Uhlenbeck or Einstein-Ornstein-Uhlenbeck's Brownian motion was initially introduced by Langevin. The classical Langevin equation and its generalization, were reviewed with a fractional treatment a few years ago by Mainardi and Pironi [23], where it was discussed Langevin equation associated only with velocity. In order to discuss the Langevin equation associated with the displacement, an integration in the time variable is still necessary.

The generalized Langevin equation, in the absence of a deterministic field, is

$$D_t^2 x(t) + \int_0^t \mu(t - \xi) D_\xi x(\xi) \, d\xi = F(t) \tag{6.43}$$

where $D_y \equiv d/dy$ and $y = t, \xi$; $\mu(t)$ the dissipative memory kernel and $F(t)$ a random force. We consider Eq. (6.43) with two deterministic conditions, $x(0) = x_0$, initial displacement, and $\dot{x}(0) = v_0$, initial velocity of a particle of unit mass. To solve Eq. (6.43) we use the Laplace transform methodology. Taking the Laplace transform on both members of Eq. (6.43) and using the convolution theorem we get

$$[s^2 + s\mu(s)]x(s) = x_0[s + \mu(s)] + v_0 + F(s) \tag{6.44}$$

in which $\mathscr{L}[\mu(t)] \equiv \mu(s)$, $\mathscr{L}[x(t)] \equiv x(s)$ and $\mathscr{L}[F(t)] \equiv F(s)$ denote, respectively, the Laplace transform of $\mu(t)$, $x(t)$ and $F(t)$, with s the parameter of Laplace transform.

Introducing the called relaxation function, denoted by $H(t)$, as the inverse Laplace transform of $H(s)$,

$$H(s) = \frac{1}{s^2 + s\mu(s)}$$

we can write Eq. (6.44) in the following form

$$x(s) = H(s)\{x_0[s + \mu(s)] + v_0 + F(s)\},$$

whose respective inverse Laplace transform is given by

$$x(t) = x_0 + v_0 H(t) + \int_0^t H(t - \xi) F(\xi) d\xi$$

which can still be rewritten as follows

$$x(t) = \langle x(t) \rangle + \int_0^t H(t - \xi) F(\xi) d\xi \tag{6.45}$$

where we introduce the notation $\langle x(t) \rangle = x_0 + v_0 H(t)$.

The first derivative, with respect to the variable t, of Eq. (6.45) is given by

$$\dot{x}(t) = \langle \dot{x}(t) \rangle + \int_0^t h(t - \xi) F(\xi) d\xi \tag{6.46}$$

where $\langle \dot{x}(t) \rangle = v_0 h(t)$, the relaxation function, $h(t)$, is the derivative of $H(t)$, and $h(t) = \dot{H}(t)$. Then, we have

$$h(s) = \frac{1}{s + \mu(s)} \equiv s H(s)$$

being $H(0) = 0$ and $h(0) = 1$, obtained by Eq. (6.45) and Eq. (6.46), respectively.

With the help of Eqs. (6.45) and (6.46), we can discuss the explicit equations for the variables associated with Eq. (6.43), in particular the anomalous diffusion and the probability distribution [24]. However, here, our intention is to discuss the fractional Langevin equation and, for that, we will follow the same one that was presented for the generalized Langevin equation. Then, we introduce the fractional version of the generalized Langevin equation and using a particular correlation function we obtain the relaxation function. Finally, we explicitly calculate the kernels in terms of the Mittag-Leffler function with three parameters. We have several ways of explaining phenomena involving anomalous diffusion, in particular, extending the definition of entropy in the conventional Gibbs-Boltzmann statistics as well as, through the Fokker-Planck equation with fractional derivatives or with coefficients associated with diffusion varying in time [25]. Here, we present and discuss the called Langevin equation considering the Laplace transform methodology in a similar way to that used for the generalized Langevin equation.

The fractional (generalized) Langevin equation, the fractional version of the generalized Langevin equation, is defined by

$$D_t^\alpha x(t) + \int_0^t \mu(t - \xi) \, D_\xi^\beta x(\xi) \, d\xi = F(t) \tag{6.47}$$

with the parameters $1 < \alpha \leq 2$ and $0 < \beta \leq 1$. The operator D denotes the fractional derivative considered in the Caputo sense. Note that we are using another symbol for the Caputo derivative, exclusively, to simplify notation.

In Eq. (6.47), $x(t)$ represents the position of a particle, $\mu(t - t')$ the dissipative memory kernel, which is taken in terms of a Mittag-Leffler function. We mention that this form seems natural, a generalization of the exponential, using a Mittag-Leffler function, since the stationary noise is exponentially correlated, and $F(t)$ is a Gaussian force. It is important to note that, considering the limits $\alpha \to 2$ and $\beta \to 1$, the results involving the generalized Langevin equation are recovered.

We mention that in a recent paper [26] a class of generalized Langevin equations with fractional derivatives was presented, also in the Caputo sense. A system involving diffusion processes with a correlation function of two types was discussed: the exponential, the law of force and the terms associated with the deterministic field equal to zero, that is, the Eq. (6.47) in the case $0 < \alpha < 1$ and $\beta = 1$. The same author [27] discusses the fractional Langevin equation with the fractional derivative in the Riemann-Liouville sense. These results are compared with those obtained in [26] considering the null deterministic term. Further, also recent, [28], is the discussion of the anomalous diffusion induced by a noise correlation function given by a Mittag-Leffler function. The exact expression for the principal values, variance and diffusion coefficients for a particle in terms of the Mittag-Leffler function with two parameters and their derivatives is obtained. We point out that the equation discussed is a generalized Langevin equation, unlike the equation presented in [27], which is a fractional generalized Langevin equation.

After this brief review, here, we are interested in obtaining analytical solutions of Eq. (6.47) through the Laplace transform. Thus, by applying the Laplace transform in Eq. (6.47) and using the relation involving the Laplace transform and its derivatives [19], we obtain the following algebraic equation

$$[s^\alpha + s^\beta \mu(s)]x(s) = s^{\alpha-1}x_0 + s^{\alpha-2}v_0 + x_0 s^{\beta-1}\mu(s) + F(s)$$

that, in the particular case, $\alpha = 2$ and $\beta = 1$, recovers Eq. (6.44).

By introducing a relaxation function, denoted by $H(t)$, we can write the previous equation, already solving for $x(s)$, in the following way

$$x(s) = H(s)\left\{\frac{x_0}{s}[s^\alpha + s^\beta \mu(s)] + v_0 s^{\alpha-2} + F(s)\right\}$$

whose inverse Laplace transform, denoted by $\mathscr{L}^{-1}[f(s)]$, can be written as

$$x(t) = x_0 + v_0 \mathscr{L}^{-1}\left[H(s)s^{\alpha-2}\right] + \mathscr{L}^{-1}\left[H(s)F(s)\right].$$

Finally, using the convolution theorem associated with the Laplace transform, we obtain the solution

$$x(t) = x_0 + v_0 \mathscr{L}^{-1}\left[H(s)s^{\alpha-2}\right] + \int_0^t H(t-\xi)F(\xi)d\xi$$

where $H(s)$ is the Laplace transform of the $H(t)$ function. From this point, we must distinguish a particular relaxation function, that is, given this relaxation function we can explain the displacement.

Then, as already mentioned, we must explain the calculation of the kernels $H(t)$ and $h(t)$, which appear in the expressions involving variances, given as follows

$$\sigma_{xx}(t) = k_B T \left[2I(t) - H^2(t)\right]$$
$$\sigma_{vv}(t) = k_B T \left[1 - h^2(t)\right]$$
$$\sigma_{xv}(t) = k_B T H(t)\left[1 - h(t)\right] \qquad (6.48)$$

where

$$I(t) = \int_0^t d\xi\, H(\xi),$$

k_B is the Boltzmann constant and T is the absolute temperature of the environment [24].

The Eq. (6.43) represents a steady state system, the functions $\mu(t)$, the dissipative memory kernel and $C(t)$, the correlation function, are related by known the floatation-dissipation theorem

$$C(t) = k_B T \mu(t).$$

We consider the case where the dissipative memory kernel is given by [28]

$$\mu(t) = \gamma_\lambda E_\lambda[-(t/\tau)^\lambda]/\tau^\lambda \tag{6.49}$$

with $0 < \lambda < 2$, being determined by a dynamic mechanism associated to the physical process. $E_\lambda(\cdot)$ is the classical Mittag-Leffler function depending on a parameter, and for $0 < \lambda < 1$, is a strictly monotonous function and for $1 < \lambda \leq 2$ can be decomposed into a completely monotonic function added to an oscillatory contribution, as in Application (16), γ_λ is a constant that depends on λ but has no temporal dependence and τ is memory temporal characteristic. Using Eq. (6.49), the correlation function is given by

$$C(t) = C_0(\lambda)\tau^{-\lambda} E_\lambda[-(t/\tau)^\lambda]$$

where τ acts as a characteristic temporal memory and $C_0(\lambda) = \gamma_\lambda k_B T$ is the coefficient of proportionality depending on the parameter λ.

Proceeding in an analogous way to the previous one, that is to say, using the dissipative memory kernel, given by Eq. (6.49), we can write the respective Laplace transform in the following form

$$\mu(s) = \gamma_\lambda \frac{s^{\lambda-1}}{1 + s^\lambda \tau^\lambda}$$

which provides the Laplace transform corresponding to the relaxation function

$$H(s) = H_0(s) + H_1(s)$$

with $H_1(s) = \tau^{-\lambda} s^{-\lambda} H_0(s)$ in which $H_0(s)$ is given by the expression

$$\widehat{H}_0(s) = \frac{s^{\lambda-\alpha}}{s^\lambda + (\gamma_\lambda/\tau^\lambda) s^{\lambda+\beta-\alpha-1} + (1/\tau^\lambda)} \tag{6.50}$$

which, for $\alpha = 2 = 2\beta$, recover the results obtained in [28].

Since, from now on, what is to be done is the calculation of the inverse Laplace transform and which, in this particular type of function, has already been discussed, according to Exercise (33) of Chap. 4, let's not repeat the calculations here [19].   ◇

30. *Changing the order of integration* [29, 30]. Let us return, by exchanging the integration order, by writing an iterated integral from the integers n.

Consider a function of two variables $f(x, y)$. If $f(x, y)$ is measurable, in particular continuous, in the quadrant $(0, \infty) \times (0, \infty)$ and the integrals, in the Riemann sense,

$$\int_0^\infty dx \int_0^\infty f(x, y)\, dy \quad \text{and} \quad \int_0^\infty dy \int_0^\infty f(x, y)\, dx$$

exist and are both absolutely convergent, then these two integrals are equal,

$$\int_0^\infty dx \int_0^\infty f(x, y)\, dy = \int_0^\infty dy \int_0^\infty f(x, y)\, dx \cdot$$

In particular, the result that interests us, specifically for the case of iterated integrals, is such that equality

$$\int_a^b dx \int_a^x f(x, y)\, dy = \int_a^b dy \int_y^b f(x, y)\, dx$$

with $a, b \geq 0$ and $b > a$. Even more, let us explain two of them, namely:

$$\int_a^x dx_1 \int_a^{x_1} f(\xi)\, d\xi$$

that can be written, changing the order of integrals

$$\int_a^x dx_1 \int_a^{x_1} f(\xi)\, d\xi = \int_a^x f(\xi)\, d\xi \int_\xi^x dx_1$$

which, after the integration in x_1, allows to write

$$\int_a^x dx_1 \int_a^{x_1} f(\xi)\, d\xi = \int_a^x f(\xi)\, d\xi \int_\xi^x dx_1 = \int_a^x (x - \xi) f(\xi)\, d\xi \cdot$$

And now, the second case, with three integrals

$$\int_a^x dx_1 \int_a^{x_1} dx_2 \int_a^{x_2} f(\xi)\, d\xi = \int_a^x dx_1 \left\{ \int_a^{x_1} dx_2 \int_a^{x_2} f(\xi)\, d\xi \right\} \cdot$$

By changing the order of integrals in brackets, we can write

$$\int_a^x dx_1 \left\{ \int_a^{x_1} dx_2 \int_a^{x_2} f(\xi)\, d\xi \right\} = \int_a^x dx_1 \left\{ \int_a^{x_1} f(\xi)\, d\xi \int_\xi^{x_1} dx_2 \right\}$$

which, after integration in x_2, allows writing

$$\int_a^x dx_1 \left\{ \int_a^{x_1} f(\xi)\,d\xi \int_\xi^{x_1} dx_2 \right\} = \int_a^x dx_1 \left[\int_a^{x_1} (x_1 - \xi) f(\xi)\,d\xi \right].$$

An analogous procedure, changing the order of the remaining two integrals, we get

$$\int_a^x dx_1 \left[\int_a^{x_1} (x_1 - \xi) f(\xi)\,d\xi \right] = \int_a^x f(\xi)\,d\xi \int_\xi^x (x_1 - \xi)\,dx_1$$

whose integration in the variable x_1 provides

$$\int_a^x f(\xi)\,d\xi \int_\xi^x (x_1 - \xi)\,dx_1 = \int_a^x f(\xi)\,d\xi \left. \frac{(x_1 - \xi)^2}{2} \right|_{x_1 = \xi}^{x_1 = x}$$

from which, finally, the result

$$\int_a^x dx_1 \int_a^{x_1} dx_2 \int_a^{x_2} f(\xi)\,d\xi = \int_a^x \frac{(x - \xi)^2}{2!} f(\xi)\,d\xi.$$

This result can be extended in order to obtain the desired expression, that is, the iterated integrals written in terms of a single integral.   ◇

31. *Abel equation.* Show that the solution of the second kind Abel integral equation

$$x(t) - \frac{\lambda}{\Gamma(\nu)} \int_0^t \frac{x(\tau)}{(t - \tau)^{1-\nu}}\,d\tau = f(t)$$

with $0 < t < 1$ and $\nu > 0$, is given by

$$x(t) = \frac{d}{dt} \int_0^t E_\nu[\lambda(t - \tau)^\nu] f(\tau)\,d\tau$$

where $E_\nu(\cdot)$ is a Mittag-Leffler function. Using Exercise (19) from Chap. 3, with $\beta = 1 = \gamma$, to get $x(t)$ in the particular case where $f(t) = 1 \cdot$
Writing the integral with the fractional operator

$$_0\mathcal{J}_t^\nu x(t) = \frac{1}{\Gamma(\nu)} \int_0^t (t - \tau)^{\nu-1} x(\tau)\,d\tau$$

we obtain, operating with the Riemann-Liouville fractional derivative

$$_0\mathcal{D}_t^\nu x(t) - \lambda \,_0\mathcal{D}_t^\nu \,_0\mathcal{J}^\nu x(t) = \,_0\mathcal{D}_t^\nu f(t) \cdot$$

Since the Riemann-Liouville fractional derivative on the left is the inverse operator of the Riemann-Liouville fractional integral, we can write

$$_0\mathcal{D}_t^\nu x(t) - \lambda x(t) = \,_0\mathcal{D}_t^\nu f(t) \cdot$$

Taking the Laplace transform on both members of the preceding one, we obtain

$$\mathscr{L}[_0\mathscr{D}_t^\nu x(t)] - \lambda X(s) = \mathscr{L}[_0\mathscr{D}_t^\nu f(t)]$$

with $X(s) = \mathscr{L}[x(t)]$. Writing the expression for the Laplace transform of the Riemann-Liouville fractional derivative, we have

$$s^\nu X(s) - \lambda X(s) = s^\nu F(s)$$

with $F(s) = \mathscr{L}[f(t)]$. This is an algebraic equation with solution

$$X(s) = s\frac{s^{\nu-1}}{s^\nu - \lambda}F(s).$$

Let's calculate the inverse Laplace transform. For this, we use the result

$$\mathscr{L}^{-1}\left[s\frac{s^{\nu-1}}{s^\nu - \lambda}\right] = \frac{\mathrm{d}}{\mathrm{d}t}E_\alpha(\lambda t^\alpha)$$

as well as the definition of the convolution product, from where it follows

$$x(t) = \frac{\mathrm{d}}{\mathrm{d}t}\int_0^t E_\nu[\lambda(t-\tau)^\nu]f(\tau)\,\mathrm{d}\tau.$$

For the particular case where $f(t) = 1$, we must evaluate the integral

$$x(t) = \frac{\mathrm{d}}{\mathrm{d}t}\int_0^t E_\nu[\lambda(t-\tau)^\nu]\,\mathrm{d}\tau,$$

which, using the result of Exercise (19) in Chap. 3, allows us to write

$$x(t) = \frac{\mathrm{d}}{\mathrm{d}t}\left[tE_{\nu,2}(\lambda t^\nu)\right].$$

By explicitly calculating the derivative, we obtain

$$x(t) = E_{\nu,2}(\lambda t^\nu) + t\frac{\mathrm{d}}{\mathrm{d}t}\left[E_{\nu,2}(\lambda t^\nu)\right].$$

From the equality involving the Mittag-Leffler functions,

$$t\frac{\mathrm{d}}{\mathrm{d}t}E_{\nu,\beta}(\lambda t^\nu) = E_{\nu,\beta-1}(\lambda t^\nu) + (1-\beta)E_{\nu,\beta}(\lambda t^\nu)$$

we can write, already simplifying,

$$x(t) = E_\nu(\lambda t^\nu)$$

which is the desired result.   ◇

32. *Non-constant coefficients.* The study of fractional differential equations with non-constant coefficients is a separate chapter, since the Leibniz rule for the product of two functions and/or the chain rule for composite functions is not always valid. Here, we are going to discuss a seemingly simple fractional ordinary differential equation with nonconstant coefficients in order to verify that the problem becomes a little more complicated.

Use the Laplace transform methodology to solve the following fractional differential equation with non-constant coefficients [31]

$$_0\mathcal{D}_t^{1/2}x(t) = \frac{x(t)}{t}$$

with $t > 0$ and the fractional derivative considered in the Riemann-Liouville sense.

Writing the fractional differential equation in the form

$$t\,_0\mathcal{D}_t^{1/2}x(t) = x(t)$$

introducing the notation $\mathcal{L}[x(t)] = F(s)$ and taking the Laplace transform of both members of the differential equation, we can write

$$\mathcal{L}\left[t\,_0\mathcal{D}_t^{1/2}x(t)\right] = F(s).$$

Using the relation

$$\mathcal{L}[tg(t)] = -\frac{\mathrm{d}}{\mathrm{d}s}G(s) \qquad (6.51)$$

with $\mathcal{L}[g(t)] = G(s)$ and s the parameter of Laplace transform, we can write, in this case

$$G(s) \equiv \mathcal{L}\left[_0\mathcal{D}_t^{1/2}x(t)\right]$$

that, from the expression for the Laplace transform of the Riemann-Liouville derivative with the order $1/2$, which entails $n = 1$,

$$\mathcal{L}\left[_0\mathcal{D}_t^{1/2}x(t)\right] = s^{1/2}F(s) - x(0)$$

with $x(0)$ is a constant.

Returning to the ordinary differential equation and using Eq. (6.51), we obtain

$$-\frac{\mathrm{d}}{\mathrm{d}s}\left\{\sqrt{s}F(s) - x(0)\right\} = F(s)$$

or, after evaluate the derivative and rearranging, as follows

$$\frac{d}{ds}F(s) = -\frac{2\sqrt{s}+1}{2s}$$

which is an ordinary differential equation of first order and separable. Solving this ordinary differential equation we obtain

$$F(s) = \frac{C}{\sqrt{s}}\exp(-2\sqrt{s})$$

where C is an integration constant. Thus, we recover the solution of the fractional differential equation through the inverse Laplace transform,

$$x(t) = C\,\mathscr{L}^{-1}\left[\frac{e^{-2\sqrt{s}}}{\sqrt{s}}\right].$$

As already mentioned, more than once, the problem of the integral transform methodology lies in the calculation of its inverse, that is, the initial problem is led to another auxiliary problem whose solution requires the calculation of the inverse. Here, in our case, we must use integration in the complex plane, using the modified Bromwich contour, since we have a branch point at $s = 0$ [32]. Going through the boundary of Bromwich and using the residue theorem (we do not have contributions, because we do not have poles and the branching points is outside the contour) we can write

$$\mathscr{L}^{-1}\left[\frac{e^{-2\sqrt{s}}}{\sqrt{s}}\right] = -\frac{1}{2\pi i}\left\{\int_{\infty}^{0}e^{-xt}\frac{e^{-2\sqrt{x}i}}{i\sqrt{x}}(-dx) + \int_{0}^{\infty}e^{-xt}\frac{e^{2\sqrt{x}i}}{-i\sqrt{x}}(-dx)\right\}$$

with $x \in \mathbb{R}$ and $t > 0$. Manipulating these two integrals, using the Euler relation involving exponentials and simplifying, we obtain

$$\mathscr{L}^{-1}\left[\frac{e^{-2\sqrt{s}}}{\sqrt{s}}\right] = \frac{1}{\pi}\int_{0}^{\infty}e^{-xt}\cos(2\sqrt{x})\frac{dx}{\sqrt{x}}.$$

To evaluate the remaining integral, we change the variable $x = y^2/4$ and consider a perfect square to lead the integral into the form

$$\mathscr{L}^{-1}\left[\frac{e^{-2\sqrt{s}}}{\sqrt{s}}\right] = \frac{1}{\pi}e^{-1/t}\int_{0}^{\infty}e^{-\frac{1}{4}z^2}dz$$

where, we have introduced the notation $z = y \pm 2i/t$. Using the result

$$\int_0^\infty e^{-px^2} dx = \frac{1}{2}\sqrt{\frac{\pi}{p}}$$

with $p > 0$ and simplifying, we obtain the inverse Laplace transform

$$\mathcal{L}^{-1}\left[\frac{e^{-2\sqrt{s}}}{\sqrt{s}}\right] = \frac{1}{\sqrt{\pi t}}e^{-1/t}.$$

Finally, the solution of the fractional differential equation is given by

$$x(t) = \frac{C}{\sqrt{\pi t}}e^{-1/t}$$

where C is a constant. We mentioned that in Ref. [31] we can find the resolution of this fractional differential equation by another methodology, in particular by appending the Leibniz rule, as well as showing that the obtained solution is really solution of the starting problem   ◇

33. *Fractional differential equation. General case.* Let $\alpha > 0$ be the order of the fractional differential equation

$$_C\mathcal{D}_t^\alpha x(t) = \mathcal{D}^\alpha\left[x(t) - \sum_{k=0}^{m-1}\frac{t^k}{k!}x^{(k)}(0^+)\right] = -x(t) + q(t) \qquad (6.52)$$

with $t > 0$ and the derivative considered in the Caputo sense. Here, m is an integer positive such that $m - 1 < \alpha \le m$, which relates the number of initial conditions

$$x^{(k)}(0^+) = c_k$$

with $k = 0, 1, 2, \ldots, m - 1$, which must be provided. In particular, for $m = 1$ we have what is known as fractional relaxation and $m = 2$ for fractional oscillations. Further, $\alpha = m$, the equation is led to an ordinary differential equation with constant coefficients whose solution is the sum of m linearly independent solutions of the homogeneous equation ($q(t) = 0$) and a particular solution of the respective non-homogeneous ordinary differential equation ($q(t) \ne 0$).

In the case where the order is an integer, the general solution has the form

$$x(t) = \sum_{k=0}^{m-1} c_k x_k(t) + \int_0^t q(t - \tau)x_\delta(\tau)\,d\tau$$

where

$$x_k(t) = \mathcal{J}^k x_0(t), \quad \text{and} \quad x_\delta(t) = -x_0'(t) \qquad (6.53)$$

being $\mathcal{J}^k$ the repeated integral k times, and

$$\mathcal{J}^1 x(t) = \mathcal{J} x(t) = \int_0^t x(\tau) \, d\tau.$$

Thus, the m functions $x_k(t)$ represent fundamental solutions of the ordinary differential equation of order m, that is, linearly independent solutions of the homogeneous differential equation, satisfying the conditions Eq. (6.53). The function $x_\delta(t)$ with which the convoluted term appears is a particular solution of the nonhomogeneous ordinary differential equation such that all $c_k = 0$ with $k = 0, 1, 2, \ldots, m - 1$ and $q(t) = \delta(t)$, the Dirac delta 'function' (actually a distribution). In particular, in the case of ordinary relaxation, we have

$$x_0(t) = e^{-t} = x_\delta(t)$$

while, in the ordinary oscillations, we have

$$x_0(t) = \cos t, \quad x_1(t) = \mathcal{J} x_0(t) = \sin t = x_\delta(t).$$

Let us return to the fractional differential equation, Eq. (6.52) in order to solve it through the Laplace transform methodology. First, we apply the fractional integral operator in both members of the equation,

$$x(t) = \sum_{k=0}^{m-1} c_k \frac{t^k}{k!} - \mathcal{J}^\alpha x(t) + \mathcal{J}^\alpha q(t)$$

since $\mathcal{J}^\alpha \mathcal{D}^\alpha$ is the identity and $x^{(k)}(0^+) = c_k$.

Taking the Laplace transform in the previous one, we have an algebraic equation

$$F(s) = \sum_{k=0}^{m-1} \frac{c_k}{s^{k+1}} - \frac{F(s)}{s^\alpha} + \frac{Q(s)}{s^\alpha}$$

with $F(s) = \mathcal{L}[x(t)]$, $Q(s) = \mathcal{L}[q(t)]$ and s the parameter of the Laplace transform.

Solving the algebraic equation, we can write

$$F(s) = \sum_{k=0}^{m-1} c_k \frac{s^{\alpha-k-1}}{s^\alpha + 1} + \frac{Q(s)}{s^\alpha + 1}.$$

We must now calculate the inverse Laplace transform. For this, we use the convolution theorem, then

$$x(t) = \sum_{k=0}^{m-1} c_k x_k(t) - \int_0^t q(t - \tau) x_0'(\tau) \, d\tau$$

where we use, from the condition $x_0(0^+) = E_\alpha(0^+) = 1$, and the notation

$$\frac{1}{s^\alpha + 1} = -\left(s\frac{s^{\alpha-1}}{s^\alpha + 1} - 1\right) \div -x_0'(t)$$

being $E_\alpha(\cdot)$ the classical Mittag-Leffler function and $\div$ denotes the pair of Laplace transform, direct and inverse.

Note that $m = 1$ represents the integer part of α, denoted by $[\alpha]$ and m the number of initial conditions necessary and sufficient to guarantee the uniqueness of solution $x(t)$. Thus, m functions

$$x_k(t) = \mathcal{J}^k E_\alpha(-t^\alpha)$$

with $k = 0, 1, 2, \ldots, m - 1$ represent the solutions of the homogeneous equation satisfying the conditions

$$x_k^{(h)}(0^+) = \delta_{kh}, \qquad h, k = 0, 1, 2, \ldots$$

that is, they represent the fundamental solutions of the fractional differential equation, in complete analogy to the integer case $\alpha = m$.

A complete discussion of this problem is found in the Ref. [11] where, also, vast material can be found, dedicated to the applications of the fractional calculus. $\diamond$

34. *Fractional partial differential equation.* Let us present what is known as the Nigmatullin fractional diffusion Eq. [33]. Let $0 < \alpha < 1$ be the order of the derivative, considered in the Riemann-Liouville sense and $\lambda > 0$ a constant. Let's solve the fractional diffusion equation

$$_0\mathcal{D}_t^\alpha u(x, t) = \lambda^2 \frac{\partial^2}{\partial x^2} u(x, t)$$

for $t > 0$ and $-\infty < x < \infty$, satisfying the boundary conditions $\lim\limits_{x \to \pm\infty} u(x, t) = 0$ and initial conditions

$$_0\mathcal{D}_t^{\alpha-1} u(x, t)\big|_{t=0} = \varphi(x)$$

through the methodology of the juxtaposition of transforms, that is, the Fourier transform in the spatial part (the problem is one-dimensional of the spatial variable) and the Laplace transform in the temporal part. Let ω be the Fourier transform parameter. Taking the Fourier transform on both sides of the equation, taking into account the boundary conditions and transforming the initial conditions, we conduct our starting problem (equation + conditions) to the following problem, composed by the fractional differential equation

$$_0\mathcal{D}_t^\alpha U(\omega, t) + \lambda^2 \omega^2 U(\omega, t) = 0$$

satisfying the initial conditions

$$_0\mathcal{D}_t^\alpha U(\omega, t)\big|_{t=0} = \phi(\omega)$$

where $U(\omega, t) = \mathcal{F}[u(x, t)]$ and $U(\phi) = \mathcal{F}[\varphi(x)]$ the respective Fourier transforms.

Now, we apply the Laplace transform in the preceding problem, that is, in both members of the differential equation and using the initial conditions, we obtain an algebraic equation with solution is given by

$$F(\omega, s) = \frac{\phi(\omega)}{s^\alpha + \lambda^2 \omega^2}$$

with notation $F(\omega, s) = \mathcal{L}[U(\omega, t)]$ being s the parameter of Laplace transform.

In order to proceed with the inversion, we begin with the inverse Laplace transform, that is, to recover the function $U(\omega, t)$, we must calculate

$$U(\omega, t) = \mathcal{L}^{-1}\left[\frac{\phi(\omega)}{s^\alpha + \lambda^2 \omega^2}\right] = \phi(\omega)\mathcal{L}^{-1}\left[\frac{1}{s^\alpha + \lambda^2 \omega^2}\right]$$

which, using the result of Exercise (14) of Chap. 4, allows us to write

$$U(\omega, t) = \phi(\omega) t^{\alpha-1} E_{\alpha,\alpha}(-\lambda^2 \omega^2 t^\alpha)$$

with $E_{\alpha,\alpha}(\cdot)$ a Mittag-Leffler function with two parameters. In order to proceed with the inversion of the Fourier transform, we introduce the called Green function, obtained when $\varphi(x) = \delta(x)$ from where we can write to the inverse Fourier transform, using the Fourier convolution product

$$\mathcal{F}^{-1}[U(\omega, t)] = u(x, t) = \int_{-\infty}^{\infty} G(x - \xi, t)\varphi(\xi)\,d\xi$$

and because the Green function is a even function in x [34] we can write

$$G(x, t) = \frac{1}{\pi}\int_0^{\infty} t^{\alpha-1} E_{\alpha,\alpha}(-\lambda^2 \omega^2 t^\alpha)\cos \omega x\,d\omega.$$

In [17], this Green function is obtained in terms of a Wright function, by integrating in the complex plane, the same result obtained in [34].   ◇

35. *Green function.* Let $a > 0$ and $\alpha > 0$ be the order of the fractional derivative. Admit that the Laplace transform of the Dirac delta 'function' is equal to unity. Get the fractional Green function, denoted by $\mathcal{G}(t)$, satisfying the fractional differential equation with constant coefficients

$$c\mathcal{D}_t^\alpha \mathcal{G}(t) + a\mathcal{G}(t) = \delta(t)$$

being the derivative in the Caputo sense and the homogeneous conditions $f^{(k)}(0^+) = 0$.

Taking the Laplace transform on both members of the non-homegeneous fractional differential equation, we have

$$s^\alpha F(s) - \sum_{k=0}^{n-1} s^{\mu-k-1} f^{(k)}(0^+) + a F(s) = 1$$

where $F(s) = \mathcal{L}[\mathcal{G}(t)]$ which, by imposing the homogeneous conditions and solving the algebraic equation for $F(s)$, allows us to write

$$F(s) = \frac{1}{s^\alpha + a}.$$

In order to calculate the inverse Laplace transform and recover the solution of the fractional differential equation, we use Exercise (14) from Chap. 4, then

$$\mathcal{G}(t) = \mathcal{L}^{-1}\left[\frac{1}{s^\alpha + a}\right] = t^{\alpha-1} E_{\alpha,\alpha}(-a\, t^\alpha),$$

which is the desired result.   ◇

36. *Fractional time difusion equation.* Let $0 < \alpha \le 1$, $x \in \mathbb{R}$, $t > 0$ and $D > 0$ be a constant. The fractional time diffusion equation (only the temporal part is fractional) is given by

$$\frac{\partial^\alpha}{\partial t^\alpha} u(x, t) = D \frac{\partial^2}{\partial x^2} u(x, t)$$

being the fractional derivative considered in the Caputo sense. Let us assume the conditions (initial and boundary)

$$u(x, 0) = \delta(x) \quad \text{and} \quad \lim_{x \to \pm\infty} u(x, t) = 0.$$

Note that in the particular case where $\alpha = 1$ we retrieve the standard diffusion equation. In order to solve the problem (equation + conditions), we will use the juxtaposition of integral transforms, that is, the Laplace transform, denoted by $u(x, s)$, in the time variable and the Fourier transform, denoted by $u(k, s)$, in the spatial variable, where s and k are the parameters associated with the Laplace and Fourier transforms, respectively.

Recalling that the Laplace transform of the Caputo derivative is

$$\mathcal{L}[c\mathcal{D}_t^\alpha u(x, t)] = s^\alpha u(x, s) - s^{\alpha-1} u(x, 0)$$

and the Fourier transform of the second derivative

$$\mathscr{F}\left[\frac{\partial^2}{\partial x^2}u(x,t)\right] = -k^2 u(k,t)$$

and taking the integral transforms of the non-homogeneous fractional partial differential equation, we obtain an algebraic equation

$$s^\alpha u(k,s) - s^{\alpha-1} = -Dk^2 u(k,s)$$

where we use the result: the Fourier transform of the Dirac delta function is unit, that is, $\mathscr{F}[\delta(x)] = 1$. The solution of the algebraic equation is

$$u(k,s) = \frac{s^{\alpha-1}}{s^\alpha + Dk^2}.$$

In order to proceed with the inversion and recover the solution of the starting problem, we begin with the inversion of the Laplace transform,

$$u(k,t) = \mathscr{L}^{-1}[u(k,s)] = \mathscr{L}^{-1}\left[\frac{s^{\alpha-1}}{s^\alpha + Dk^2}\right] = E_\alpha(-Dk^2 t^\alpha)$$

where $E_\alpha(\cdot)$ is the classical Mittag-Leffler function.
In order to invert the Fourier transform, we use the result [35]

$$\int_0^\infty \cos(kt)\, E_{\alpha,\beta}(-at^2)\, dt = \frac{\pi}{k} H_{1,1}^{1,0}\left[\frac{k^2}{a}\,\bigg|\,\begin{matrix}\beta,\alpha\\(1,2)\end{matrix}\right]$$

where $H_{c,d}^{a,b}(\cdot)$ is a Fox's H-function from where it follows for the solution

$$u(x,t) = \frac{1}{|x|} H_{1,1}^{1,0}\left[\frac{k^2}{a}\,\bigg|\,\begin{matrix}\beta,\alpha\\(1,2)\end{matrix}\right]$$

which is the desired result.   ◇

37. *Mittag-Leffler function with three parameters.* Let's show the following result [12]. Consider the parameters $\alpha > 0$, $\beta > 0$ and $x,y \in \mathbb{C}$. Then, for the Mittag-Leffler function with three parameters, denoted by $E_{a,b}^c(\cdot)$, we have

$$\sum_{r=0}^\infty (x+y)^r E_{2\alpha,\alpha r+\beta}^{r+1}(-xy) = \sum_{k=0}^\infty (-xy)^k E_{\alpha,2\alpha k+\beta}^{k+1}(x+y).$$

Denoting by Λ the first member of the preceding,

$$\Lambda = \sum_{r=0}^{\infty} (x+y)^r E_{2\alpha, \alpha r+\beta}^{r+1}(-xy),$$

we will introduce the series representation of the Mittag-Leffler function with three parameters and change the order of integration, admitting the absolutely convergent series, then

$$\Lambda = \sum_{k=0}^{\infty} \frac{(-xy)^k}{k!} \sum_{r=0}^{\infty} \frac{(r+1)_k}{\Gamma(2\alpha k, \alpha r + \beta)} (x+y)^r$$

with $(\cdot)_n$ a Pochhammer symbol.
Using equality involving the Pochhammer symbols

$$\frac{(r+1)_k}{k!} = \frac{(k+1)_r}{r!}$$

we can write the precedent in the following way

$$\Lambda = \sum_{k=0}^{\infty} (-xy)^k \sum_{r=0}^{\infty} \frac{(k+1)_r}{\Gamma(\alpha r + 2\alpha k + \beta)} \frac{(x+y)^r}{r!}$$

which, by identifying the series with the Mittag-Leffler function with three parameters, allows us to write

$$\Lambda = \sum_{k=0}^{\infty} (-xy)^k E_{\alpha, 2\alpha k+\beta}^{k+1}(x+y)$$

which is exactly the second member of equality.   ◇

38. *Volterra integral equation.* By means of the Laplace transform methodology, we will discuss and solve a Volterra integral equation in the particular case in which the kernel is a Mittag-Leffler function with three parameters.
Consider the kernel of the second-order Volterra integral equation

$$\Phi(x) = f(x) + \lambda \int_0^x k(x|t)\Phi(t)\, dt$$

of the form $k(x-t)$, a continuous function on both variables, for $t \leq x$, and $F(s)$ and $K(s)$ the Laplace transform of the functions $f(x)$ and $k(x|t)$, respectively, we then show

$$\Phi(x) = \frac{1}{2\pi i} \int_{\gamma_0 - i\infty}^{\gamma_0 + i\infty} \frac{F(s)}{1 - \lambda K(s)} e^{xs}\, ds.$$

Let us consider the Volterra integral equation whose kernel is a Mittag-Leffler function with three parameters, and let x and y real constants, independent of t, and α, β, γ, μ complex parameters with Re $(\alpha) > 0$ and Re $(\gamma) > 0$,

$$z(t) = t^{\beta-1} E_{\alpha,\beta}(xt^\alpha) + y \int_0^t (t-\tau)^{\mu-1} E_{\alpha,\mu}^\gamma [x(t-\tau)^\alpha] z(\tau) \, d\tau \qquad (6.54)$$

in which $E_{\alpha,\beta}(\cdot)$ and $E_{\alpha,\beta}^\gamma(\cdot)$ are the Mittag-Leffler function with two and three parameters, respectively.

In order to solve the integral equation we will restrict ourselves to the functions that are continuous in $[0, \infty)$. Thus, according to Lerch's theorem [36], the solution is unique. Then follows the theorem [37]:

Theorem 6.3 *Let $E_{\alpha,\beta}^\gamma(\cdot)$ be a Mittag-Leffler function with three parameters with* Re$(\alpha) > 0$, Re$(\beta) > 0$ *and* Re$(\gamma) > 0$, $x, y \in \mathbb{R}$ *be constants, the function*

$$z(t) = \sum_{k=0}^\infty (yt^\mu)^k t^{\beta-1} E_{\alpha,\mu k+\beta}^{\gamma k+1}(xt^\alpha)$$

is the unique solutions of Eq. (6.54).

Proof Taking the Laplace transform of Eq. (6.54) we obtain an algebraic equation whose solution is given by

$$F(s) = \frac{\mathscr{L}[t^{\beta-1} E_{\alpha,\beta}(xt^\alpha)]}{1 - y\mathscr{L}[t^{\mu-1} E_{\alpha,\mu}^\gamma(xt^\alpha)]}.$$

where $F(s) \equiv \mathscr{L}[z(t)]$ is the Laplace transform with parameter s.

Using the relation [19]

$$\mathscr{L}[t^{\nu-1} E_{\alpha,\nu}^\xi(at^\alpha)] = \frac{s^{\alpha\xi-\nu}}{(s^\alpha - a)^\xi}$$

we can write for the precedent

$$F(s) = \frac{s^{\alpha-\beta}/(s^\alpha - x)}{1 - ys^{\alpha\gamma-\mu}/(s^\alpha - x)^\gamma} = \frac{s^{\alpha-\beta}}{s^\alpha - x} \frac{1}{1 - y\dfrac{s^{\alpha\gamma-\mu}}{(s^\alpha - x)^\gamma}}.$$

Let us now consider the particular case in which the condition

$$\left| y \frac{s^{\alpha\gamma-\mu}}{(s^\alpha - x)^\gamma} \right| < 1.$$

is valid. Then, using the geometric series, we have

$$F(s) = \sum_{k=0}^{\infty} y^k \frac{s^{\alpha\gamma k - \mu k + \alpha - \beta}}{(s^\alpha - x)^{\gamma k + 1}}.$$

Taking the inverse Laplace transform and using the relation

$$\mathscr{L}^{-1}\left[\frac{s^{\alpha\xi - \nu}}{(s^\alpha - a)^\xi}\right] = t^{\nu-1} E_{\alpha,\nu}^\xi(at^\alpha)$$

we obtain, after some algebraic manipulations,

$$z(t) = \sum_{k=0}^{\infty} (yt^\mu)^k t^{\beta-1} E_{\alpha,\mu k+\beta}^{\gamma k+1}(xt^\alpha), \tag{6.55}$$

which is the solution of the second kind Volterra integral equation whose kernel is a Mittag-Leffler function with three parameters. □

39. *Sums involving a Mittag-Leffler function.* In a recent paper the following result was shown as a theorem [12]

$$\sum_{k=0}^{\infty} (-xy)^k E_{\alpha,2\alpha k+\beta}^{k+1}(x+y) = \frac{xE_{\alpha,\beta}(x) - yE_{\alpha,\beta}(y)}{x - y} \tag{6.56}$$

that is, a sum containing the Mittag-Leffler function with three parameters, resulting in a quotient involving two Mittag-Leffler functions with two parameters, where $x \neq y$. Let us here discuss the case $y \to x$ and recover some particular sums.

Taking the limit $y \to x$ and using the l'Hôpital role in the second member, follows,

$$\sum_{k=0}^{\infty} (-x^2)^k E_{\alpha,2\alpha k+\beta}^{k+1}(2x) = E_{\alpha,\beta}(x) + x \frac{d}{dx} E_{\alpha,\beta}(x)$$

which is the desired result. Considering $\alpha = 1$ in Eq. (6.56) we get

$$\sum_{k=0}^{\infty} (-xy)^k E_{1,2k+\beta}^{k+1}(x+y) = \frac{xE_{1,\beta}(x) - yE_{1,\beta}(y)}{x - y}.$$

Using the result

$$\Gamma(\mu) E_{1,\mu}^\rho(x) = {}_1F_1(\rho; \mu; x)$$

with ${}_1F_1(\rho; \mu; x)$ a confluent hypergeometric function, we have,

$$\sum_{k=0}^{\infty} \frac{(-xy)^k}{\Gamma(2k+\beta)} {}_1F_1(k+1; 2k+\beta; x+y) = \frac{xE_{1,\beta}(x) - yE_{1,\beta}(y)}{x - y}.$$

Using the precedent result and considering $\beta = 1$ we get, for $x \neq y$,

$$\sum_{k=0}^{\infty} \frac{(-x^2)^k}{(2k)!} \, {}_1F_1(k+1; 2k+1; x+y) = \frac{xe^x - ye^y}{x - y},$$

a particular series involving the confluent hypergeometric function. In the particular case $y \to x$ we get, using the l'Hôpital role

$$\sum_{k=0}^{\infty} \frac{(-x^2)^k}{(2k)!} \, {}_1F_1(k+1; 2k+1; 2x) = (1+x)\, e^x.$$

Finally, in the case $\beta = 2$, we can write

$$\sum_{k=0}^{\infty} (-xy)^k \frac{{}_1F_1(k+1; 2k+2; x+y)}{\Gamma(2k+2)} = \frac{e^x - e^y}{x - y},$$

which, in the limit $y \to x$ and, also, using the l'Hôpital, provides

$$\sum_{k=0}^{\infty} (-x^2)^k \frac{{}_1F_1(k+1; 2k+2; 2x)}{\Gamma(2k+2)} = e^x$$

which is the desired result   ◇

40. *On the classical Mittag-Leffler function* [38]. Let us answer the following question: is there another function, different from the exponential function, possibly dependent on parameters, for which its Laplace transform is equal to

$$\mathcal{L}[f(t)] = \frac{1}{s - \mu} = \frac{1/s}{1 - \mu/s} \qquad (6.57)$$

with $\mu \in \mathbb{R}$?

Consider the well-known result, the Laplace transform of the function $e^{\mu t}$ is

$$\mathcal{L}[e^{\mu t}] = \int_0^{\infty} e^{\mu t} e^{-st} \, dt = \frac{1}{s - \mu} \qquad (6.58)$$

for $|\mu/s| < 1$.

Deriving the precedent, with respect to the parameter μ, k times, we obtain

$$\int_0^{\infty} e^{\mu t} t^k e^{-st} \, dt = \frac{k!}{(s - \mu)^{k+1}}$$

with $k \geq 0$. Note that the particular case $k = 0$ retrieves Eq. (6.57). Moreover, the preceding equality can be interpreted as the Laplace transform of $f(t) = t^k e^{\mu t}$ which, using the inverse theorem, allow us to write

$$\mathcal{L}^{-1}\left[\frac{1}{(s-\mu)^{k+1}}\right] = \frac{t^k e^{\mu t}}{k!}. \qquad (6.59)$$

Consider the function $g(t) = \sum_{n=0}^{\infty} a_n (\mu t^\alpha)^n$, where $\alpha \in \mathbb{R}_+^*$ is a parameter and a_n are coefficients to be determined. In particular, in the case $\alpha = 1$ and $a_n = 1/n!$ we obtain $g(t) = e^{\mu t}$, that is, the exponential function is recovered.

Let's look for the convenient α, a parameter and a_n, a coefficient, in order to answer the question. For this, we consider $s = 1$ and $k = 0$ in Eq. (6.59) and we exchange the integral signal with the sum signal. Thus, we must find the values of the parameter and the coefficients, for which equality is valid

$$\sum_{n=0}^{\infty} a_n \mu^n \int_0^{\infty} e^{-t} t^{\alpha n} \, dt = \frac{1}{1-\mu}.$$

Evaluating the integral and using the definition of the gamma function, we can write

$$\sum_{n=0}^{\infty} a_n \mu^n \Gamma(\alpha n + 1) = \frac{1}{1-\mu}.$$

From the preceding expression, imposing the condition $a_n \Gamma(\alpha n + 1) = 1$, we get for the function $g(t)$

$$g(t) = \sum_{n=0}^{\infty} \frac{(\mu t^\alpha)^n}{\Gamma(\alpha n + 1)}$$

with $\alpha \in \mathbb{R}_+^*$ a free parameter.

This function $g(t)$, denoted in the literature by $E_\alpha(\mu t^\alpha)$ is the classical Mittag-Leffler function, introduced several years ago [39].

Then, rearranging these results, we can write to the integral

$$\int_0^{\infty} e^{-t} E_\alpha(\mu t^\alpha) \, dt = \frac{1}{1-\mu}$$

and, by introducing a variable change $t \to st$, with $\mu \to \mu s^{-\alpha}$, we obtain

$$\int_0^{\infty} e^{-st} E_\alpha(\mu t^\alpha) \, dt = \frac{s^{\alpha-1}}{s^\alpha - \mu}$$

interpreted as the Laplace transform of the classical Mittag-Leffler function.

Finally, in answering the question, we conclude that in the case $s = 1$ there is a function whose integral gives the value equal to the value of the second member of Eq. (6.58), and for $\mathrm{Re}(s) > 0$, we obtain a general result, in which the case $\alpha = 1$ recovers the Laplace transform of the exponential function   ◇

41. *Fractional differential equation with nonconstant coefficient—I.* Using the inverse operator, solve the fractional differential equation with nonconstant coefficient, and the derivative considered in the Caputo sense

$$(1 + x)\,{}_x\mathbb{D}_0^\alpha y(x) = 1 \tag{6.60}$$

with $0 < \alpha \leq 1$, satisfying the initial condition $y(0) = 0$.
This fractional differential equation can be written as

$$ {}_x\mathbb{D}_0^\alpha y(x) = \frac{1}{1 + x} $$

and operating with the corresponding integral operator (inverse operator) we get

$$ \mathcal{J}^\alpha \left[{}_x\mathbb{D}_0^\alpha y(x) \right] = \mathcal{J}^\alpha \left[\frac{1}{1 + x} \right]. $$

Using the property $\mathcal{J}^\alpha \left[{}_x\mathbb{D}_0^\alpha y(x) \right] = y(x)$ and the expression for the integral operator, we can write

$$ y(x) = \frac{1}{\Gamma(\alpha)} \int_0^x (x - \xi)^{\alpha-1} \frac{d\xi}{1 + \xi}. $$

Introducing $\xi = xt$ in the previous integral we have

$$ y(x) = \frac{x^\alpha}{\Gamma(\alpha)} \int_0^1 (1 - t)^{\alpha-1} \frac{dt}{1 + xt}. $$

This is a known integral [40], then

$$ y(x) = \frac{x^\alpha}{\Gamma(\alpha)} B(1, \alpha)\, {}_2F_1(1, 1; 1 + \alpha; -x) $$

where $B(1, \alpha)$ is a beta function and ${}_2F_1(1, 1; 1 + \alpha; -x)$ a hypergeometric function.
Using the relation involving gamma and beta functions, and simplifying we have

$$ y(x) = \frac{x^\alpha}{\Gamma(1 + \alpha)}\, {}_2F_1(1, 1; 1 + \alpha; -x) $$

which is the solution of the fractional differential equation, Eq. (6.60), satisfying the initial condition   ◇

42. *Fractional differential equation with nonconstant coefficient—II.* We discuss the same problem (equation + condition) involving the fractional differential equation with nonconstant coefficient presented in precedent exercise, by means of the Laplace transform methodology. We discuss also the particular case $\alpha = 1$. Taking the Laplace transform with parameter s in both sides of Eq. (6.60) and using the linearity of the Laplace transform, we have

$$\mathscr{L}[_x\mathbb{D}_0^\alpha y(x)] + \mathscr{L}[x \ _x\mathbb{D}_0^\alpha y(x)] = \mathscr{L}[1]$$

which can be written as

$$\mathscr{L}[_x\mathbb{D}_0^\alpha y(x)] - \frac{d}{dx}\mathscr{L}[_x\mathbb{D}_0^\alpha y(x)] = \frac{1}{s}.$$

Using the relation involving the Laplace transform of the Caputo derivative, Eq. (5.12), we obtain

$$s^\alpha F(s) - s^{\alpha-1}y(0) - \frac{d}{dx}\left[s^\alpha F(s) - s^{\alpha-1}y(0)\right] = \frac{1}{s}$$

where $F(s) = \mathscr{L}[y(x)]$. Introducing the initial condition and rearranging we get

$$\frac{d}{ds}F(s) + \left(\frac{\alpha}{s} - 1\right)F(s) = -\frac{1}{s^{\alpha+1}}$$

which is a nonhomogeneous ordinary differential equation. To solve this ordinary differential equation we use the variation of parameters, then

$$F(s) = C\,e^s s^{-\alpha} - e^s s^{-\alpha}\mathsf{Ei}(-s)$$

with C a constant and $\mathsf{Ei}(\cdot)$ the exponential integral [40]. Taking the corresponding inverse Laplace transform, we obtain

$$y(x) = C\mathscr{L}^{-1}\left[e^s s^{-\alpha}\right] - \mathscr{L}^{-1}\left[e^s s^{-\alpha}\mathsf{Ei}(-s)\right].$$

These inverse Laplace transforms are tabulated [41], then

$$y(x) = C\frac{(x+1)^{\alpha-1}}{\Gamma(\alpha)} + \frac{x^\alpha}{\Gamma(\alpha+1)}\,_2F_1(1, 1; 1+\alpha; -x)$$

where $_2F_1(1, 1; 1+\alpha; -x)$ is a hypergeometric function. Using initial condition we have $C = 0$, then

$$y(x) = \frac{x^\alpha}{\Gamma(1+\alpha)} \, {}_2F_1(1, 1; 1+\alpha; -x)$$

which is the same result as obtained in the previous exercise.
For the particular case $\alpha = 1$ (integer case) we have

$$y(x) = x \, {}_2F_1(1, 1; 2; -x)$$

which can be written as

$$y(x) = -\sum_{k=0}^{\infty} \frac{(-x)^{k+1}}{k+1}.$$

This is a known series associated with the logarithm, then

$$y(x) = -[-\ln(1+x)] = \ln(1+x)$$

which is the solution of the nonhomogeneous first order ordinary differential
equation

$$(1+x)\frac{\mathrm{d}}{\mathrm{d}x}y(x) = 1$$

satisfying $y(0) = 0$.   ◇

43. *Riemann-Liouville fractional derivative.* Show that the Riemann-Liouville frac-
    tional derivative of order $\alpha \in \mathbb{R}$ of an analytic function f is given by

$$(\mathcal{D}^\alpha f)(t) = \sum_{n=0}^{\infty} \binom{\alpha}{n} \frac{f^{(n)}(t)t^{n-\alpha}}{\Gamma(n-\alpha+1)} \tag{6.61}$$

where $\binom{\alpha}{n}$ is the binomial coefficient.

First, we consider $\alpha < 0$. Thus, $\mathcal{D}^\alpha = \mathcal{J}^{-\alpha}$ being $\mathcal{J}^{-\alpha}$ the Riemann-Liouville
fractional integral operator, we can write

$$\mathcal{D}^\alpha f(t) = \mathcal{J}^{-\alpha} f(t) = \mathcal{J}^\beta f(t) = \frac{1}{\Gamma(\beta)} \int_0^t f(\tau)(t-\tau)^{\alpha-1}\mathrm{d}\tau$$

with $\beta = -\alpha$.

As f is an analitic function, we obtain

$$f(\tau) = \sum_{n=0}^{\infty} \frac{f^{(n)}(t)(\tau-t)^n}{n!}.$$

Substituting this expansion for f in the definition of Riemann-Liouville fractional derivative and changing the order of the sum with the integral we have

$$\mathcal{D}^\alpha f(t) = \frac{1}{\Gamma(\beta)} \sum_{n=0}^{\infty} \frac{f^{(n)}(t)(-1)^n}{n!} \int_0^t (t-\tau)^{\beta-1+n} d\tau.$$

Evaluating the resultant integral we get

$$\mathcal{D}^\alpha f(t) = \frac{1}{\Gamma(\beta)} \sum_{n=0}^{\infty} \frac{f^{(n)}(t)(-1)^n}{n!} \frac{t^{\beta+n}}{\beta+n}$$

which can be written as follows

$$\mathcal{D}^\alpha f(t) = \frac{1}{\Gamma(\beta)} \sum_{n=0}^{\infty} \frac{f^{(n)}(t)(-1)^n}{n!} \frac{t^{\beta+n} \Gamma(\beta+n)}{\Gamma(\beta+n+1)}.$$

Using the reflection formula we obtain

$$\mathcal{D}^\alpha f(t) = \sum_{n=0}^{\infty} \binom{-\beta}{n} \frac{f^{(n)}(t) t^{n+\beta}}{\Gamma(n+\beta+1)} = \sum_{n=0}^{\infty} \binom{\alpha}{n} \frac{f^{(n)}(t) t^{n-\alpha}}{\Gamma(n-\alpha+1)}$$

which is the desired result.

The case $\alpha > 0$. Let $m \in \mathbb{N}$ and $m - 1 < \alpha \le m$ and using the previous result, associated with $\alpha < 0$, we can write

$$\mathcal{D}^\alpha f(t) = \frac{d^m}{dt^m} \mathcal{J}^{m-\alpha} f(t) = \frac{d^m}{dt^m} \sum_{n=0}^{\infty} \binom{\alpha-m}{n} \frac{f^{(n)}(t) t^{n+m-\alpha}}{\Gamma(n+m-\alpha+1)}.$$

Considering the last series as a uniformly convergent, using the Leibniz rule for the integer case and rearranging, we obtain

$$\mathcal{D}^\alpha f(t) = \sum_{n=0}^{\infty} \sum_{k=0}^{\infty} \binom{m}{k} f^{(n+k)}(t) \frac{t^{n+k-\alpha}}{\Gamma(n+k-\alpha+1)} \binom{\alpha-m}{n}.$$

Introducing the change $k \to j - n$ in the previous expression, we get

$$\mathcal{D}^\alpha f(t) = \sum_{j=0}^{\infty} \sum_{n=0}^{\infty} \binom{m}{j-n} \binom{\alpha-m}{n} f^{(j)}(t) \frac{t^{j-\alpha}}{\Gamma(j-\alpha+1)}.$$

Using the relation

$$\sum_{n=0}^{\infty} \binom{\alpha}{m-n} \binom{\beta}{n} = \binom{\alpha+\beta}{m}$$

we obtain

$$\mathcal{D}^{\alpha} f(t) = \sum_{j=0}^{\infty} \binom{\alpha}{j} f^{(j)}(t) \frac{t^{j-\alpha}}{\Gamma(j-\alpha+1)}$$

which is the desired result. We conclude that Eq. (6.61) for the Riemann-Liouville fractional derivative is valid for $\alpha \in \mathbb{R}$.   ◇

44. *Riemann-Liouville derivative and Leibniz rule.* Show that the Riemann-Liouville fractional derivative satisfies the generalized Leibniz rule, i.e., the expression

$$\mathcal{D}^{\alpha}(fg)(t) = \sum_{k=0}^{\infty} \binom{\alpha}{k} f^{(k)}(t) \mathcal{D}^{\alpha-k} g(t)$$

for two analitic functions f and g, is valid.

Using the result presented in the previous exercise, the Leibniz rule for integer order and rearranging, we can write the following expression

$$\mathcal{D}^{\alpha}(fg)(t) = \sum_{k=0}^{\infty} f^{(k)}(t) \sum_{n=k}^{\infty} \binom{\alpha}{n} \binom{n}{k} \frac{t^{n-\alpha}}{\Gamma(n-\alpha+1)} g^{(n-k)}(t).$$

Introducing the change of index $n \to n+k$, we get

$$\mathcal{D}^{\alpha}(fg)(t) = \sum_{k=0}^{\infty} f^{(k)}(t) \sum_{n=0}^{\infty} \binom{\alpha}{n+k} \binom{n+k}{k} \frac{t^{n+k-\alpha}}{\Gamma(n+k-\alpha+1)} g^{(n)}(t).$$

Using the known relation, involving binomial coefficients,

$$\binom{\alpha}{n+k} \binom{n+k}{k} = \binom{\alpha}{k} \binom{\alpha-k}{n}$$

we obtain for the Riemann-Liouville fractional derivative

$$\mathcal{D}^{\alpha}(fg)(t) = \sum_{k=0}^{\infty} \binom{\alpha}{k} f^{(k)}(t) \sum_{n=0}^{\infty} \binom{\alpha-k}{n} \frac{t^{n+k-\alpha}}{\Gamma(n+k-\alpha+1)} g^{(n)}(t).$$

Finally, using Eq. (6.61) we obtain

$$\mathcal{D}^{\alpha}(fg)(t) = \sum_{k=0}^{\infty} \binom{\alpha}{k} f^{(k)}(t) \mathcal{D}^{\alpha-k} g(t)$$

which is the desired result.   ◇

45. *Nonhomogeneous fractional differential equation.* Let $A, B \in \mathbb{R}$ be constants, $\alpha(1 < \alpha \leq 2)$ and $\beta(0 < \beta \leq 1)$. Considering the derivative operator in the Caputo sense, solve the nonhomogeneous fractional differential equation

$$\mathcal{D}^{\alpha} y(x) + \mathcal{D}^{\beta} y(x) = 1$$

satisfying the initial conditions $y(0) = A$ and $y'(0) = B$. Discuss the particular cases: (i) $A = 0 = B$ and (ii) $A = 0 = B$ and $\alpha = 2 = 2\beta$.

As the coefficients are constants, the Laplace transform methodology is adequate. Thus, taking the Laplace transform in both sides and using the linearity of the Laplace transform, we obtain

$$\mathcal{L}\left[\frac{d^{\alpha}}{dx^{\alpha}} y(x)\right] + \mathcal{L}\left[\frac{d^{\beta}}{dx^{\beta}} y(x)\right] = \mathcal{L}[1].$$

Using the relation involving the Laplace transform and the Caputo fractional derivative twice, we get

$$s^{\alpha} F(s) - s^{\alpha-1} y(0) - s^{\alpha-2} y'(0) + s^{\beta} F(s) - s^{\beta-1} y(0) = \frac{1}{s},$$

where $F(s) = \mathcal{L}[y(x)]$ is the Laplace transform with parameter s. Substituting the initial conditions and solving the algebraic equation, we obtain

$$F(s) = \frac{s^{-1}}{s^{\alpha} + s^{\beta}} + \left(\frac{s^{\alpha-1}}{s^{\alpha} + s^{\beta}} + \frac{s^{\beta-1}}{s^{\alpha} + s^{\beta}}\right) A + \frac{s^{\alpha-2}}{s^{\alpha} + s^{\beta}} B$$

which can be written in the following form

$$F(s) = \frac{s^{-1-\beta}}{1 + s^{\alpha-\beta}} + \left(\frac{s^{\alpha-\beta-1}}{1 + s^{\alpha-\beta}} + \frac{s^{-1}}{1 + s^{\alpha-\beta}}\right) A + \frac{s^{\alpha-\beta-2}}{1 + s^{\alpha-\beta}} B.$$

To recover the solution of the nonhomogeneous fractional differential equation we must take the corresponding inverse Laplace transform

$$y(x) = \mathcal{L}^{-1}[F(s)].$$

Using the result obtained in Exercise (14) of Chap. 4, and the linearity of the inverse Laplace transform, we can write

$$y(x) = x^{\alpha} E_{\alpha-\beta,\alpha+1}(-x^{\alpha-\beta})$$
$$+ A\left[E_{\alpha-\beta}(-x^{\alpha-\beta}) + x^{\alpha-\beta} E_{\alpha-\beta,\alpha-\beta+1}(-x^{\alpha-\beta})\right] + B x E_{\alpha-\beta,\alpha+2}(-x^{\alpha-\beta})$$

where $E_\mu(\cdot)$ and $E_{\mu,\nu}(\cdot)$ are the Mittag-Leffler functions with one and two parameters, respectively. (i) Putting $A = 0 = B$ in the precedent, we have

$$y(x) = x^\alpha E_{\alpha-\beta,\alpha+1}(-x^{\alpha-\beta}).$$

(ii) In this case we put $\alpha = 2$ and $\beta = 1$ in the previous equation to get

$$y(x) = x^2 E_{1,3}(-x).$$

Using the result obtained in the Proposed exercise (15) of Chap. 3, and simplifying we can write

$$y(x) = e^x - x - 1$$

which is the solution of the second order ordinary differential equation

$$\frac{d^2}{dx^2}y(x) + \frac{d}{dx}y(x) = 1$$

satisfying the initial conditions $y(0) = 0 = y'(0)$.   ◇

46. *Fractional integrodifferential equation.* Let $\alpha(0 < \alpha \leq 1)$. Use the Laplace transform methodology to solve the fractional integrodifferential equation

$$\int_0^x \mu(x - \xi)\mathcal{D}_\xi^\alpha y(\xi)\,d\xi = x$$

satisfying $y(0) = 1$, considering the derivative in the Caputo sense and the kernel being the classical Mittag-Leffler function, i.e.,

$$\mu(x) = E_\beta(-t^\beta)$$

with $\beta > 0$. Discuss the particular case $\alpha = 1 = \beta$.

Taking the Laplace transform with parameter s in both sides and using the convolution theorem, we get

$$\mathcal{L}[\mu(x)]\mathcal{L}[\mathcal{D}_x^\alpha y(x)] = \frac{1}{s^2}.$$

By means of the relation involving the Laplace transform and the Caputo fractional derivative we can write

$$\mathcal{L}[\mu(x)][s^\alpha F(s) - s^{\alpha-1}] = \frac{1}{s^2}$$

where $F(s) = \mathcal{L}[y(x)]$. The Laplace transform of the classical Mittag-Leffler function is a known result and using the initial condition we obtain

$$\frac{s^{\beta-1}}{s^\beta+1}[s^\alpha F(s) - s^{\alpha-1}] = \frac{1}{s^2}$$

which is an algebraic equation whose solution is given by

$$F(s) = \frac{1}{s} + \frac{1}{s^{\alpha+1}} + \frac{1}{s^{\alpha+\beta+1}}.$$

To recover the solution of the integrodifferential equation we must take the corresponding inverse Laplace transform

$$y(x) = \mathscr{L}^{-1}[F(s)],$$

with the linearity of the inverse Laplace transform, furnishes

$$y(x) = \mathscr{L}^{-1}\left[\frac{1}{s}\right] + \mathscr{L}^{-1}\left[\frac{1}{s^{\alpha+1}}\right] + \mathscr{L}^{-1}\left[\frac{1}{s^{\alpha+\beta+1}}\right].$$

These three inverse Laplace transform are known, then

$$y(x) = 1 + \frac{x^\alpha}{\Gamma(\alpha+1)} + \frac{x^{\alpha+\beta}}{\Gamma(\alpha+\beta+1)}$$

where $\Gamma(\cdot)$ is a gamma function.
The particular case $\alpha = 1 = \beta$ furnishes

$$y(x) = 1 + x + \frac{x^2}{2}$$

which is the solution of the integer order integrodifferential equation

$$\int_0^x e^{-(x-\xi)}\frac{d}{d\xi}y(\xi)\,d\xi = x$$

satisfying the condition $y(0) = 1$.   ◇

47. *Fractional integral equation.* Consider $x > 0$ and $\alpha > 0$. Use the Laplace transform methodology to solve the integral equation

$$\int_0^x E_\alpha[-(x-\xi)^\alpha]y(\xi) = \frac{x^2}{2}$$

satisfying the condition $y(0) = 0$ and $E_\alpha(\cdot)$ is the classical Mittag-Leffler function.
Taking the Laplace transform in both sides and using the convolution theorem, we can write

$$\mathcal{L}[E_\alpha(-x^\alpha)]\mathcal{L}[y(x)] = \frac{1}{2}\mathcal{L}[x^2].$$

Using the known relation

$$\mathcal{L}[E_\alpha(-x^\alpha)] = \frac{s^{\alpha-1}}{s^\alpha + 1}$$

we obtain an algebraic equation whose solution is given by

$$F(s) = \frac{1}{s^2} + \frac{1}{s^{\alpha+2}}$$

where $F(s) = \mathcal{L}[y(x)]$ is the Laplace transform of parameter s. Taking the inverse Laplace transform we have

$$y(x) = x + \frac{x^{\alpha+1}}{\Gamma(\alpha + 1)}$$

where $\Gamma(\cdot)$ is the gamma function. In the particular case $\alpha = 1$ we obtain

$$y(x) = x + \frac{x^2}{2}$$

which is the solution of the integer order integral equation

$$\int_0^x e^\xi y(\xi)\, d\xi = \frac{x^2}{2} e^x$$

satisfying $y(0) = 0$.   ◇

48. *Fractional differential equation.* Let $\gamma (0 < \gamma \le 1)$ and $\alpha, \beta > 0$. Use the Laplace transform to solve the fractional differential equation

$$\mathcal{D}_x^\alpha y(x) = \int_0^x (x - \xi)^{\beta-1} E_{\alpha,\beta}[-(x - \xi)^\alpha]\, d\xi$$

satisfying $y(0) = 0$ with $\mathcal{D}_x^\alpha$ is considered in the Caputo sense and $E_{\alpha,\beta}(\cdot)$ is the Mittag-Leffler function with two parameters.

Note that, the integral in the second member can be evaluated using the series for the Mittag-Leffler function with two parameters and after by means of the inverse integral operator, we obtain the solution, but as the kernel has a particular function, the Laplace transform methodology seems to be more adequate.

Thus, taking the Laplace transform and use the convolution theorem, we get

$$s^\gamma F(s) - s^{\gamma-1} y(0) = \mathcal{L}[x^{\beta-1} E_{\alpha,\beta}(-x^\alpha)]\mathcal{L}[1]$$

where $F(s) = \mathcal{L}[y(x)]$ is the Laplace transform with parameter s. Using the known relation, Eq. (30) of Chap. 4,

$$\mathcal{L}\left[x^{b-1}E_{a,b}(-x^a)\right] = \frac{s^{a-b}}{s^a + 1}$$

and rearranging we get

$$F(s) = \frac{s^{a-b-\gamma-1}}{s^a + 1}.$$

Taking the corresponding inverse Laplace transform, we obtain

$$y(x) = x^{\beta+\gamma}E_{\alpha,\beta+\gamma+1}(-x^\alpha)$$

which is the desired result. Taking $\alpha = 1 = \beta$ we get

$$y(x) = x^{\gamma+1}E_{1,2+\gamma}(-x)$$

which is the solution of the fractional differential equation

$$\mathcal{D}_x^\gamma y(x) = 1 - e^{-x}.$$

Also, taking $\alpha = \beta = \gamma = 1$ we have

$$y(x) = x^2 E_{1,3}(-x).$$

Using the known relation, Proposed exercise (15) of Chap. 3,

$$x^2 E_{1,3}(-x) = e^{-x} + x - 1,$$

we get the expression

$$y(x) = e^{-x} + x - 1$$

which is the solution of the ordinary differential equation

$$\frac{d}{dx}y(x) = 1 - e^{-x}$$

satisfying the condition $y(0) = 0$. ◇

49. *An important property of the Fox's H-function.* Show that [35]

$$H_{p,q}^{m,n}\left[z \,\middle|\, \begin{matrix} (a_p, \alpha_p) \\ (b_q, \beta_q) \end{matrix}\right] = k\, H_{p,q}^{m,n}\left[z^k \,\middle|\, \begin{matrix} (a_p, k\alpha_p) \\ (b_q, k\beta_q) \end{matrix}\right]$$

with k a real positive parameter.

First, we write the Fox's H-function in terms of an integral in the complex plane

$$H_{p,q}^{m,n}(z) = \frac{1}{2\pi i} \int_\gamma \mathcal{H}_{p,q}^{m,n}(s) z^{-s} \, ds \qquad (6.62)$$

where we have introduced

$$\mathcal{H}_{p,q}^{m,n}(s) = \frac{\displaystyle\prod_{j=1}^{n} \Gamma(1 - a_j + \alpha_j s) \quad \prod_{j=1}^{m} \Gamma(b_j - \beta_j s)}{\displaystyle\prod_{j=n+1}^{p} \Gamma(a_j - \alpha_j s) \quad \prod_{j=m+1}^{q} \Gamma(1 - b_j + \beta_j s)}$$

with $0 \le n \le p$ and $1 \le m \le q$. Without loss of generality, we work with $\mathcal{H}_{p,q}^{m,n}(s)$. Taking $s \to ks$ in previous expression, we can write

$$\mathcal{H}_{p,q}^{m,n}(ks) = \frac{\displaystyle\prod_{j=1}^{n} \Gamma(1 - a_j + k\alpha_j s) \quad \prod_{j=1}^{m} \Gamma(b_j - k\beta_j s)}{\displaystyle\prod_{j=n+1}^{p} \Gamma(a_j - k\alpha_j s) \quad \prod_{j=m+1}^{q} \Gamma(1 - b_j + k\beta_j s)}$$

also in Eq. (6.62) we obtain the expression

$$H_{p,q}^{m,n}(z) = k \, \frac{1}{2\pi i} \int_\gamma \mathcal{H}_{p,q}^{m,n}(ks)(z^k)^{-s} \, ds.$$

Identifying these last two expression we get

$$H_{p,q}^{m,n}\left[z \, \middle| \, \begin{matrix} (a_p, \alpha_p) \\ (b_q, \beta_q) \end{matrix} \right] = k \, H_{q,p}^{m,n}\left[z^k \, \middle| \, \begin{matrix} (a_p, k\alpha_p) \\ (b_q, k\beta_q) \end{matrix} \right]$$

which is the desired result.   ◇

50. *Mellin transform and Fox's H-function.* In a recent study [42] was discussed a fractional differential equation associated with the diffusion with the diffusion coefficient being time dependent. Using the similarity method and the Mellin transform methodology was found an explicit solution of the fractional differential equation expressing it in terms of a Fox's H-function.

Let $\alpha, \gamma \in \mathbb{R}$ and $\ell \notin \mathbb{N}$. Using the same notation, the integral in the complex plane was obtained in the following form

$$F(\eta) = \frac{1}{2\pi i} \int_{c-i\infty}^{c+i\infty} \frac{(\rho^{-\gamma}k)^{\frac{s}{\alpha}} \Gamma(s - \ell)}{\Gamma\left(\frac{\gamma}{\alpha}s + 1\right)} \eta^{-s} \, ds$$

where $\rho > 0$, $k > 0$ and s is the parameter of the Mellin transform. (i) Express $F(\eta)$ as a Fox's H-function; (ii) $F(\eta)$ can be express as a Meijer's G-function? and (iii) Discuss the particular case $\ell = -\alpha/\gamma$ and $\alpha = 2$ and $\gamma = 1$.

(i) First, we write $F(\eta)$ as follows

$$F(\eta) = \frac{1}{2\pi i} \int_{c-i\infty}^{c+i\infty} \frac{\Gamma(s-\ell)}{\Gamma\left(\frac{\gamma}{\alpha}s + 1\right)} z^{-s} \, ds$$

with $z = \frac{\eta}{(\rho^{-\gamma}k)^{\frac{1}{\alpha}}}$.

Using Definition (2.4.4) we can identify the parameters: $m = 1$, $n = 0$ and $p = 1 = q$, then we obtain $\alpha_1 = \gamma/\alpha$ and $a_1 = 1$; $\beta_1 = 1$ and $b_1 = -\ell$. Then, we can write

$$F(\eta) = H_{1,1}^{1,0}\left[z \,\middle|\, \begin{matrix} (1, \gamma/\alpha) \\ (-\ell, 1) \end{matrix}\right].$$

(ii) By means of this last expression, $F(\eta)$ can be written as a Meijer's G-function if, and only if, $\gamma = \alpha$, which furnishes

$$F(\eta) = G_{1,1}^{1,0}\left[z^{-1} \,\middle|\, \begin{matrix} (1, 1) \\ (-\ell, 1) \end{matrix}\right].$$

(iii) Using the result obtained in item (i), putting $\ell = -\alpha/\gamma$ with $\alpha = 2$ and $\gamma = 1$, we can write

$$F(\eta) = H_{1,1}^{1,0}\left[z \,\middle|\, \begin{matrix} (1, 1/2) \\ (2, 1) \end{matrix}\right]$$

with $z = \eta\sqrt{\rho/k}$.

Writing this expression in terms of an integral in the complex plane, we get

$$F(\eta) = \frac{1}{2\pi i} \int_\gamma \frac{\Gamma(2+s)}{\Gamma(1+s/2)} z^{-s} \, ds.$$

In this expression we consider the change $s \to 2s$, then

$$F(\eta) = \frac{1}{\pi i} \int_\gamma \frac{\Gamma(2+2s)}{\Gamma(1+s)} (z^2)^{-s} \, ds.$$

Using the duplication formula for the gamma function and rearranging, we have

$$F(\eta) = \frac{z^3}{2\sqrt{\pi}} \underbrace{\frac{1}{2\pi i} \int_\gamma \Gamma(s) \left(\frac{z^2}{4}\right)^{-s} \, ds}.$$

The last integral is given by an exponential function, then

$$F(\eta) = \frac{(\eta\sqrt{\rho/k})^3}{2\sqrt{\pi}} \exp\left(-\frac{\rho}{4k}\eta^2\right)$$

which is the desired result.   ◇

51. *Gamma function and its inverse Mellin transform.* As we have already said, the exponential function has the Mellin transform given by

$$\mathcal{M}[e^{-x}] = \Gamma(s)$$

where s is the parameter of the Mellin transform (see Chap. 2, Exercise (4.3)). Here, we will show the inverse problem, i.e., the inverse Mellin transform of gamma function is the exponential function

$$\mathcal{M}^{-1}[\Gamma(s)] = \frac{1}{2\pi i}\int_{\gamma} x^{-s}\Gamma(s)\,ds \qquad (6.63)$$

where the contour γ in the complex plane is a rectangular contour as in Fig. 6.2. To evaluate this integral we recall the Jordan's lemma and the residue theorem. Thus, taking $R \to \infty$ we can write for the inverse Mellin transform

$$\mathcal{M}^{-1}[\Gamma(s)] = \sum_{k=0}^{\infty} \text{Res}\left[\Gamma(s)x^{-s}\right]_{s=-k}.$$

As we know (see Chap. 2, Exercise (2.11)) the residue of $\Gamma(s)$ is given by

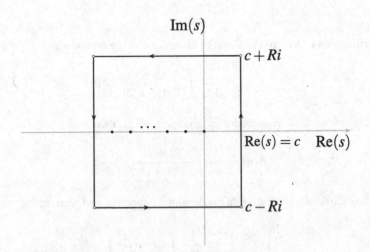

Im(s)

$c+Ri$

$Re(s) = c$ $Re(s)$

$c - Ri$

Fig. 6.2 Contour for the integral in Eq. (6.63)

$$\text{Res}\left[\Gamma(s)\right]_{s=-k} = \frac{(-1)^k}{k!}$$

thus, expanding the product $\Gamma(s)x^{-s}$, we have for the non null term

$$\Gamma(s)x^{-s} = \frac{1}{s+k}\left\{\frac{(-1)^k}{k!}x^k + \cdots\right\}$$

Then, we have

$$\text{Res}[\Gamma(s)x^{-s}]_{s=-k} = \frac{(-x)^k}{k!}.$$

Substituting this result in Eq. (6.63) we obtain

$$\mathscr{M}^{-1}[\Gamma(s)] = \sum_{k=0}^{\infty}\frac{(-x)^k}{k!} = e^{-x}$$

which is the desired result ◇

52. *Nonhomogeneous differential equation with nonconstant coefficients.* Let $0 < \alpha \leq 1, 0 < \beta \leq 1, 0 < \gamma \leq 1, 0 < \mu \leq 1, |x| < 1$ and $x \neq 0$. Solve the following fractional differential equation with nonconstant coefficients

$$\mathscr{D}^{\alpha}[x^{\beta}(1-x)^{\gamma}\mathscr{D}^{\mu}y(x)] = 1$$

with $\mathscr{D} \equiv d/dx$ and the derivative considered in the Caputo/Riemann-Liouville sense. Discuss particular cases involving the parameters.
Using the identity $\mathscr{D}^{-\alpha}\mathscr{D}^{\alpha} = \mathbb{I}$ (identity operator) we obtain

$$x^{\beta}(1-x)^{\gamma}\mathscr{D}^{\mu}y(x) = \mathscr{D}^{-\alpha}[1]$$

which can be written by means of the Riemann-Liouville integral operator in the following form

$$x^{\beta}(1-x)^{\gamma}\mathscr{D}^{\mu}y(x) = \frac{1}{\Gamma(\alpha)}\int_0^x (x-\tau)^{\alpha-1}\,d\tau.$$

To evaluate the integral, we introduce an adequate change of variable $\tau = xt$ and performing the remaining integral, we get

$$x^{\beta}(1-x)^{\gamma}\mathscr{D}^{\mu}y(x) = \frac{x^{\alpha}}{\Gamma(\alpha+1)}$$

which can be written as follows

$$\mathscr{D}^{\mu} y(x) = \frac{x^{\alpha-\beta}}{\Gamma(\alpha+1)} (1-x)^{-\gamma}.$$

Proceeding as above we have

$$y(x) = \frac{1}{\Gamma(\alpha+1)} \mathscr{D}^{-\mu} [x^{\alpha-\beta} (1-x)^{-\gamma}]$$

which can be written in the following form

$$y(x) = \frac{1}{\Gamma(\alpha+1)\Gamma(\mu)} \int_0^x (x-\tau)^{\mu-1} \tau^{\alpha-\beta} (1-\tau)^{-\gamma} d\tau.$$

To evaluate the remaining integral we introduce an adequate change of variable $\tau = xt$ and then

$$y(x) = \frac{x^{\mu+\alpha-\beta}}{\Gamma(\alpha+1)\Gamma(\mu)} \int_0^1 (1-t)^{\mu-1} t^{\alpha-\beta} (1-xt)^{-\gamma} dt.$$

The resulting integral is well-known ([40], p. 286)

$$y(x) = \frac{x^{\mu+\alpha-\beta}}{\Gamma(\alpha+1)\Gamma(\mu)} B(\alpha+1-\beta, \mu) \, {}_2F_1(\gamma, \alpha+1-\beta; \mu+\alpha+1-\beta; x)$$

where $B(\cdot, \cdot)$ is a beta function and $_2F_1(a, b; c; \cdot)$ is a hypergeometric function. Using the relation between gamma and beta functions and simplifying, we obtain

$$y(x) = \frac{x^{\mu+\alpha-\beta}}{\Gamma(\alpha+1)} \frac{\Gamma(\alpha+1-\beta)}{\Gamma(\alpha+1-\beta+\mu)} \, {}_2F_1(\gamma, \alpha+1-\beta; \mu+\alpha+1-\beta; x)$$

which is the desired result.

As particular cases we will discuss: (i) $\alpha = \beta = \gamma = \mu$, expressing the result in terms of the α parameter, and (ii) $\alpha = \beta = \gamma = \mu = 1$.

(i) Substituting $\beta = \gamma = \mu = \alpha$ in the previous result, after simplification, we obtain

$$y(x) = \frac{x^{\alpha}}{\Gamma(\alpha+1)\Gamma(\alpha+1)} \, {}_2F_1(\alpha, 1; \alpha+1; x)$$

We note that, this result can be written in terms of the $\Phi(x, 1, \alpha)$ function [43] defined by

$$\sum_{k=0}^{\infty} \frac{x^k}{k+\alpha} = \int_0^{\infty} \frac{e^{-\alpha t}}{1-x \, e^{-t}} dt \equiv \Phi(x, 1, \alpha).$$

To this end, we express the hypergeometric function in terms of the hypergeometric series.

(ii) Taking $\alpha = 1$ in the previous expression, we get

$$y(x) = x \, {}_2F_1(1, 1; 2; x)$$

which can be written in terms of a logarithm function, i.e.,

$$y(x) = x \frac{\ln(1 - x)}{x} = -\ln(1 - x)$$

which is the solution of the integer order differential equation

$$\frac{d}{dx}\left[x(1 - x)\frac{d}{dx}y(x)\right] = 1$$

as can be verified by direct substitution   ◇

53. *Generalized hypergeometric function.* Let $0 < \alpha \le 1, 0 < \beta \le 1 \ x \in \mathbb{R}$. Solve the following fractional differential equation with nonconstant coefficients

$$\mathscr{D}^{\alpha}[(1 + x^2)\mathscr{D}^{\beta}y(x)] = 1 \tag{6.64}$$

with $\mathscr{D} \equiv d/dx$ and the derivative considered in the Caputo/Riemann-Liouville sense. Discuss particular cases involving the parameters.
Using the identity $\mathscr{D}^{-\alpha}\mathscr{D}^{\alpha} = \mathbb{I}$ (identity operator) we obtain

$$(1 + x^2)\mathscr{D}^{\beta}y(x) = \mathscr{D}^{-\alpha}[1]$$

which can be written by means of the Riemann-Liouville integral operator in the following form

$$(1 + x^2)\mathscr{D}^{\beta}y(x) = \frac{1}{\Gamma(\alpha)}\int_0^x (x - \tau)^{\alpha-1}\, d\tau = \frac{x^{\alpha}}{\Gamma(\alpha + 1)}.$$

The second identity was obtained in the same way as in the previous exercise. Using the same procedure we obtain

$$y(x) = \frac{1}{\Gamma(\alpha + 1)\Gamma(\beta)}\int_0^x (x - \tau)^{\beta-1}\tau^{\alpha}(1 + \tau^2)^{-1}\, d\tau.$$

This is a well-known integral ([40], p. 297, 3.251-1) and using the relation between gamma and beta functions we can write

$$y(x) = \frac{x^{\alpha+\beta}}{\Gamma(\alpha + \beta + 1)}{}_3F_2\left(1, \frac{\alpha + 1}{2}, \frac{\alpha + 2}{2}; \frac{\alpha + \beta + 1}{2}, \frac{\alpha + \beta + 2}{2}; -x^2\right)$$

where $_3F_2(a, b, c; d, e; \cdot)$ is a generalized hypergeometric function with three parameters in numerator and two in denominator. $y(x)$ as given above is the solution of the fractional differential equation with nonconstant coefficients, Eq. (6.64), satisfying $y(0) = 0$ which permit us use the derivative in the Caputo sense or in the Riemann-Liouville sense.

As a particular case we will discuss Eq. (6.64) with $\alpha = 1 = \beta$. To this end, first we substitute these values in the previous expression, use the series representation for the generalized hypergeometric function, simplifying and rearranging, we have

$$y(x) = \frac{x^2}{2} \sum_{k=0}^{\infty} \frac{(-x^2)^k}{k+1} = -\frac{1}{2} \sum_{k=0}^{\infty} \frac{(-x^2)^{k+1}}{k+1}.$$

This sum is well-known ([43], p. 698) and then we finally get

$$y(x) = \frac{1}{2} \ln(1 + x^2)$$

valid for $0 < x \leq 1$, which is the solution of the classical integer order ordinary differential equation

$$\frac{d}{dx}\left[(1 + x^2)\frac{d}{dx}y(x)\right] = 1$$

which is the desired result.   ⋄

54. *General Caputo fractional derivative.* Using the Laplace transform methodology in the study of the fractional harmonic oscillator with the general Caputo fractional derivative, whose kernel is a Mittag-Leffler function with three parameter, also known as Prabhakar-like kernel, we must evaluate a specific inverse Laplace transform, as follows.

Consider $\alpha, \beta, \gamma \in \mathbb{R}$ with $0 < \alpha \leq 1, 0 < \beta \leq 1$ and $0 < \gamma \leq 1$, satisfying the inequality $\beta \geq \alpha\gamma$. Let μ and ω be positive constants. (a) Evaluate the inverse Laplace transform

$$f(t) = \mathscr{L}^{-1}\left[\frac{s^{\beta-1}(1 - \mu s^{-\alpha})^\gamma}{s^\beta(1 - \mu s^{-\alpha})^\gamma + \omega^\beta}\right]$$

where s is the parameter of the Laplace transform. (b) Discuss the limit case $\alpha \to 1, \beta \to 1$ and $\gamma \to 1$.

(a) Rearranging we can rewrite the above expression in the following form

$$f(t) = \mathscr{L}^{-1}\left[\frac{s^{-1}}{1 + \dfrac{\omega^\beta s^{-\beta}}{(1 - \mu s^{-\alpha})^\gamma}}\right].$$

Imposing the restriction $|\omega^\beta s^{\alpha\gamma-\beta}/(s^\alpha - \mu)^\gamma| < 1$, using the geometric series and rearranging, we obtain

$$f(t) = \sum_{k=0}^{\infty} (-\omega^\beta)^k \mathscr{L}^{-1} \left[\frac{s^{\alpha\gamma k - \beta k - 1}}{(s^\alpha - \mu)^{\gamma k}} \right].$$

To evaluate the inverse Laplace transform we use the Proposed exercise (14) whose identification permit us obtain the parameters $\lambda = \mu$, $c = \alpha$, $\rho = \gamma k$ and $b = \beta k + 1$. Substituting these parameters we have

$$f(t) = \sum_{k=0}^{\infty} (-\omega^\beta)^k t^{\beta k} E_{\alpha,\beta k+1}^{\gamma k}(\mu t^\alpha)$$

where $E_{a,b}^c(\cdot)$ is a Mittag-Leffler function with three parameters.

(b) In what follow we will consider $\alpha = \beta = \gamma = 1$. In this case we have

$$f(t) = \sum_{k=0}^{\infty} (-\omega t)^k E_{1,k+1}^k(\mu t).$$

Using Exercise (6) Chap. 3, we obtain

$$f(t) = \sum_{k=0}^{\infty} \frac{(-\omega t)^k}{\Gamma(k+1)} {}_1F_1(k; k+1; \mu t)$$

where ${}_1F_1(a; b; \cdot)$ is a confluent hypergeometric function.

By means of an adequate integral representation for the confluent hypergeometric function [44], using the relation involving the beta and gamma functions, and simplifying, we can write

$$f(t) = \sum_{k=1}^{\infty} \frac{(-\omega t)^k}{\Gamma(k)} \int_0^1 e^{\mu t \xi} \xi^{k-1} \, d\xi.$$

Note that, the index was changed $k = 0 \to k = 1$ because for $k = 0$ we get $f(t) = 0$. Thus, changing the order of the sum with the integral and resetting the index, we get

$$f(t) = -\omega t \int_0^1 e^{\mu t \xi} \, d\xi \underbrace{\sum_{k=0}^{\infty} \frac{(-\omega t \xi)^k}{k!}}.$$

The highlighted expression is the representation of an exponential function, then, rearranging we get

$$f(t) = -\omega t \int_0^1 e^{(\mu t - \omega t)\xi}\, d\xi$$

whose integration furnishes, after simplification,

$$f(t) = \frac{\omega}{\omega - \mu}\left[e^{-(\omega - \mu)t} - 1\right]$$

which is the desired result result   ◊

References

1. Wang, Z.X., Guo, D.R.: Special Functions. World Scientific, Singapore (1989)
2. de Oliveira, E.C.: Analytical Methods of Integration. Editora Livraria da Física, São Paulo (2012). (in Portuguese)
3. Costa, F.S., Oliveira, D.S., Rodrigues, F.G., Capelas de Oliveira, E.: The fractional space-time radial diffusion equation in terms of the Fox's H-function. Phys. A Stat. Mech. Appl. (2018). https://doi.org/10.1016/j.physa.2018.10.002
4. Duan, J.S., Guo, A.P., Yun, W.Z.: Similarity solution for fractional diffusion equation. Abs. Appl. Anal. **2014**, 548126 (2014)
5. Lenzi, E.K., Vieira, D.S., Lenzi, M.K., Gonçalves, G., Leitoles, D.P.: Solutions for a fractional diffusion equation with radial symmetry and integrodifferential boundary conditions. Therm. Sci. **19**, S1–S6 (2015)
6. Muslih, S.I., Agrawal, O.P.: Solutions of wave equation in fractional dimensional space. In: Baleanu, D. et al. (eds.) Fractional Dynamics and Control, pp. 217–228 (2012)
7. Grigoletto, E.C., Figueiredo Camargo, R., de Oliveira, E.C.: Three-parameter Mittag-Leffler function with an integral representation on the positive real axis. Proc. Ser. Braz. Soc. Comput. Appl. Math. **6**, 010331-1–010331-7 (2018)
8. Inizan, P.: Homogeneous fractional embeddings. J. Math. Phys. **49**, 082901 (2008)
9. Diethelm, K.: The analysis of fractional differential equations. Lecture Notes in Mathematics. Springer, Heidelberg (2010)
10. Herrmann, R.: Fractional Calculus: An Introduction for Physicis. World Scientific, New Jersey (2011)
11. Gorenflo, R., Kilbas, A.A., Mainardi, F., Rogosin, S.V.: Mittag-Leffler Functions, Related Topics and Applications. Springer, Heidelberg (2014)
12. Shoubia, A.L., Figueiredo Camargo, R., de Oliveira, E.C., Vaz, Jr, J.: Theorem for series in three-parameter Mittag-Leffler function. Fract. Cal. Appl. Anal. **13**, 9–20 (2010)
13. Wittaker, E.T., Watson, G.N.: A Course of Modern Analysis, 4th edn. Cambridge Mathematical Library, Cambridge (1996)
14. Mainardi, F., Gorenflo, R.: On Mittag-Leffler-type functions in fractional evolution processes. J. Comput. Appl. Math. **118**, 283–299 (2000)
15. de Oliveira, E.C., Mainardi, F., Vaz, Jr, J.: Models based on Mittag-Leffler functions for anomalous relaxation in dielectrics. Eur. Phys. J. **193**, 161–171 (2011)
16. de Oliveira, E.C., Mainardi, F., Vaz, Jr, J.: Fractional models of anomalous relaxation based on the Kilbas and Saigo function. Meccanica **49**, 2049–2060 (2014)
17. Podlubny, I.: Fractional Differential Equations. Mathematical in Sciences and Engineering, vol. 198. Academic Press, San Diego (1999)
18. Matlob, M.A., Jamali, Y.: The concepts and applications of fractional order differential calculus in modelling of viscoelastic systems: a primer (2017). arXiv:1706.06446v2

19. Camargo, R.F., de Oliveira, E.C.: Fractional Calculus. Editora Livraria da Física, São Paulo (2015). (in Portuguese)
20. Teodoro, G.S.: Fractional derivatives: types and criteria. Ph.D. thesis (2019), Imecc-Unicamp, Campinas. (in Portuguese)
21. Hilfer, R. (ed.): Applications of Fractional Calculus in Physics. World Scientific, Singapore (2000)
22. Tarasov, V.E.: No violation of the Leibniz rule. No fractional derivative. Commun. Nonlinear Sci. Numer. Simul. **18**, 2945–2948 (2013)
23. Mainardi, F., Pironi, P.: The fractional Langevin equation: Brownian motion revisited. Extr. Math. **11**, 140–154 (1996)
24. Porrà, J.M., Wand, K.G., Masoliver, J.: Generalized Langevin equation: anomalous diffusion and probability distributions. Phys. Rev. E. **53**, 5872–5881 (1996)
25. Camargo, R.F, Chiacchio, A.O., Charnet, R., de Oliveira, E.C.: Solution of the fractional Langevin equation and the Mittag-Leffler functions. J. Math. Phys. **50**, 063507 (2009)
26. Sau Fa, K.: Generalized Langevin equation with fractional derivative and long-time correlation function. Phys. Rev. E. **73**, 061104 (2006)
27. Sau Fa, K.: Fractional Langevin equation and Riemann-Liouville fractional derivative. Eur. Phys. J. E. **24**, 139–143 (2007)
28. Viñales, A.D., Despósito, M.A.: Anomalous diffusion induced by a Mittag-Leffler correlated noise. Phys. Rev. E. **75**, 042102 (2007)
29. Körner, T.W.: Fourier Analysis. Cambridge University Press, Cambridge (1990)
30. Love, E.R.: Changing the order of integration. J. Austral. Math. Soc. **9**, 421–432 (1970)
31. Miller, K.S., Ross, B.: An Introduction to the Fractional Calculus and Fractional Differential Equations. Wiley, New York (1993)
32. de Oliveira, E.C., Rodrigues Jr., W.A.: Analytical Functions with Applications. Editora Livraria da Física, São Paulo (2005). (in Portuguese)
33. Nigmatullin, R.R.: The realization of the generalized transfer equation in a medium with fractal geometry. Phys. Sta. Sol. B **133**, 425–430 (1986)
34. Mainardi, F.: The fundamental solutions for the fractional diffusion-wave equation. Appl. Math. Lett. **9**, 23–28 (1996)
35. Mathai, A.M., Saxena, R.K., Haubold, H.J.: The H-Function: Theory and Applications. Springer, New York (2010)
36. Arfken, G.B., Weber, H.J.: Mathematical Methods for Physicists. Academic Press, New York (1995)
37. Oliveira, D.S., de Oliveira, E.C., Deif, S.: On a sum with a three-parameter Mittag-Leffler function. Int. Transf. Spec. Funct. **27**, 639–652 (2016)
38. Teodoro, G.S, de Oliveira, E.C.: Laplace transform and the Mittag-Leffler function. Int. J. Math. Educ. Sci. Technol. **45**, 595–604 (2014)
39. Mittag-Leffler, G.M.: Sur la nouvelle fonction $E_\alpha(x)$. CR Acad. Sci. Paris **137**, 554–558 (1903)
40. Gradshteyn, I.S., Ryzhik, I.M.: Table of Integrals, Series and Products. Academic Press Inc, New York (1980)
41. Prudnikov, A.P., Brychkov, YuA, Marichev, O.I.: Integral and Series (Inverse Laplace Transforms). Gordon and Breach Science Publishers, New York (1992)
42. Costa, F.S., Plata, A.R.G., de Oliveira, E.C.: Fractional Spacetime Diffusion with Time Dependent Diffusion Coefficients. Submited for Publication (2019)
43. Prudnikov, A.P., Brychkov, YuA, Marichev, O.I.: Integral and Series (Elementary Functions). Gordon and Breach Science Publishers, London (1986)
44. de Oliveira, E.C.: Special Functions and Applications, 2nd edn. Livraria Editora da Física, São Paulo (2012). (in Portuguese)

Appendix A
Mellin-Barnes Integrals

In analogy to integer-order calculus, fractional calculus also presents classes of functions that can be called special. In solving a fractional differential equation with constant coefficients, the classical Mittag-Leffler function and its particular cases emerge naturally. The most general special function appearing in problems of fractional calculus is the Fox's H-function, which is defined in terms of the Mellin-Barnes integral. Here we present Mellin-Barnes integrals associated with Fox's H-function and some particulares cases, among them, the classical hypergeometric function.

In this appendix we will present the called Mellin-Barnes integrals [1, 2], because of their importance in the inversion of the Laplace and Mellin transforms. The particularity of these integrals lies in the fact that they are integral in the complex plane and, therefore, the integration contour plays a fundamental rule [3].

First, before presenting the definition of the Mellin-Barnes integral, it should be pointed out that Pincherle [4], in 1888, therefore, before Mellin [2] and Barnes [1], obtained the following formula

$$\Psi(t) = \int_{\gamma} \frac{\displaystyle\prod_{i=1}^{m} \Gamma(x - \rho_i)}{\displaystyle\prod_{i=1}^{m-1} \Gamma(x - \sigma_i)} e^{xt} \, dx$$

whose convergence was proved using an asymptotic formula for the gamma function.

We note the similarity of this expression with the inverse Laplace transform, since by defining the quotient involving the gamma functions, by $f(x, \rho_i, \sigma_i)$ for example, we have exactly the integral, in the complex plane, that recovers the function through the inverse Laplace transform. Thus, this integral can be considered as the first example in the literature of what is now known as the Mellin-Barnes integral [5].

In order to make explicit the calculations, we will discuss, through examples, involving the Mittag-Leffler function with three parameters and the classical hyper-

© Springer Nature Switzerland AG 2019

E. Capelas de Oliveira, *Solved Exercises in Fractional Calculus*, Studies in Systems, Decision and Control 240, https://doi.org/10.1007/978-3-030-20524-9

geometric function, how to present the function as an integral in the complex plane and vice versa, that is, from the integral in complex plane, to recover the particular function.

Definition A.1 (*Mellin-Barnes integral*) Let γ be a contour in the complex plane starting at $c - i\infty$ and ending at $c + i\infty$, with $\mathrm{Re}(s) = c > 0$. It is called Mellin-Barnes integral to any integral in the complex plane whose integrand contemplates at least one gamma function, given by

$$I(z) = \frac{1}{2\pi i} \int_\gamma f(s) z^{-s} \, \mathrm{d}s \qquad (A.1)$$

where the density function, $f(s)$, in general, solution of a differential equation with polynomial coefficients, is given by a quotient of products of gamma functions depending on parameters. The choice of the contour γ, as we have already mentioned, plays a crucial rule, and will be explained through examples, involving a Mittag-Leffler function with three parameters, the hypergeometric function as well as, the confluent hypergeometric function, because with these functions we can discuss the case of the called 'empty product', expressing these functions as a Meijer's G-function or a Fox's H-function.

Moreover, we will direct this appendix to be able to use it, when appropriate, in the inversion of integrals, in particular, involving the Meijer and Fox functions. Therefore, in this sense, we will consider the density function in the following form

$$f(s) = \frac{A(s)B(s)}{C(s)D(s)}$$

with $A(s)$, $B(s)$, $C(s)$ and $D(s)$ are products of gamma functions, as in Sect. 2.4.3.

Example A.1 (Mittag-Leffler function with three parameters) Let $\alpha \in \mathbb{R}_+$; $\beta, \rho \in \mathbb{C}$, $\beta \neq 0$ and $\mathrm{Re}(\rho) > 0$. Show that the integral

$$\Lambda \equiv \frac{1}{2\pi i} \int_{\mathcal{L}} \frac{\Gamma(s)\Gamma(\rho - s)}{\Gamma(\beta - \alpha s)} (-z)^{-s} \, \mathrm{d}s$$

for a convenient integration contour, $\mathcal{L}$, in the complex plane, is a less than adequate gamma function, a Mittag-Leffler function with three parameters.

We consider $|\arg(z)| < \pi$ and the contour $\mathcal{L}$, as in Fig. A.1, starting at $c - i\infty$ going to $c + i\infty$, being $0 < c < \mathrm{Re}(\rho)$, separating all the poles $s = -k$, with $k = 0, 1, 2, \ldots$ to the left and all the poles $s = \rho + n$, with $n = 0, 1, 2, \ldots$ to the right (do not contribute to the residue theorem) relatively to the line $\mathrm{Re}(s) = c$.

To compute the integral in the complex plane, we close the integration contour so that only the poles at $s = -k$, with $k = 0, 1, 2, \ldots$ contribute, and we use the residue theorem, that is, the sum of the residues at the poles $s = 0, -1, -2, \ldots$, from where follows

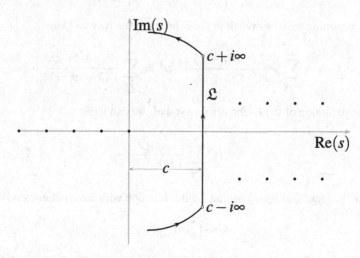

Fig. A.1 Contour $\mathfrak{L}$ for Mittag-Leffler with three parameters

$$\Lambda = \sum_{k=0}^{\infty} \lim_{s \to -k} \left\{ (s+k) \left[\frac{\Gamma(s)\Gamma(\rho - s)}{\Gamma(\beta - \alpha s)}(-z)^{-s} \right] \right\}$$

that, using the fact that the limit of the product is equal to the product of the limits,

$$\Lambda = \sum_{k=0}^{\infty} \frac{\Gamma(k + \rho)}{\Gamma(\alpha k + \beta)}(-z)^k \lim_{s \to -k}(s+k)\Gamma(s) \cdot$$

Note that we have an indetermination of type $0 \cdot \infty$. In order to raise this indetermination, we will calculate the remaining limit separately,

$$\lim_{s \to -k}(s+k)\Gamma(s) = \lim_{s \to -k}(s+k)\frac{(s+k-1)\cdots s\Gamma(s)}{(s+k-1)\cdots s} = \lim_{s \to -k}\frac{\Gamma(s+k+1)}{(s+k-1)\cdots s}$$

$$= \frac{\Gamma(1)}{(-1)(-2)\cdots(-k)} = \frac{1}{(-1)^k(1 \cdot 2 \cdots k)} = \frac{(-1)^k}{k!}.$$

This result is the residue of the gamma function at the poles $s = -k$ with $k = 0, 1, 2, \ldots$ Another way to obtain this result is through the called Mittag-Leffler expansion, according to Exercises (2) and (3) from Chap. 2.

Thus, returning with this result in the expression for Λ, we obtain

$$\Lambda = \sum_{k=0}^{\infty} \frac{(-1)^k}{k!} \frac{\Gamma(\rho+k)}{\Gamma(\beta+\alpha k)} (-z)^k = \sum_{k=0}^{\infty} \frac{\Gamma(\rho+k)}{\Gamma(\beta+\alpha k)} \frac{z^k}{k!}.$$

Using the definition of the Pochhammer symbol, we can write

$$\Lambda = \Gamma(\rho) \sum_{k=0}^{\infty} \frac{(\rho)_k}{\Gamma(\alpha k+\beta)} \frac{z^k}{k!}$$

which can be identified with a Mittag-Leffler function with three parameters, that is,

$$\Lambda = \Gamma(\rho) E_{\alpha,\beta}^{\rho}(z)$$

from which it follows that the integral is the product of a gamma function by the Mittag-Leffler function with three parameters

$$\frac{1}{2\pi i} \int_{\mathcal{L}} \frac{\Gamma(s)\Gamma(\rho-s)}{\Gamma(\beta-\alpha s)} (-z)^{-s} \, \mathrm{d}s = \Gamma(\rho) E_{\alpha,\beta}^{\rho}(z) \qquad (A.2)$$

which is the desired result   ◇

Example A.2 (Particular cases) Since the Mittag-Leffler function with three parameters contains as particular cases Mittag-Leffler functions with two and one parameter, we will express such functions in terms of an integral in the complex plane. Using the previous result, taking $\rho = 1$ in Eq. (A.2) we obtain

$$\frac{1}{2\pi i} \int_{\mathcal{L}} \frac{\Gamma(s)\Gamma(1-s)}{\Gamma(\beta-\alpha s)} (-z)^{-s} \, \mathrm{d}s = E_{\alpha,\beta}(z)$$

an integral representation for the Mittag-Leffler function with two parameters. Also, taking $\rho = 1 = \beta$ in Eq. (A.2) we obtain

$$\frac{1}{2\pi i} \int_{\mathcal{L}} \frac{\Gamma(s)\Gamma(1-s)}{\Gamma(1-\alpha s)} (-z)^{-s} \, \mathrm{d}s = E_{\alpha}(z)$$

an integral representation for the classical Mittag-Leffler function. Finally, for $\rho = 1 = \beta = \alpha$ in Eq. (A.2) we obtain

$$\frac{1}{2\pi i} \int_{\mathcal{L}} \Gamma(s)(-z)^{-s} \, \mathrm{d}s = e^z$$

an integral representation for the exponential function. The last three expressions can be interpreted as inverse Mellin transforms   ◇

Example A.3 (Mittag-Leffler function as a Fox's *H*-function) Here we will address the inverse problem, that is, we have the integral in the complex plane and we want to determine the function, that is, to calculate the integral explicitly.

Let us show that this function can not be written in terms of a Meijer function, which, as particular cases, contains all classical hypergeometric functions. Just to mention, the particular Mittag-Leffler function that can be written in terms of a confluent hypergeometric function, yes, can be written as a Meijer function.

In order to write the Mittag-Leffler function with three parameters, let us first identify its representation in terms of a Mellin-Barnes integral, Eq. (A.2), with the Mellin-Barnes integral defining Fox's *H*-function, that is,

$$\frac{1}{2\pi i} \int_{\mathcal{L}} \frac{\prod_{i=1}^{m} \Gamma(b_j + \beta_j s) \prod_{j=1}^{n} \Gamma(1 - a_j - \alpha_j s)}{\prod_{i=m+1}^{q} \Gamma(1 - b_j - \beta_j s) \prod_{j=n+1}^{p} \Gamma(a_j + \alpha_j s)} (-z)^s \, ds \cdot$$

In order to identify, we start with the numerator, that is, the integrand numerator in Eq. (A.2), has two gamma functions, where we consider $m = 1 = n$ while in the denominator we have only one gamma function, then we consider $q = 2$ and $p = 1$, the smallest that does not contribute, that is, it will generate an 'empty product', since the initial index begins with $j = n + 1$ and, like $n = 1$, the beginning will be $j = 2$, so it is equal to unity.

Let's explain these products individually, identifying them as follows:

$$\prod_{j=1}^{1} \Gamma(b_j + \beta_j s) = \Gamma(b_1 + \beta_1 s) \qquad \text{numerator,}$$

$$\prod_{j=1}^{1} \Gamma(1 - a_j - \alpha_j s) = \Gamma(1 - a_1 - \alpha_1 s) \qquad \text{numerator,}$$

$$\prod_{j=2}^{2} \Gamma(1 - b_j - \beta_j s) = \Gamma(1 - b_2 - \beta_2 s) \qquad \text{denominator,}$$

$$\prod_{j=2}^{1} \Gamma(a_j + \alpha_j s) = 1 (\text{not contributes}) \qquad \text{denominator.}$$

Using the integrand in Eq. (A.2) and identifying with the products,

$$\frac{\Gamma(s)\Gamma(\rho - s)}{\Gamma(\beta - \alpha s)} = \frac{\Gamma(b_1 + \beta_1 s)\Gamma(1 - a_1 - \alpha_1 s)}{\Gamma(1 - b_2 - \beta_2 s)}$$

we have $a_1 = 1 - \rho, \alpha_1 = 1, b_1 = 0, \beta_1 = 1, b_2 = 1 - \beta, \beta_2 = \alpha$ and $z \to -z$. Identifying the symbols for the Fox's H-function, we have

$$H_{1,2}^{1,1}\left[-z \left| \begin{array}{l} (a_1, \alpha_1) \\ (b_1, \beta_1), (b_2, \beta_2) \end{array} \right. \right] = H_{1,2}^{1,1}\left[-z \left| \begin{array}{l} (1 - \rho, 1) \\ (0, 1), (1 - \beta, \alpha) \end{array} \right. \right]$$

which, finally, allows us to write

$$\Gamma(\rho)E_{\alpha,\beta}^{\rho}(z) = H_{1,2}^{1,1}\left[-z \left| \begin{array}{l} (1 - \rho, 1) \\ (0, 1), (1 - \beta, \alpha) \end{array} \right. \right]$$

which is the desired result   ◇

Example A.4 (A particular case. Meijer's G-function) From the above expression it is clear that Fox's H-function can not be reduced to a Meijer's function, except if $\alpha = 1$, a particular Mittag-Leffler function $E_{1,\beta}^{\rho}(\cdot)$, that is,

$$\Gamma(\rho)E_{1,\beta}^{\rho}(z) = G_{1,2}^{1,1}\left[-z \left| \begin{array}{l} 1 - \rho \\ 0, 1 - \beta \end{array} \right. \right]$$

with $G_{1,2}^{1,1}\left[\cdot \left| \begin{array}{l} \cdot \\ \cdot \end{array} \right. \right]$ is a Meijer's G-function.

Since this particular Mittag-Leffler function with three parameters is related to a confluent hypergeometric function, through the expression

$$\Gamma(\beta)E_{1,\beta}^{\rho}(z) = {}_1F_1(\rho; \beta; z)$$

we can write this confluent hypergeometric function in terms of a Meijer's G-function,

$$_1F_1(\rho; \beta; x) = G_{1,2}^{1,1}\left[-z \left| \begin{array}{l} 1 - \rho \\ 0, 1 - \beta \end{array} \right. \right]$$

which is the desired result   ◇

Example A.5 (Hypergeometric function) Here, we star with the calculation of the Mellin transform, denoted by $\mathscr{M}$, of the classical hypergeometric function $_2F_1(a, b; c; -x)$, that is

$$\Omega \equiv \mathscr{M}[_2F_1(a, b; c; -x)] = \int_0^\infty x^{s-1} {}_2F_1(a, b; c; -x)\, dx \cdot$$

By introducing the integral representation for the hypergeometric function, given by Eq. (2.10) and changing the integration orders, we can write

$$\Omega = \frac{\Gamma(c)}{\Gamma(b)\Gamma(c - b)} \int_0^1 t^{b-1}(1 - t)^{c-b-1} \int_0^\infty \frac{x^{s-1}}{(1 + xt)^a}\, dx\, dt \cdot$$

To calculate the integral in the variable x, we introduce the change of variable $xt = \xi$ from where we obtain, only for the integral in x

$$\int_0^\infty \frac{x^{s-1}}{(1+xt)^a}\, dx = t^{-s} \int_0^\infty \frac{\xi^{s-1}}{(1+\xi)^a}\, d\xi.$$

In order to calculate the integral in the variable ξ, we introduce another change of variable, η, given by $\xi = \eta + \xi\eta$, from where it follows to the integral, already simplifying

$$t^{-s} \int_0^\infty \frac{\xi^{s-1}}{(1+\xi)^a}\, d\xi = t^{-s} \int_0^1 \eta^{s-1}(1-\eta)^{-s+a-1}\, d\eta.$$

The remaining integral is nothing more than a beta function, according to Eq. (2.2), then

$$\int_0^\infty \frac{x^{s-1}}{(1+xt)^a}\, dx = t^{-s} B(s, a-s) = t^{-s} \frac{\Gamma(s)\Gamma(a-s)}{\Gamma(a)}$$

where, in the last step, we use the relation between the gamma and beta functions.

Returning in the expression to Ω we can write

$$\Omega = \frac{\Gamma(c)}{\Gamma(b)\Gamma(c-b)} \frac{\Gamma(s)\Gamma(a-s)}{\Gamma(a)} \int_0^1 t^{b-s-1}(1-t)^{c-b-1}\, dt$$

which, also, is another beta function. Proceeding as in the previous one and expressing the result in terms of gamma functions, we obtain, already simplifying

$$\Omega = \frac{\Gamma(c)}{\Gamma(a)\Gamma(b)} \frac{\Gamma(s)\Gamma(a-s)\Gamma(b-s)}{\Gamma(c-s)}.$$

Returning with the inverse we obtain the Mellin-Barnes integral

$$\frac{\Gamma(a)\Gamma(b)}{\Gamma(c)}\, _2F_1(a, b; c; z) = \frac{1}{2\pi i} \int_{-i\infty}^{i\infty} \frac{\Gamma(s)\Gamma(a-s)\Gamma(b-s)}{\Gamma(c-s)} (-z)^{-s}\, ds \quad \text{(A.3)}$$

which is also an integral representation for the classical hypergeometric function.

In order to show, from the previous integral representation, the inverse process, that is, from the integral representation, to obtain the hypergeometric function, we proceed as in the case of the Mittag-Leffler function with three parameters.

We explain only the steps, and then write this function as a particular case of the Meijer's G-function. Moreover, just to remember, we can also write it in terms of a Fox's H-function, but, as we have already mentioned, Meijer's G-function is a particular case of Fox's H-function, so, we write only in terms of Meijer's G-function.

Then we return the integral given by Eq. (A.3) and outline the steps, in analogy to the case of the Mittag-Leffler function with three parameters. First the contour, where we admit $\arg(-z) < \pi$ and separating the poles from the $\Gamma(a - s)$ and $\Gamma(b - s)$ those of function $\Gamma(s)$, as well as considering $(-z)^{-s}$ with its principal value. Then we consider the poles of $\Gamma(s)$ such that $-s = k = 0, 1, 2, \ldots$, so we must calculate the limit

$$\sum_{k=0}^{\infty} \lim_{s \to -k} \left\{ (s + k) \left[\Gamma(s) \frac{\Gamma(-s + a)\Gamma(-s + b)}{\Gamma(-s + c)} (-z)^{-s} \right] \right\}.$$

Even more, analogous to the previous one, we have

$$\lim_{s \to -k} (k + s)\Gamma(s) = \frac{(-1)^k}{k!}$$

from where follows the result for the integral

$$\frac{\Gamma(a)\Gamma(b)}{\Gamma(c)} \, {}_2F_1(a, b; c; z),$$

which is the desired result   ◇

Example A.6 (Hypergeometric function as a Meijer's G-function) Consider the following integral in the complex plane

$$\Lambda = \frac{1}{2\pi i} \int_{-i\infty}^{i\infty} \frac{\Gamma(-s)\Gamma(a + s)\Gamma(b + s)}{\Gamma(c + s)} (-z)^s \, ds \cdot \qquad (A.4)$$

We recall that, as in all gamma functions in the integrand of the previous one, the variable s is not multiplied by any coefficient, we can directly identify with the Meijer's G-function, being the contour as described in the previous example. Then we must compare the integrants, given in Λ with that of the Meijer's G-function, given by

$$\mathscr{G}_{p,q}^{m,n}(s) = \frac{\displaystyle\prod_{i=1}^{m} \Gamma(b_i - s) \cdot \prod_{j=1}^{n} \Gamma(1 - a_j + s)}{\displaystyle\prod_{i=m+1}^{q} \Gamma(1 - b_j + s) \cdot \prod_{j=n+1}^{p} \Gamma(a_j - s)} \cdot$$

Then, since in the denominator we only have a gamma function, we take $m = 1$ and $q = 2$ and $p < n + 1$ so that the product is 'empty', identified with the unit. Let $n = 2$ (we still have two more gamma functions in the numerator) and $p = 2$.

Let's explain these products individually, identifying them as follows:

$$\prod_{i=1}^{1} \Gamma(b_j - s) = \Gamma(b_1 - s) \quad \text{numerator},$$

$$\prod_{j=1}^{2} \Gamma(1 - a_j + s) = \Gamma(1 - a_1 + s)\Gamma(1 - a_2 + s) \quad \text{numerator,}$$

$$\prod_{i=2}^{2} \Gamma(1 - b_j + s) = \Gamma(1 - b_2 + s) \quad\quad\quad\quad \text{denominator,}$$

$$\prod_{j=3}^{2} \Gamma(a_j - s) = 1 (\text{not contributes}) \quad\quad\quad \text{denominator.}$$

Using the integrand in Eq. (A.4) and identifying with the products,

$$\frac{\Gamma(-s)\Gamma(a + s)\Gamma(b + s)}{\Gamma(c + s)} = \frac{\Gamma(b_1 - s)\Gamma(1 - a_1 + s)\Gamma(1 - a_2 + s)}{\Gamma(1 - b_2 + s)}$$

we have $a_1 = 1 - a$, $a_2 = 1 - b$, $b_1 = 0$ and $b_2 = 1 - c$. Identifying with the symbols for the Meijer's G-function, we have

$$G_{2,2}^{1,2}\left[-z \,\middle|\, \begin{matrix} a_1, a_2 \\ b_1, b_2 \end{matrix}\right] = G_{2,2}^{1,2}\left[-z \,\middle|\, \begin{matrix} 1 - a, 1 - b \\ 0, 1 - c \end{matrix}\right]$$

which finally allows us to write

$$\frac{\Gamma(a)\Gamma(b)}{\Gamma(c)} \,_2F_1(a, b; c; z) = G_{2,2}^{1,2}\left[-z \,\middle|\, \begin{matrix} 1 - a, 1 - b \\ 0, 1 - c \end{matrix}\right]$$

which is the desired result   ◇

Example A.7 (Hypergeometric function. A particular case) Knowing that the relation is worth

$$(1 + x^2)^{-1/2} = \,_2F_1\left(\frac{1}{2}, \frac{1}{2}; \frac{1}{2}; -x^2\right)$$

express this function in terms of the Meijer's G-function.

Using the previous result, we can write, already simplifying

$$(1 + x^2)^{-1/2} = \frac{1}{\sqrt{\pi}} G_{2,2}^{1,2}\left[x^2 \,\middle|\, \begin{matrix} 1/2, 1/2 \\ 0, 1/2 \end{matrix}\right].$$

This expression can be simplified. To this end, we use Exercise (30) from Chap. 2, then

$$(1 + x^2)^{-1/2} = \frac{1}{\sqrt{\pi}} G_{2,2}^{1,2}\left[x^2 \,\middle|\, \begin{matrix} 1/2 \\ 0 \end{matrix}\right]$$

which is the desired result.   ◇

References

1. Barnes, E.W.: A new development in the theory of the hypergeometric functions. Proc. Lond. Math. Soc. **6**, 141–177 (1908)
2. Mellin, H.: Om definita integraler. Acta Societatis Scientiarum Fennicae **20**, 1–39 (1895)
3. Costa, F.S., Vaz Jr., J., Capelas de Oliveira, E., Figueiredo Camargo, R.: Mellin-Barnes integrals and the Fox function. Tend. Mat. Apl. Comput. **12**, 157–169 (2011)
4. Pincherle, S.: On generalized hypergeometric functions. Atti R. Accad. Lincei Rend. Cl. Sci. Fis. Natur. **4**, 694–700 and 792–799 (1888). (in Italy)
5. Mainardi, F., Pagnini, G.: Salvatore Pincherle: the pioneer of the Mellin-Barnes integral. J. Comput. Appl. Math. **153**, 331–342 (2003)

Index

© Springer Nature Switzerland AG 2019
E. Capelas de Oliveira, *Solved Exercises in Fractional
Calculus*, Studies in Systems, Decision and Control 240,
https://doi.org/10.1007/978-3-030-20524-9

Printed in the United States
By Bookmasters